研究生教材

电力工程科技英语教程

Dianli Gongcheng
Keji Yingyu Jiaocheng

编著　胡　钋　华小梅
主审　查晓明

中国电力出版社
CHINA ELECTRIC POWER PRESS

内 容 提 要

本书围绕电力系统及其自动化、高电压与绝缘技术、电力电子技术以及新能源四大专题集中讨论了其中的若干基本知识模块、科技前沿以及一些重要的通用专业英语语法知识等。

本书可作为电力类相关专业的硕士研究生或博士研究生的专业英语教学用书，也可作为这些专业本科高年级学生的教学或学习用书，还可用于电力企业工程技术和管理人员学习专业英语的培训教材。

图书在版编目（CIP）数据

电力工程科技英语教程/胡钋，华小梅编著．—北京：中国电力出版社，2019.6

研究生教材

ISBN 978-7-5198-2574-4

Ⅰ.①电… Ⅱ.①胡… ②华… Ⅲ.①电力工程－英语－研究生－教材 Ⅳ.①TM7

中国版本图书馆 CIP 数据核字（2018）第 241524 号

出版发行：中国电力出版社

地　　址：北京市东城区北京站西街 19 号（邮政编码 100005）

网　　址：http：//www.cepp.sgcc.com.cn

责任编辑：牛梦洁　（010-63412528）

责任校对：黄　蓓　闫秀英

装帧设计：郝晓燕

责任印制：钱兴根

印　　刷：三河市百盛印装有限公司

版　　次：2018 年 10 月第一版　2019 年 6 月第二版

印　　次：2019 年 6 月北京第三次印刷

开　　本：787 毫米×1092 毫米　16 开本

印　　张：19.5

字　　数：479 千字

定　　价：45.00 元

前言

随着中国和世界电力科技的飞速发展，国际科技交流变得十分频繁，这是加快我国包括特高压在内智能电网的现代电力技术进步的“催化剂”之一。广大学生和科技工作者只有通过比较系统的专业英语学习，熟练掌握本专业的专业英语，才能实现有效的科技交流，从而提升自己的科学研究和技术创新水平。

本书是作者在长期从事电气工程专业本科生、研究生专业英语教学研究与实践的基础上编写而成的，其主要着眼于高效培养、训练和提高本科生、硕士或博士研究生快速准确阅读电力专业英语的能力并使他们深入牢固地掌握通用专业英语语法和一些实用、重要的惯用法，达到科技英语书写自如的目的。

本书由电力系统及其自动化、高电压与绝缘技术、电力电子技术以及新能源四篇组成，在选材原则上注重这些专业知识的点与面的有机结合，尽可能在较为有限的篇幅里反映这些专业的基本和最新知识。课文内容全部选自欧美国家出版的相关专业书刊，以确保所读材料的“原汁原味”。

本书第1篇和第4篇由华小梅撰写，第2篇和第3篇由胡钋撰写，全书由胡钋负责统稿。武汉大学电气工程学院专业英语课程组林涛、张慧、陶劲松、关伟民等全体教师以及唐炬、陈红坤、查晓明、刘开培、徐箭、常湧等有关方面专家及学者在本书的编写过程中提出了许多宝贵建议，在此一并表示衷心的感谢。

限于作者的专业英语水平，书中疏漏之处在所难免，敬请读者批评指正，以便使本书能更加完善地飨于读者。您可以请中国电力出版社转达您的赐教或直接给我们写邮件表达您对这本书的关心和指正（电子邮箱：phu1126@126.com）。

编　者

2019年3月　于珞珈山

前言

目　录

第3篇 电力电子与专业英语语法

第4篇 新能源与专业英语语法

第1篇

电力系统及其自动化与专业英语语法

1

电力系统及其自动化

1.1 Definition and Classification of Power System Stability

1.1.1 Introduction

Power system stability has been recognized as an important problem for secure system operation since the 1920s. Many major blackouts caused by power system instability have illustrated the importance of this phenomenon. Historically, transient instability has been the dominant stability problem on most systems, and has been the focus of much of the industry's attention concerning system stability. As power systems have evolved through continuing growth in interconnections, use of new technologies and controls, and the increased operation in highly stressed conditions, different forms of system instability have emerged. For example, voltage stability, frequency stability and interarea oscillations have become greater concerns than in the past. This has created a need to review the definition and classification of power system stability. A clear understanding of different types of instability and how they are interrelated is essential for the satisfactory design and operation of power systems. As well, consistent use of terminology is required for developing system design and operating criteria, standard analytical tools, and studying procedures.

The problem of defining and classifying power system stability is an old one, and there have been several previous reports on the subject by CIGRE and IEEE Task Forces. These, however, do not completely reflect current industry needs, experiences, and understanding. In particular, definitions are not precise and the classifications do not encompass all practical instability scenarios.

This report is the result of long deliberations of the Task Force set up jointly by the CIGRE Study Committee 38 and the IEEE Power System Dynamic Performance Committee. Our objectives are to:

- Define power system stability more precisely, inclusive of all forms.
- Provide a systematic basis for classifying power system stability, identifying and defining different categories, and providing a broad picture of the phenomena.
- Discuss linkages to related issues such as power system reliability and security.

Power system stability is similar to the stability of any dynamic system, and has fundamental mathematical underpinnings. Precise definitions of stability can be found in the literature dealing with the rigorous mathematical theory of stability of dynamic systems. Our intent here is to provide a physically motivated definition of power system stability which in broad terms conforms to precise mathematical definitions.

The report is organized as follows. In Section 1. 1. 2 the definition of Power System Stability is provided. A detailed discussion and elaboration of the definition are presented. The conformance of this definition with the system theoretic definitions is established. Section 1. 1. 3 provides a detailed classification of power system stability. In Section 1. 1. 4 of the report the relationship between the concepts of power system reliability, security, and stability is discussed. A description of how these terms have been defined and used in practice is also provided. Finally, in Section 1. 1. 5 definitions and concepts of stability from mathematics and control theory are reviewed to provide background information concerning stability of dynamic systems in general and to establish theoretical connections.

The analytical definitions presented in Section 1. 1. 5 constitute a key aspect of the report. They provide the mathematical underpinnings and bases for the definitions provided in the earlier sections. These details are provided at the end of the report so that interested readers can examine the finer points and assimilate the mathematical rigor.

1. 1. 2　Definition of Power System Stability

In this section, we provide a formal definition of power system stability. The intent is to provide a physically based definition which, while conforming to definitions from system theory, is easily understood and readily applied by power system engineering practitioners.

A. Proposed Definition

- *Power system stability is the ability of an electric power system, for a given initial operating condition, to regain a state of operating equilibrium after being subjected to a physical disturbance, with most system variables bounded so that practically the entire system remains intact.*

B. Discussion and Elaboration

The definition applies to an interconnected power system as a whole. Often, however, the stability of a particular generator or group of generators is also of interest. A remote generator may lose stability (synchronism) without cascading instability of the main system. Similarly, stability of particular loads or load areas maybe of interest; motors may lose *stability*(run down and stall) without cascading instability of the main system.

The power system is a highly nonlinear system that operates in a constantly changing environment; loads, generator outputs and key operating parameters change continually. When subjected to a disturbance, the stability of the system depends on the initial operating condition as well as the nature of the disturbance.

Stability of an electric power system is thus a property of the system motion around an equilibrium set, i. e. , the initial operating condition. In an equilibrium set, the various opposing forces that exist in the system are equal instantaneously (as in the case of equilibrium points) or over a cycle (as in the case of slow cyclical variations due to continuous small fluctuations in loads or aperiodic attractors).

Power systems are subjected to a wide range of disturbances, small and large. Small disturbances in the form of load changes occur continually; the system must be able to adjust to the changing conditions and operate satisfactorily. It must also be able to survive numerous disturbances of a severe nature, such as a short circuit on a transmission line or loss of a large generator. A large disturbance may lead to structural changes due to the isolation of the faulted elements.

At an equilibrium set, a power system may be stable for a given (large) physical disturbance, and unstable for another. It is impractical and uneconomical to design power systems to be stable for every possible disturbance. The design contingencies are selected on the basis they have a reasonably high probability of occurrence. Hence, large-disturbance stability always refers to a specified disturbance scenario. A stable equilibrium set thus has a finite region of attraction; the larger the region, the more robust the system with respect to large disturbances. The region of attraction changes with the operating condition of the power system.

The response of the power system to a disturbance may involve much of the equipment. For instance, a fault on a critical element followed by its isolation by protective relays will cause variations in power flows, network bus voltages, and machine rotor speeds; the voltage variations will actuate both generator and transmission network voltage regulators; the generator speed variations will actuate prime mover governors; and the voltage and frequency variations will affect the system loads to varying degrees depending on their individual characteristics. Further, devices used to protect individual equipment may respond to variations in system variables and cause tripping of the equipment, thereby weakening the system and possibly leading to system instability.

If following a disturbance the power system is stable, it will reach a new equilibrium state with the system integrity preserved i. e. , with practically all generators and loads connected through a single contiguous transmission system. Some generators and loads may be disconnected by the isolation of faulted elements or intentional tripping to preserve the continuity of operation of bulk of the system. Interconnected systems, for certain severe disturbances, may also be intentionally split into two or more "islands" to preserve as much of the generation and load as possible. The actions of automatic controls and possibly human operators will eventually restore the system to normal state. On the other hand, if the system is unstable, it will result in a run-away or run-down situation; for example, a progressive increase in angular separation of generator rotors, or a progressive decrease in bus voltages. An unstable system condition could lead to cascading outages and a shutdown of a major portion of the power system.

Power systems are continually experiencing fluctuations of small magnitudes. However, for assessing stability when subjected to a specified disturbance, it is usually valid to assume that the system is initially in a true steady-state operating condition.

C. Conformance with System—Theoretic Definitions

In Section1. 1. 2-A, we have formulated the definition by considering a given operating condition and the system being subjected to a physical disturbance. Under these conditions we require the system to either regain a new state of operating equilibrium or return to the original operating condition (if no topological changes occurred in the system). These requirements are directly correlated to the system-theoretic definition of asymptotic stability given in Section 1. 1. 5-C-I. It should be recognized here that this definition requires the equilibrium to be (a) stable in the sense of Lyapunov, i. e. , all initial conditions starting in a small spherical neighborhood of radius δ result in the system trajectory remaining in a cylinder of radius ε for all time $t \geqslant t_0$, the initial time which corresponds to all of the system state variables being bounded, and (b) at time $t \to \infty$ the system trajectory approaches the equilibrium point which corresponds to the equilibrium point being attractive. As a result, one observes that the analytical definition directly correlates to the expected behavior in a physical system.

1. 1. 3　Classification of Power System Stability

A typical modern power system is a high-order multivariable process whose dynamic response is influenced by a wide array of devices with different characteristics and response rates. Stability is a condition of equilibrium between opposing forces. Depending on the network topology, system operating condition and the form of disturbance, different sets of opposing forces may experience sustained imbalance leading to different forms of instability. In this section, we provide a systematic basis for classification of power system stability.

A. Need for Classification

Power system stability is essentially a single problem; however, the various forms of instabilities that a power system may undergo cannot be properly understood and effectively dealt with by treating it as such. Because of high dimensionality and complexity of stability problems, it helps to make simplifying assumptions to analyze specific types of problems using an appropriate degree of detail of system representation and appropriate analytical techniques. Analysis of stability, including identifying key factors that contribute to instability and devising methods of improving stable operation, is greatly facilitated by classification of stability into appropriate categories. Classification, therefore, is essential for meaningful practical analysis and resolution of power system stability problems. As discussed in Section1. 1. 5C. 1, such classification is entirely justified theoretically by the concept of partial stability.

B. Categories of Stability

The classification of power system stability proposed here is based on the following considerations:

- The physical nature of the resulting mode of instability as indicated by the main system variable in which instability can be observed.
- The size of the disturbance considered, which influences the method of calculation and prediction of stability.
- The devices, processes, and the time span that must be taken into consideration in order to assess stability.

Figure 1.1 gives the overall picture of the power system stability problem, identifying its categories and subcategories. The followings are descriptions of the corresponding forms of stability phenomena.

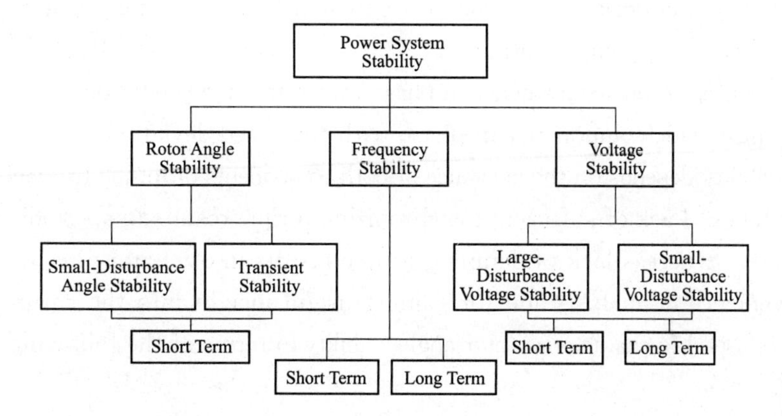

Figure 1.1 Classification of power system stability.

B.1 Rotor Angle Stability

Rotor angle stability refers to the ability of synchronous machines of an interconnected power system to remain in synchronism after being subjected to a disturbance. It depends on the ability to maintain/restore equilibrium between electromagnetic torque and mechanical torque of each synchronous machine in the system. Instability that may result occurs in the form of increasing angular swings of some generators leading to their loss of synchronism with other generators.

The rotor angle stability problem involves the study of the electromechanical oscillations inherent in power systems. A fundamental factor in this problem is the manner in which the power outputs of synchronous machines vary as their rotor angles change. Under steady-state conditions, there is equilibrium between the input mechanical torque and the output electromagnetic torque of each generator, and the speed remains constant. If the system is perturbed, this equilibrium is upset, resulting in acceleration or deceleration of the rotors of the machines according to the laws of motion of a rotating body. If one generator temporarily runs faster than another, the angular position of its rotor relative to that of the slower machine will advance. The resulting angular difference transfers part of the load from the slow machine to the fast machine, depending on the power-angle relationship. This tends to reduce the speed difference and hence the angular separation. The power-angle relationship is highly nonlinear. Beyond a certain limit, an increase in angular separation is accompanied by a decrease in power transfer such that the angular separation is increased further. Instability results if the system cannot absorb the kinetic energy corresponding to these rotor speed differences. For any given situation, the stability of the system depends on whether or not the deviations in angular positions of the rotors result in sufficient restoring torques. Loss of synchronism can occur between one machine and the rest of the system, or between groups of machines, with synchronism maintained within each group after separating from each other.

The change in electromagnetic torque of a synchronous machine following a perturbation can be resolved into two components:

- Synchronizing torque component, in phase with rotor angle deviation.
- Damping torque component, in phase with the speed deviation.

System stability depends on the existence of both components of torque for each of the synchronous machines. Lack of sufficient synchronizing torque results in aperiodic or nonoscillatory instability, whereas lack of damping torque results in oscillatory instability.

For convenience in analysis and for gaining useful insight into the nature of stability problems, it is useful to characterize rotor angle stability in terms of the following two subcategories:

- Small-disturbance (or small-signal) rotor angle stability is concerned with the ability of the power system to maintain synchronism under small disturbances. The disturbances are considered to be sufficiently small that linearization of system equations is permissible for purposes of analysis.

— Small-disturbance stability depends on the initial operating state of the system. Instability that may result can be of two forms: i) increase in rotor angle through a nonoscillatory or aperiodic mode due to lack of synchronizing torque, or ii) rotor oscillations of increasing amplitude due to lack of sufficient damping torque.

— In today's power systems, small-disturbance rotor angle stability problem is usually associated with insufficient damping of oscillations. The aperiodic instability problem has been largely eliminated by use of continuously acting generator voltage regulators; however, this problem can still occur when generators operate with constant excitation when subjected to the actions of excitation limiters (field current limiters).

— Small-disturbance rotor angle stability problems may be either local or global in nature. Local problems involve a small part of the power system, and are usually associated with rotor angle oscillations of a single power plant against the rest of the power system. Such oscillations are called local plant mode oscillations. Stability (damping) of these oscillations depends on the strength of the transmission system as seen by the power plant, generator excitation control systems and plant output.

— Global problems are caused by interactions among large groups of generators and have widespread effects. They involve oscillations of a group of generators in one area swinging against a group of generators in another area. Such oscillations are called interarea mode oscillations. Their characteristics are very complex and significantly differ from those of local plant mode oscillations. Load characteristics, in particular, have a major effect on the stability of interarea modes.

— The time frame of interest in small-disturbance stability studies is on the order of 10 to 20 seconds following a disturbance.

• Large-disturbance rotor angle stability or transient stability, as it is commonly referred to, is concerned with the ability of the power system to maintain synchronism when subjected to a severe disturbance, such as a short circuit on a transmission line. The resulting system response involves large excursions of generator rotor angles and is influenced by the nonlinear power-angle relationship.

— Transient stability depends on both the initial operating state of the system and the severity of the disturbance. Instability is usually in the form of aperiodic angular separation due to insufficient synchronizing torque, manifesting as first swing instability. However, in large power systems, transient instability may not always occur as first swing instability associated with a single mode; it could be a result of superposition of a slow interarea swing mode and a local-plant swing mode causing a large excursion of rotor angle beyond the first swing. It could also be a result of nonlinear effects affecting a single mode causing instability beyond the first swing.

— The time frame of interest in transient stability studies is usually 3 to 5 seconds following the disturbance. It may extend to 10-20 seconds for very large systems with dominant inter-area swings.

As identified in Figure. 1, small-disturbance rotor angle stability as well as transient stability are categorized as short term phenomena.

The term dynamic stability also appears in the literature as a class of rotor angle stability. However, it has been used to denote different phenomena by different authors. In the North American literature, it has been used mostly to denote small-disturbance stability in the presence of automatic controls (particularly, the generation excitation controls) as distinct from the classical "steady-state stability" with no generator controls. In the European literature, it has been used to denote transient stability. Since much confusion has resulted from the use of the term dynamic stability, we recommend against its usage, as did the previous IEEE and CIGRE Task Forces.

B. 2　Voltage Stability

Voltage stability refers to the ability of a power system to maintain steady voltages at all buses in the system after being subjected to a disturbance from a given initial operating condition. It depends on the ability to maintain/restore equilibrium between load demand and load supply from the power system. Instability that may result occurs in the form of a progressive fall or rise of voltages of some buses. A possible outcome of voltage instability is loss of load in an area, or tripping of transmission lines and other elements by their protective systems leading to cascading outages. Loss of synchronism of some generators may result from these outages or from operating conditions that violate field current limit.

Progressive drop in bus voltages can also be associated with rotor angle instability. For example, the loss of synchronism of machines as rotor angles between two groups of machines approach 180°causes rapid drop in voltages at intermediate points in the network close to the electrical center. Normally, protective systems operate to separate the two groups of machines and the voltages recover to levels depending on the post-separation conditions. If, however, the system is not so separated, the voltages near the electrical center rapidly oscillate between high and low values as a result of repeated "pole slips" between the two groups of machines. In contrast, the type of sustained fall of voltage that is related to voltage instability involves loads and may occur where rotor angle stability is not an issue.

The term voltage collapse is also often used. It is the process by which the sequence of events accompanying voltage instability leads to a blackout or abnormally low voltages in a significant part of the power system. Stable (steady) operation at low voltage may continue after transformer tap changers reach their boost limit, with intentional and/or unintentional tripping of some load. Remaining load tends to be voltage sensitive, and the connected demand at normal voltage is not met.

The driving force for voltage instability is usually the loads; in response to a disturbance, power consumed by the loads tends to be restored by the action of motor slip adjustment, distribution voltage regulators, tap-changing transformers, and thermostats. Restored loads increase the stress on the high voltage network by increasing the reactive power consumption and causing further voltage reduction. A run-down situation causing voltage instability occurs when load dynamics attempt to restore power consumption beyond the capability of the transmission network and the connected generation.

A major factor contributing to voltage instability is the voltage drop that occurs when active and reactive power flow through inductive reactances of the transmission network; this limits the capability of the transmission network for power transfer and voltage support. The power transfer and voltage support are further limited when some of the generators hit their field or armature current time-overload capability limits. Voltage stability is threatened when a disturbance increases the reactive power demand beyond the sustainable capacity of the available reactive power resources.

While the most common form of voltage instability is the progressive drop of bus voltages, the risk of overvoltage instability also exists and has been experienced at least on one system. It is caused by a capacitive behavior of the network (EHV transmission lines operating below surge impedance loading) as well as by under excitation limiters preventing generators and/or synchronous compensators from absorbing the excess reactive power. In this case, the instability is associated with the inability of the combined generation and transmission system to operate below some load level. In their attempt to restore this load power, transformer tap changers cause long-term voltage instability.

Voltage stability problems may also be experienced at the terminals of HVDC links used for either long distance or back-to-back applications. They are usually associated with HVDC links connected to weak ac systems and may occur at rectifier or inverter stations, and are associated with the unfavorable reactive power "load" characteristics of the converters. The HVDC link control strategies have a very significant influence on such problems, since the active and reactive power at the AC/DC junction are determined by the controls. If the resulting loading on the AC transmission stresses it beyond its capability, voltage instability occurs. Such a phenomenon is relatively fast with the time frame of interest being in the order of one second or less. Voltage instability may also be associated with converter transformer tap-changer controls, which is a considerably slower phenomenon. Recent developments in HVDC technology (voltage source converters and capacitor commutated converters) have significantly increased the limits for stable operation of HVDC links in weak systems as compared with the limits for line commutated converters.

One form of voltage stability problem that results in uncontrolled overvoltages is the self-excitation of synchronous machines. This can arise if the capacitive load of a synchronous machine is too large. Examples of excessive capacitive loads that can initiate self-excitation are open ended high voltage lines and shunt capacitors and filter banks from HVDC stations. The overvoltages that result when generator load changes to capacitive are characterized by an instantaneous rise at the instant of change followed by a more gradual rise. This latter rise depends on the relation between the capacitive load component and machine reactances together with the excitation system of the synchronous machine. Negative field current capability of the exciter is a feature that has a positive influence on the limits for self-excitation.

As in the case of rotor angle stability, it is useful to classify voltage stability into the following subcategories:

- Large-disturbance voltage stability refers to the system's ability to maintain steady voltages following large disturbances such as system faults, loss of generation, or circuit contingencies. This ability is determined by the system and load characteristics, and the interactions of both continuous and discrete controls and protections. Determination of large-disturbance voltage stability requires the examination of the nonlinear response of the power system over a period of time sufficient to capture the performance and interactions of such devices as motors, underload transformer tap changers, and generator field-current limiters. The study period of interest may extend from a few seconds to tens of minutes.

- Small-disturbance voltage stability refers to the system's ability to maintain steady voltages when subjected to small perturbations such as incremental changes in system load. This form of stability is influenced by the characteristics of loads, continuous controls, and discrete controls at a given instant of time. This concept is useful in determining, at any instant, how the system voltages will respond to small system changes. With appropriate assumptions, system equations can be linearized for analysis thereby allowing computation of valuable sensitivity information useful in identifying factors influencing stability. This linearization, however, cannot account for nonlinear effects such as tap changer controls (deadbands, discrete tap steps, and time delays). Therefore, a combination of linear and nonlinear analyzes is used in a complementary manner.

As noted above, the time frame of interest for voltage stability problems may vary from a few seconds to tens of minutes. Therefore, voltage stability may be either a short-term or a long-term phenomenon as identified in Figure 1.

- Short-term voltage stability involves dynamics of fast acting load components such as induction motors, electronically controlled loads, and HVDC converters. The study period of interest is in the order of several seconds, and analysis requires solution of appropriate system differential equations; this is similar to analysis of rotor angle stability. Dynamic modeling of loads is often essential. In contrast to angle stability, short circuits near loads are important. It is recommended that the term transient voltage stability not be used.
- Long-term voltage stability involves slower acting equipment such as tap-changing transformers, thermostatically controlled loads, and generator current limiters. The study period of interest may extend to several or many minutes, and long-term simulations are required for analysis of system dynamic performance. Stability is usually determined by the resulting outage of equipment, rather than the severity of the initial disturbance. Instability is due to the loss of long-term equilibrium (e. g. , when loads try to restore their power beyond the capability of the transmission network and connected generation), post-disturbance steady-state operating point being small-disturbance unstable, or a lack of attraction toward the stable post-disturbance equilibrium (e. g. , when a remedial action is applied too late). The disturbance could also be a sustained load buildup (e. g. , morning load increase). In many cases, static analysis can be used to estimate stability margins, identify factors influencing stability, and screen a wide range of system conditions and a large

number of scenarios. Where timing of control actions is important, this should be complemented by quasi-steady-state time-domain simulations.

B. 3 Basis for Distinction between Voltage and Rotor Angle Stability

It is important to recognize that the distinction between rotor angle stability and voltage stability is not based on weak coupling between variations in active power/angle and reactive power/voltage magnitude. In fact, coupling is strong for stressed conditions and both rotor angle stability and voltage stability are affected by pre-disturbance active power as well as reactive power flows. Instead, the distinction is based on the specific set of opposing forces that experience sustained imbalance and the principal system variable in which the consequent instability is apparent.

B. 4 Frequency Stability

Frequency stability refers to the ability of a power system to maintain steady frequency following a severe system upset resulting in a significant imbalance between generation and load. It depends on the ability to maintain/restore equilibrium between system generation and load, with minimum unintentional loss of load. Instability that may result occurs in the form of sustained frequency swings leading to tripping of generating units and/or loads.

Severe system upsets generally result in large excursions of frequency, power flows, voltage, and other system variables, thereby invoking the actions of processes, controls, and protections that are not modeled in conventional transient stability or voltage stability studies. These processes may be very slow, such as boiler dynamics, or only triggered for extreme system conditions, such as volts/hertz protection tripping generators. In large interconnected power systems, this type of situation is most commonly associated with conditions following splitting of systems into islands. Stability in this case is a question of whether or not each island will reach a state of operating equilibrium with minimal unintentional loss of load. It is determined by the overall response of the island as evidenced by its mean frequency, rather than relative motion of machines. Generally, frequency stability problems are associated with inadequacies in equipment responses, poor coordination of control and protection equipment, or insufficient generation reserve. Examples of such problems are reported in references. In isolated island systems, frequency stability could be of concern for any disturbance causing a relatively significant loss of load or generation.

During frequency excursions, the characteristic times of the processes and devices that are activated will range from fraction of seconds, corresponding to the response of devices such as underfrequency load shedding and generator controls and protections, to several minutes, corresponding to the response of devices such as prime mover energy supply systems and load voltage regulators. Therefore, as identified in Figure. 1. 1, frequency stability may be a short-term phenomenon or a long-term phenomenon. An example of short-term frequency instability is the formation of an undergenerated island with insufficient underfrequency load shedding such that frequency decays rapidly causing blackout of the island within a few seconds. On the other hand, more complex situations in which frequency instability is caused by steam turbine overspeed controls or boiler/reactor protection and controls are longer-term phenomena with the time frame of interest ranging from tens of seconds to several minutes.

During frequency excursions, voltage magnitudes may change significantly, especially for islanding conditions with underfrequency load shedding that unloads the system. Voltage magnitude changes, which may be higher in percentage than frequency changes, affect the load-generation imbalance. High voltage may cause undesirable generator tripping by poorly designed or coordinated loss of excitation relays or volts/Hertz relays. In an overloaded system, low voltage may cause undesirable operation of impedance relays.

B. 5　Comments on Classification

We have classified power system stability for convenience in identifying causes of instability, applying suitable analysis tools, and developing corrective measures. In any given situation, however, any one form of instability may not occur in its pure form. This is particularly true in highly stressed systems and for cascading events; as systems fail one form of instability may ultimately lead to another form. However, distinguishing between different forms is important for understanding the underlying causes of the problem in order to develop appropriate design and operating procedures.

While classification of power system stability is an effective and convenient means to deal with the complexities of the problem, the overall stability of the system should always be kept in mind. Solutions to stability problems of one category should not be at the expense of another. It is essential to look at all aspects of the stability phenomenon, and at each aspect from more than one viewpoint.

1. 1. 4　Relationship Between Reliability, Security, and Stability

In this section, we discuss the relationship between the concepts of power system reliability, security, and stability. We will also briefly describe how these terms have been defined and used in practice.

A. Conceptual Relationship

Reliability of a power system refers to the probability of its satisfactory operation over the long run. It denotes the ability to supply adequate electric service on a nearly continuous basis, with few interruptions over an extended time period.

Security of a power system refers to the degree of risk in its ability to survive imminent disturbances (contingencies) without interruption of customer service. It relates to robustness of the system to imminent disturbances and, hence, depends on the system operating condition as well as the contingent probability of disturbances.

Stability of a power system, as discussed in Section 1. 1. 3, refers to the continuance of intact operation following a disturbance. It depends on the operating condition and the nature of the physical disturbance.

The following are the essential differences among the three aspects of power system performance:

1) Reliability is the overall objective in power system design and operation. To be reliable, the power system must be secure most of the time. To be secure, the system must be stable but must also be secure against other contingencies that would not be classified as stability problems e. g., damage to equipment such as an explosive failure of a cable, fall of transmission towers due to ice loading or sabotage. As well, a system may be stable following a contingency, yet insecure due to post-fault system conditions resulting in equipment overloads or voltage violations.

2) System security may be further distinguished from stability in terms of the resulting consequences. For example, two systems may both be stable with equal stability margins, but one may be relatively more secure because the consequences of instability are less severe.

3) Security and stability are time-varying attributes which can be judged by studying the performance of the power system under a particular set of conditions. Reliability, on the other hand, is a function of the time-average performance of the power system; it can only be judged by consideration of the system's behavior over an appreciable period of time.

B. NERC Definition of Reliability

NERC (North American Electric Reliability Council) defines power system reliability as follows.

- *Reliability, in a bulk power electric system, is the degree to which the performance of the elements of that system results in power being delivered to consumers within accepted standards and in the amount desired. The degree of reliability may be measured by the frequency, duration, and magnitude of adverse effects on consumer service.*

Reliability can be addressed by considering two basic functional aspects of the power systems:

Adequacy—the ability of the power system to supply the aggregate electric power and energy requirements of the customer at all times, taking into account scheduled and unscheduled outages of system components.

Security—the ability of the power system to withstand sudden disturbances such as electric short circuits or nonanticipated loss of system components.

The above definitions also appear in several IEEE and CIGRE Working Group/Task Force documents.

Other alternative forms of definition of power system security have been proposed in the literature. For example, in reference, security is defined in terms of satisfying a set of inequality constraints over a subset of the possible disturbances called the "next contingency set."

C. Analysis of Power System Security

The analysis of security relates to the determination of the robustness of the power system relative to imminent disturbances. There are two important components of security analysis. For a power system subjected to changes (small or large), it is important that when the changes are completed, the system settles to new operating conditions such that no physical constraints are violated. This implies that, in addition to the next operating conditions being acceptable, the system must survive the transition to these conditions.

The above characterization of system security clearly highlights two aspects of its analysis:

- Static security analysis—This involves steady-state analysis of post-disturbance system conditions to verify that no equipment ratings and voltage constraints are violated.
- Dynamic security analysis—This involves examining different categories of system stability described in Section 1. 1. 3.

Stability analysis is thus an integral component of system security and reliability assessment.

The general industry practice for security assessment has been to use a deterministic approach. The power system is designed and operated to withstand a set of contingencies referred to as "normal contingencies" selected on the basis that they have a significant likelihood of occurrence. In practice, they are usually defined as the loss of any single element in a power system either spontaneously or preceded by a single-, double-, or three-phase fault. This is usually referred to as the *N*-1 criterion because it examines the behavior of an *N*-component grid following the loss of any one of its major components. In addition, loss of load or cascading outages may not be allowed for multiple-related outages such as loss of a double-circuit line. Consideration may be given to extreme contingencies that exceed in severity the normal design contingencies. Emergency controls, such as generation tripping, load shedding, and controlled islanding, may be used to cope with such events and prevent widespread blackouts.

The deterministic approach has served the industry reasonably well in the past—it has resulted in high security levels and the study effort is minimized. Its main limitation, however, is that it treats all security-limiting scenarios as having the same risk. It also does not give adequate consideration as to how likely or unlikely various contingencies are.

In today's utility environment, with a diversity of participants with different business interests, the deterministic approach may not be acceptable. There is a need to account for the probabilistic nature of system conditions and events, and to quantify and manage risk. The trend will be to expand the use of risk-based security assessment. In this approach, the probability of the system becoming unstable and its consequences are examined, and the degree of exposure to system failure is estimated. This approach is computationally intensive but is possible with today's computing and analysis tools.

1.1.5 System-Theoretic Foundations of Power System Stability

A. Preliminaries

In this section, we address fundamental issues related to definitions of power system stability from a system-theoretic viewpoint. We assume that the model of a power system is given in the form of explicit first-order differential equations (i. e. , a state-space description). While this is quite common in the theory of dynamical systems, it may not always be entirely natural for physical systems such as power systems. First-principle models that are typically used to describe power systems are seldom in this form, and transformations required to bring them to explicit first-order form may, in general, introduce spurious solutions.

More importantly, there often exist algebraic (implicit) equations that constrain various quantities, and a set of differential-algebraic equations (DAE) is often used in simulations of power system transients. The algebraic part often arises from a singular perturbation-type reasoning that uses time separation between subsets of variables to postulate that the fast variables have already reached their steady state on the time horizon of interest.

Proving the existence of solutions of DAE is a very challenging problem in general. While local results can be derived from the implicit function theorem that specifies rank conditions for the Jacobian of the algebraic part, the non-local results are much harder to obtain. One general approach to non-local study of stability of DAE systems that is based on differential geometry is presented (for sufficient conditions). The surfaces on which rank conditions for the Jacobian of the algebraic part do not hold are commonly denoted as impasse surfaces, and in the analysis of models of power systems, it is typically assumed that equilibrium sets of interest in stability analysis are disjoint from such surfaces.

An often useful approximation of the fast dynamics is based on the concept of dynamic (time-varying) phasors and dynamic symmetrical components. It is also typically assumed that distributed nature of some elements of a power system (e. g. , transmission lines) can be approximated with lumped parameter models without a major loss of model fidelity. This is mostly dictated by the fundamental intractability of models that include partial differential equations, and by satisfactory behavior of lumped parameter models (when evaluated on the level of single element—e. g. , the use of multiple "π" section models for a long transmission line). Some qualitative aspects of fault propagation in spatially extended power systems can, however, be studied effectively with distributed models.

Power systems are also an example of constrained dynamical systems, as their state trajectories are restricted to a particular subset in the state space (phase space in the language of dynamical systems) denoted as the feasible (or technically viable, or permitted) operating region. The trajectories that exit this desired region may either lead to structural changes (e. g. , breaker tripping in a power system), or lead to unsafe operation. This type of consideration will introduce restricted stability regions in power system stability analysis.

Several additional issues are raised by the fact that the power system interacts with its (typically unmodeled) environment, making the power system model nonautonomous (or time-varying). Examples include load variations and network topology changes due to switching in substations. Additional interactions with the environment include disturbances whose physical description may include outages of system elements, while a mathematical description may involve variations in the system order, or the number of variables of interest.

Finally, a power system is a controlled (or forced) system with numerous feedback loops, and it is necessary to include the effects of control inputs (including their saturation), especially on longer time horizons.

The outlined modeling problems are typically addressed in a power system analysis framework in the following way:

1) The problem of defining stability for general nonautonomous systems is very challenging even in the theoretical realm, and one possible approach is to say that a system to which the environment delivers square integrable signals as inputs over a time interval is stable if variables of interest (such as outputs) are also square integrable. In a more general setup, we can consider signals truncated in time, and denote the system as well-posed if it maps square integrable truncated signals into signals with the same property. In a power system setting, one typically assumes that the variables at the interface with the environment are known (or predictable)—e. g. , that mechanical inputs to all generators are constant, or that they vary according to the known response of turbine regulators.

2) The disturbances of interest will fall into two broad categories—event-type (typically described as outages of specific pieces of equipment) and norm-type (described by their size e. g. , in terms of various norms of signals); we will return to this issue shortly. We also observe that in cases when event-type (e. g. , switching) disturbances occur repeatedly, a proper analysis framework is that of hybrid systems; event-type disturbances may also be initiated by human operators. Our focus is on time horizons of the order of seconds to minutes; on a longer time scale, the effects of market structures may become prominent. In that case, the relevant notion of stability needs re-examination; some leads about systems with distributed decision making may be found.

3) Given our emphasis on stability analysis, we will assume that the actions of all controllers are fully predictable in terms of known system quantities (states), or as functions of time; the dual problem of designing stabilizing controls for nonlinear systems is very challenging.

A typical power system stability study consists of the following steps:

1) Make modeling assumptions and formulate a mathematical model appropriate for

the time-scales and phenomena under study;

2) Select an appropriate stability definition;

3) Analyze and/or simulate to determine stability, typically using a scenario of events;

4) Review results in light of assumptions, compare with the engineering experience ("reality"), and repeat if necessary.

Before considering specifics about power system stability, we need to assess the required computational effort. In the case of linear system models, the stability question is decidable, and can be answered efficiently in polynomial time. In the case of nonlinear systems, the available algorithms are inherently inefficient. For example, a related problem of whether all trajectories of a system with a single scalar nonlinearity converge to the origin turns out to be very computationally intensive (i. e. , NP-hard), and it is unclear if it is decidable at all. Given the large size of power systems and the need to consider event-type perturbations that will inevitably lead to nonlinear models, it is clear that the task of determining stability of a power system will be a challenging one. It turns out, however, that our main tools in reducing the computational complexity will be our ability (and willingness) to utilize approximations, and the particular nature of event-type disturbances that we are analyzing.

We also want to point out that a possible shift in emphasis regarding various phenomena in power systems (e. g., hybrid aspects) would necessarily entail a reassessment of notions of stability.

B. A Scenario for Stability Analysis

We consider the system

$$\dot{x} = f(t, x)$$

where x is the state vector (a function of time, but we omit explicitly writing the time argument), $\dot{x}$ is its derivative, f is sufficiently differentiable and its domain includes the origin. The system described above is said to be autonomous if $f(t,x)$ is independent of t and is said to be nonautonomous otherwise.

A typical scenario for power system stability analysis involves three distinct steps.

1) The system is initially operating in a pre-disturbance equilibrium set X_n (e. g. , an equilibrium point or perhaps even a benign limit cycle in the state space); in that set, various driving terms (forces) affecting system variables are balanced (either instantaneously, or over a time interval). We use the notion of an equilibrium set to denote equilibrium points, limit sets and more complicated structures like aperiodic attractors (which may be possible in realistic models of power systems). However, in the vast majority of cases of practical interest today, the equilibrium points are the sets of interest.

In general, an equilibrium set, or an attractor, is a set of trajectories in the phase space to which all neighboring trajectories converge. Attractors therefore describe the long-term behavior of a dynamical system. A point attractor, or an equilibrium point, is an attractor consisting of a single point in the phase space. A limit cycle attractor, on the other hand, corresponds to closed curves in phase space; limit cycles imply periodic behavior. A chaotic (or aperiodic, or strange) attractor corresponds to an equilibrium set where system trajectories never converge to a point or a closed curve, but remain within the same region of phase space. Unlike limit cycles, strange attractors are non-periodic, and trajectories in such systems are very sensitive to the initial conditions.

2) Next, a disturbance acts on the system. An event-type (or incident-type) disturbance is characterized by a specific fault scenario (e. g. , short circuit somewhere in the transmission network followed by a lined is connection including the duration of the event—"fault clearing time"), while norm-type (described by their size in terms of various norms of signals—e. g. , load variations) disturbances are described by their size (norm, or signal intensity). A problem of some analytical interest is determining the maximum permissible duration of the fault (the so called "critical clearing time") for which the subsequent system response remains stable. This portion of stability analysis requires the knowledge of actions of protective relaying.

3) After an event-type disturbance, the system dynamics is studied with respect to a known post-disturbance equilibrium set X_p (which may be distinct from X_n). The system initial condition belongs to a (known) starting set X_p, and we want to characterize the system motion with respect to X_p i. e. , if the system trajectory will remain inside the technically viable set Ω_p (which includes X_p). In the case of norm-type disturbances, very often we have $X_p = X_n$. If the system response turns out to be stable (a precise definition will follow shortly), it is said that X_p (and sometimes X_n as well) are stable. A detected instability (during which system motion crosses the boundary of the technically viable set $\partial\Omega_p$—e. g. , causing line tripping or a partial load shedding) may lead to a new stability study for a new (reduced) system with new starting and viable sets, and possibly with different modeling assumptions (or several such studies, if a system gets partitioned into several disconnected parts).

The stability analysis of power systems is in general nonlocal, as various equilibrium sets may get involved. In the case of event-type disturbances, the perturbations of interest are specified deterministically (the same may apply to X_n as well), and it is assumed that the analyst has determined all X_p that are relevant for a given X_n and the disturbance. In the case of norm-type perturbations, the uncertainty structure is different—the perturbation is characterized by size (in the case of the so-called small-disturbance or small-signal analysis this is done implicitly, so that linearized analysis remains valid), and the same equilibrium set typically characterizes the system before and after the disturbance. Note, however, that norm-type perturbations could in principle be used in large-signal analyses as well.

We propose next a formulation of power system stability that will allow us to explore salient features of general stability concepts from system theory.

- "*An equilibrium set of a power system is stable if, when the initial state is in the given starting set, the system motion converges to the equilibrium set, and operating constraints are satisfied for all relevant variables along the entire trajectory.*"

The operating constraints are of inequality (and equality) type, and pertain to individual variables and their collections. For example, system connectedness is a collective feature, as it implies that there exist paths (in graph-theoretic terms) from any given bus to all other buses in the network. Note also that some of the operating constraints (e. g., voltage levels) are inherently soft i. e., the power system analyst may be interested in stability characterization with and without these constraints. Note that we assume that our model is accurate within Ω_p in the sense that there are no further system changes (e. g., relay-initiated line tripping) until the trajectory crosses the boundary $\partial\Omega_p$.

C. Stability Definitions From System Theory

In this section, we provide detailed analytical definitions of several types of stability including Lyapunov stability, input-output stability, stability of linear systems, and partial stability. Of these various types, the Lyapunov stability definitions related to stability and asymptotic stability are the ones most applicable to power system nonlinear behavior under large disturbances. The definition of stability related to linear systems finds wide use in small-signal stability analysis of power systems. The concept of partial stability is useful in the classification of power system stability into different categories.

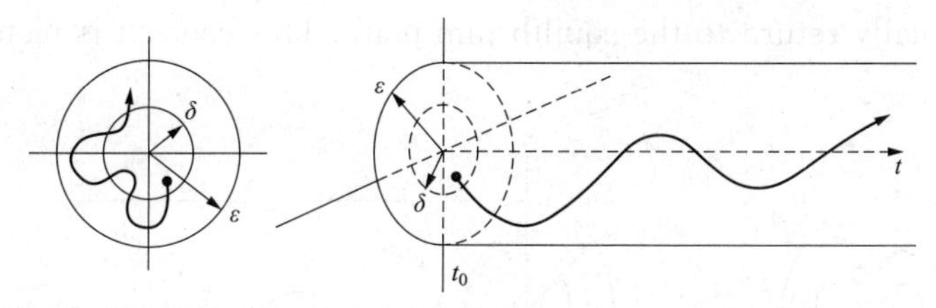

Figure 1. 2 Illustration of the definition of stability.

C. 1 Lyapunov Stability

The definitions collected here mostly follow the presentation; we present definitions for cases that are typical for power system models (e. g., assuming differentiability of functions involved), and not necessarily in the most general context. We will concentrate on the study of stability of equilibrium points; a study of more intricate equilibrium sets, like periodic orbits, can often be reduced to the study of equilibrium points of an associated system whose states are deviations of the states of the original system from the periodic orbit; another possibility is to study periodic orbits via sampled states and Poincare maps.

We again consider the nonautonomous system:

$$\dot{x} = f(t, x) \tag{1.1}$$

where x is the state vector, $\dot{x}$ is its derivative, f is sufficiently differentiable and its domain includes the origin. Note that the forcing (control input) term is not included i. e., we do not write $f(t, x, u)$. In stability analysis in the Lyapunov framework that is not a limitation, since all control inputs are assumed to be known functions of time t and/or known functions of states x. For technical reasons, we will assume that the origin is an equilibrium point (meaning that $f(t, 0)=0, \forall t \geqslant 0$). An equilibrium at the origin could be a translation of a nonzero equilibrium after a suitable coordinate transformation.

The equilibrium point $x=0$ of (1.1) is:

- stable if, for each $\varepsilon > 0$, there is $\delta = \delta(\varepsilon, t_0) > 0$ such that:

$$\|\chi(t_0)\| < \delta \Rightarrow \|\chi(t)\| < \varepsilon, \forall t \geqslant t_0 \geqslant 0 \tag{1.2}$$

Note that in (1.2) any norm can be used due to topological equivalence of all norms. In Figure 1.2, we depict the behavior of trajectories in the vicinity of a stable equilibrium for the case $\chi \in R^2$ (a two-dimensional system in the space of real variables). By choosing the initial conditions in a sufficiently small spherical neighborhood of radius δ, we can force the trajectory of the system for all time $t \geqslant t_0$ to lie entirely in a given cylinder of radius ε.

- uniformly stable if, for each $\varepsilon > 0$, there is $\delta = \delta(\varepsilon) > 0$, independent of t_0, such that (1.2) is satisfied;
- unstable if not stable;
- asymptotically stable if it is stable and in addition there is $\eta(t_0) > 0$ such that:

$$\|\chi(t_0)\| < \eta(t_0) \Rightarrow \chi(t) \to 0 \quad \text{as} \quad t \to \infty$$

It is important to note that the definition of asymptotic stability combines the aspect of stability as well as attractivity of the equilibrium. This is a stricter requirement of the system behavior to eventually return to the equilibrium point. This concept is pictorially presented in Figure 1.3.

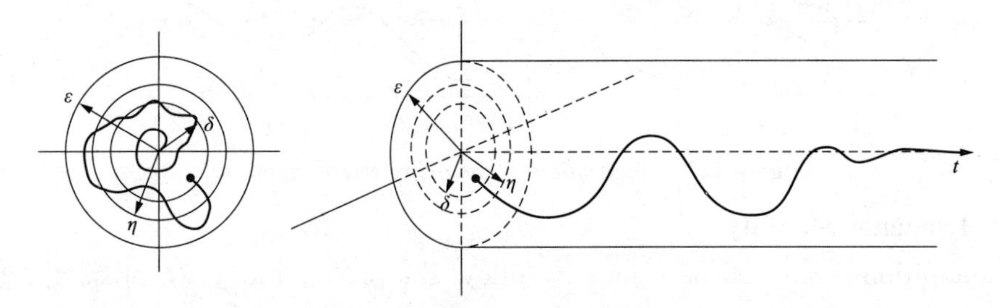

Figure 1.3 Illustration of the definition of asymptotic stability.

- uniformly asymptotically stable if it is uniformly stable and there is $\delta_0 > 0$, independent of t_0, such that for all $\|\chi(t_0)\| < \delta_0$, $\chi(t) \to 0$ as $t \to \infty$, uniformly in t_0 and $\chi(t_0)$; that is, for each $\varepsilon > 0$, there is $T = T(\varepsilon, \delta_0) > 0$ such that

$$\|\chi(t_0)\| < \delta_0 \Rightarrow \|\chi(t)\| < \varepsilon, \forall t \geqslant t_0 + T(\varepsilon, \delta_0)$$

In Figure 1.4, we depict the property of uniform asymptotic stability pictorially. By choosing the initial operating points in a sufficiently small spherical neighborhood at $t = t_0$, we can force the trajectory of the solution to lie inside a given cylinder for all $t > t_0 + T(\varepsilon, \delta_0)$.

- globally uniformly asymptotically stable if it is uniformly stable, and, for each pair of positive numbers ε and δ_0, there is $T = T(\varepsilon, \delta_0) > 0$ such that:

$$\|\chi(t_0)\| < \delta_0 \Rightarrow \|\chi(t)\| < \varepsilon, \forall t \geqslant t_0 + T(\varepsilon, \delta_0)$$

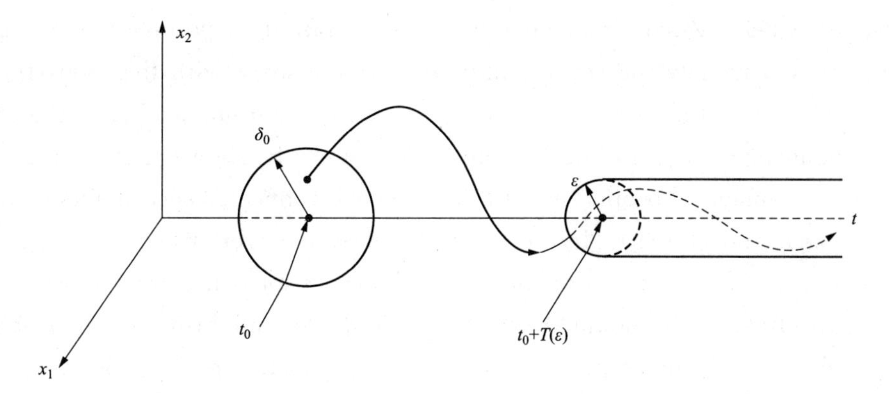

Figure 1. 4 Illustration of the definition of uniform asymptotic stability.

- exponentially stable if there are $\delta>0$, $\varepsilon>0$, $\alpha>0$. such that:

$$|| \chi(t_0) ||<\delta \Rightarrow || x(t) ||\leqslant \varepsilon || x(t_0) || \mathrm{e}^{-\alpha(t-t_0)}, \ t\geqslant t_0$$

In Figure 1. 5 the behavior of a solution in the vicinity of an exponentially stable equilibrium point is shown.

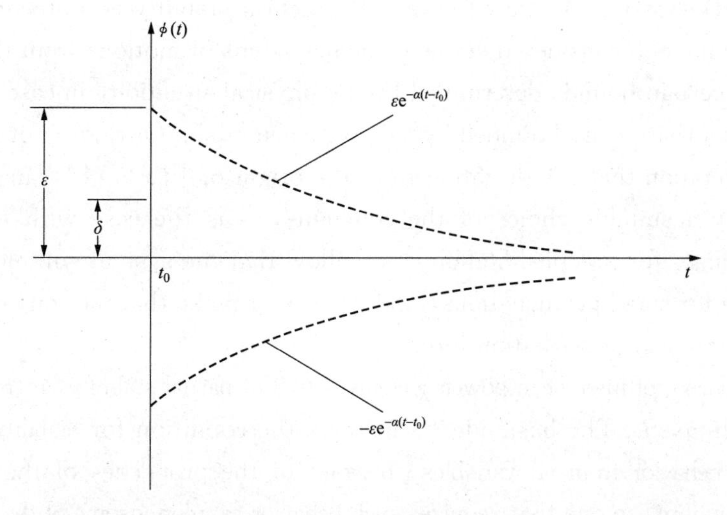

Figure 1. 5 Illustration of the definition of exponential stability.

- globally exponentially stable if the exponential stability condition is satisfied for any initial state.

These definitions form the foundation of the Lyapunov approach to system stability, and can be most naturally checked for a specific system via so called Lyapunov functions. Qualitatively speaking, i. e. , disregarding subtleties due to the nonautonomous characteristics of the system, we are to construct a smooth positive definite "energy" function whose time derivative [along trajectories of (1. 1)] is negative definite. While unfortunately there is no systematic method to generate such functions (some leads for the case of simple power systems are given), the so called converse Lyapunov theorems establish the existence of such functions if the system is stable in a certain sense.

In power systems we are interested in, the region of attraction $R(X_p)$ of a given equilibrium set X_p, namely the set of points in the state space with the property that all trajectories initiated at the points will converge to the equilibrium set X_p. If the equilibrium set is a point that is asymptotically stable, then it can be shown that the region of attraction has nice analytical properties (it is an open and connected set, and its boundary is formed by system trajectories). In the case of large scale power systems, we are naturally interested in the effects of approximations and idealizations that are necessary because of system size. Even if the nominal system has a stable equilibrium at the origin, this may not be the case for the actual perturbed system, which is not entirely known to the analyst. We cannot necessarily expect that the solution of the perturbed system will approach the origin, but could if the solution is ultimately bounded i. e., $||x(t)||$ is bounded by a fixed constant, given that the initial condition is in a ball of a fixed radius, and for sufficiently large time t. Characterization of stability in this case requires knowledge of the size of the perturbation term, and of a Lyapunov function for the nominal (non-perturbed) system. A related notion of practical stability is motivated by the idea that a system may be "considered stable if the deviations of motions from the equilibrium remain within certain bounds determined by the physical situation, in case the initial values and the perturbation are bounded by suitable constants". One does not require a more narrow interpretation that the deviation from the origin of $||x(t)||$ can be made arbitrarily small by a suitable choice of the constants, as is the case with total stability. Roughly speaking, for practical stability, we allow that the system will move away from the origin even for small perturbations, and we cannot make that motion arbitrarily small by reducing the model perturbation term.

Another concept of interest in power systems is that of partial stability, introduced already by Lyapunov himself. The basic idea is to relax the condition for stability from one that requires stable behavior from all variables [because of the properties of the norm used in (1.2) and elsewhere] to one that requires such behavior from only some of the variables. This formulation is natural in some engineered systems, and leads to substantial simplifications in others (e. g., in some adaptive systems). It has been used in the context of power system stability as well.

A power system is often modeled as an interconnection of lower-order subsystems, and we may be interested in a hierarchical (two-level) approach to stability determination. At the first step, we analyze the stability of each subsystem separately (i. e., while ignoring the interconnections). In the second step, we combine the results of the first step with information about the interconnections to analyze the stability of the overall system. In a Lyapunov framework, this results in the study of composite Lyapunov functions. An important qualitative result is that if the isolated subsystems are sufficiently stable, compared to the strength of the interconnections, then the overall system is uniformly asymptotically stable at the origin.

C. 2 Input/Output Stability

This approach considers the system description of the form:

$$y=H(u) \tag{1.3}$$

where H is an operator (nonlinear in general) that specifies the q -dimensional output vector y in terms of the m-dimensional input u. The input u belongs to a normed linear space of vector signals L'''_s—e. g. , extended bounded or square integrable signals, meaning that all truncations u_T of such signals (set to zero for $t>T$) are bounded or square integrable (this allows inclusion of "growing" signals like ramps etc. that are of interest in stability analysis).

- Definition: A continuous function α: $[0,\alpha)\rightarrow[0,\infty)$ is said to belong to class **K** if it is strictly increasing and $\alpha(0)=0$.
- Definition: A continuous function β: $[0,\alpha)\times[0,\infty)\rightarrow[0,\infty)$ is said to belong to class **KL** if, for each fixed s, the mapping $\beta(r,s)$ belongs to class **K** with respect to r and, for each fixed r, the mapping $\beta(r,s)$ is decreasing with respect to s and $\beta(r,s)\rightarrow0$ as $s\rightarrow\infty$.
- A mapping H: $\tau_e^m\rightarrow L_e^q$ is L *stable* if there exists a class-**K** function $\alpha(\cdot)$ defined on $[0,\infty)$, and a non-negative constant β such that:

$$||(Hu)||\leqslant\alpha(||u||)+\beta \tag{1.4}$$

for all $u\in L_e^m$ and $\tau\in[0,\infty)$.

It is f inite-gain L stable if there exist non-negative constants α and β such that:

$$||(Hu)_T||\ L\leqslant\gamma(||u_\tau||\Sigma)+\beta \tag{1.5}$$

for all $u\in L_e^m$ and $\tau\in[0,\infty)$.

Note that if L is the space of uniformly bounded signals, then this definition yields the familiar notion of bounded-input, bounded-output stability. The above definitions exclude systems for which inequalities (1. 4) and (1. 5) are defined only for a subset of the input space; this is allowed in the notion of small-signal τ stability, where the norm of the input signals is constrained. Let us consider a nonautonomous system with input:

$$\dot{\tau}=f(t,x,u) \tag{1.6}$$

Note the shift in analytical framework, as in input/output stability inputs are not assumed to be known functions of time, but assumed to be in a known class typically described by a norm.

A system (1. 6) is said to be locally input-to-state stable if there exists a class-**KL** function $\beta(\cdot,\cdot)$, a class-**K** function $\alpha(\cdot)$, and positive constants k_1 and k_2 such that for any initial state $x(t_0)$ with $||x(t_0)||<k_1$ and any input $u(t)$ with $\sup_{t\geqslant t_0}||u(t)||<k_2$, the solution $x(t)$ exists and satisfies:

$$||x(t)||\leqslant\beta(||x(t_0)||,t-t_0)+\alpha\left(\sup_{t_0\leqslant\tau\leqslant t}||u(\tau)||\right) \tag{1.7}$$

for all $t\geqslant t_0\geqslant0$.

It is said to be input-to-state stable if the local input-to-state property holds for the entire input and output spaces, and inequality (1. 7) is satisfied for any initial state $x(t_0)$ and any bounded input $u(t)$. This property is typically established by Lyapunov-type arguments.

Next, we consider the system (1.6) with the output y determined from:

$$y=h(t, x, u) \tag{1.8}$$

where h is again assumed smooth.

A system (1.6) is said to be locally input-to-output stable if there exists a class-**KL** function β, a class-**K** function α, and positive constants k_1 and k_2 such that for any initial state $x(t_0)$ With $||x(t_0)||<k_1$ and any input $u(t)$ with $\sup_{t\geqslant t_0}||u(t)||<k_2$, the solution $x(t)$ exists and the output $y(t)$ satisfies:

$$||y(t)||\leqslant\beta(||x(t_0)||, t-t_0)+\alpha(\sup_{t_0\leqslant\tau\leqslant t}||u(\tau)||) \tag{1.9}$$

for all $t\geqslant t_0\geqslant 0$.

It is said to be input-to-output stable if the local input-to-state property holds for entire input and output spaces, and inequality (1.9) is satisfied for any initial state $x(t_0)$ and any bounded input $u(t)$.

The first term describes the (decreasing) effects of the initial condition, while the function α in the second term bounds the "amplification" of the input through the system. In the case of square-integrable signals, the maximal amplification from a given input to a given output is denoted as the L_2 gain of the system. This gain can be easily calculated, in general, only for linear systems, where it equals the maximal singular value (the supremum of the two-norm of the transfer function evaluated along the imaginary axis, or the H_∞ norm) of the transfer function. One of main goals of control design is then to minimize this gain, if the input represents a disturbance. There exist a number of theorems relating Lyapunov and input-to-output stability, and some of the main tools for establishing input-to-output stability come from the Lyapunov approach. Note, however, that input-to-output stability describes global properties of a system, so in its standard form, it is not suitable for study of individual equilibrium sets. Input-to-output stability results are thus sometimes used in the stability analysis to establish Lyapunov stability results in a global sense. For a more sophisticated use of the input-to-output stability concept, in which the input-output properties are indexed by the operating equilibrium.

C.3　Stability of Linear Systems

The direct ways to establish stability in terms of the preceding definitions are constructive; the long experience with Lyapunov stability offers guidelines for generating candidate Lyapunov functions for various classes of systems, but no general systematic procedures. For the case of power systems, Lyapunov functions are known to exist for simplified models with special features, but again not for many realistic models. Similarly, there are no general constructive methods to establish input-to-output stability using (1.9) for nonlinear systems.

One approach of utmost importance in power engineering practice is then to try to relate stability of a nonlinear system to the properties of a linearized model at a certain operating point. While such results are necessarily local, they are still of great practical interest, especially if the operating point is judiciously selected. This is the method of choice for analytical (as contrasted with simulation-based) software packages used in the power industry today. The precise technical conditions required from the linearization procedure are given. The essence of the approach is that if the linearized system is uniformly asymptotically stable (in the nonautonomous case, where it is equivalent to exponential stability), or if all eigenvalues have negative real parts (in the autonomous case), then the original nonlinear system is also locally stable in the suitable sense. The autonomous system case when some eigenvalues have zero real parts, and others have negative real parts, is covered by the center manifold theory.

In this subsection, we consider a system of the form:

$$\dot{x} = A(t)x(t) \tag{1.10}$$

which is the linearization of (1.1) around the equilibrium at the origin. General stability conditions for the nonautonomous case are given in terms of the state transition matrix $\Phi(t,t_0)$:

$$x(t)=\Phi(t,t_0)x(t_0) \tag{1.11}$$

While such conditions are of little computational value, as it is impossible to derive an analytical expression for $\Phi(t,t_0)$ in the general case, they are of significant conceptual value.

In the case of autonomous systems (i.e., $A(t)=A$: The origin of (1.10) is (globally) asymptotically (exponentially) stable if and only if all eigenvalues of $\boldsymbol{A}$ have negative real parts. The origin is stable if and only if all eigenvalues of $\boldsymbol{A}$ have nonpositive real parts, and in addition, every eigenvalue of $\boldsymbol{A}$ having a zero real part is a simple zero of the minimal characteristic polynomial of $\boldsymbol{A}$.

In the autonomous case, an alternative to calculating eigenvalues of $\boldsymbol{A}$ is to solve a linear Lyapunov Matrix Equation for a positive definite matrix solution; if such solution exists, it corresponds to a quadratic Lyapunov function that establishes stability of the system.

D. Stability Definitions and Power Systems

D.1 Complementarity of Different Approaches

While Lyapunov and input/output approaches to defining system stability have different flavors, they serve complementary roles in stability analysis of power systems. Intuitively speaking, "input/output stability protects against noise disturbances, whereas Lyapunov stability protects against a single impulse-like disturbance".

The ability to select specific equilibrium sets for analysis is a major advantage of the Lyapunov approach; it also connects naturally with studies of bifurcations that have been of great interest in power systems, mostly related to the topic of voltage collapse. Note, however, that standard definitions like (1.2) are not directly applicable, as both the starting set χ_p and the technically viable set Ω_p are difficult to characterize with the norm-type bounds used in (1.2). An attempt to use such bounds would produce results that are too conservative for practical use. For outage of a single element (e.g., transmission line), and assuming a known post-equilibrium set x_p and an autonomous system model, the starting set is a point; for a finite list of different outages of this type, the starting set will be a collection of distinct points. The requirement that such χ_p be "covered" by a norm-type bound is not very suitable, as it would likely include many other disturbances to which the system may not be stable, and which are not of interest to the power system analyst. Note also that partial stability may be very suitable for some system models. In a very straightforward example, we are typically not interested in some states, like generator angles, but only in their differences. The concept of partial stability is hence of fundamental importance in voltage and angle stability studies. In such studies, we focus on a subset of variables describing the power system, and we assume that the disregarded variables will not influence the outcome of the analysis in a significant way. In practice, we tend to use simpler reduced models where the ignored variables do not appear, but conceptually we effectively use partial stability. The other key difficulty in analysis stems from the fact that the construction of Lyapunov functions for detailed power system models, particularly accounting for load models, is still an open question, as we commented earlier. Because of these two reasons, the stability of power systems to large disturbances is typically explored in simulations. Advances in this direction come from improved computer technology and from efficient power system models and algorithms; for a recent review of key issues in power system simulations that are related to stability analysis. In the case of power system models for which there exist energy functions, it is possible to approximate the viable set Ω_p using the so-called BCU method and related ideas; a detailed exposition with geometric and topological emphasis is presented in.

The input/output framework is a natural choice for analysis of some persistent disturbances acting on power systems (e. g. , load variations). Note, however, that conditions like (1. 9) are difficult to establish in a non-local (large signal) setting, and simulations are again the main option available today.

The two approaches of stability coalesce in the case of linear system models. The use of such models is typically justified with the assumed small size of the signals involved. A range of powerful analysis tools (like participation factors) has been developed or adapted to power system models. For noise-type disturbances, an interesting nonstochastic approach to the worst-case analysis is offered by the set-based description of noise detailed. Small signal analyses are a part of standard practice of power system operation today.

D. 2　An Illustration of a Typical Analysis Scenario

In terms of the notation introduced here, a scenario leading toward a blackout is as follows: Following a disturbance, χ_p turns out to be unstable, and the system trajectory passes through $\partial\Omega_p$. After actions of relays and line tripping, the system splits into k_2 (mutually disconnected) components. The post-fault equilibria in each component are $x_p^{2,l}, l = 1, \cdots, k_2$, and some of them again turn out to be unstable, as their boundaries $\partial\Omega_p^{2,l}$ are crossed by corresponding (sub) system trajectories. Note that up to this point k_2+1, stability analyses have been performed. Then the stability assessment process repeats on the third step and so on. In this framework under a "power system," we understand a set of elements that is supplying a given set of loads, and if it becomes disconnected (in graph-theoretic sense) at any point, we have to consider as many newly created power systems as there are connected components.

There exists a point of difference between theorists and practitioners that we want to comment upon here: Stability theorists tend to see a new system after the initial event (e. g. , a line switching), while practitioners tend to keep referring back to the original (pre-disturbance) system. This is because stability limits are specified in terms of pre-disturbance system conditions. While this is typically not a major obstacle, it points out toward the need for a more comprehensive treatment of stability theory for power systems as discussed in this section.

1.1.6 Summary

This report has addressed the issue of stability definition and classification in power systems from a fundamental viewpoint and has examined the practical ramifications of stability phenomena in significant detail. A precise definition of power system stability that is inclusive of all forms is provided. A salient feature of the report is a systematic classification of power system stability, and the identification of different categories of stability behavior. Linkages between power system reliability, security, and stability are also established and discussed. The report also includes a rigorous treatment of definitions and concepts of stability from mathematics and control theory. This material is provided as background information and to establish theoretical connections.

1.2 Computational Intelligence for the Smart Grid-history, Challenges and Opportunities

1.2.1 Introduction

Computational intelligence faces a wide variety of opportunities to help us meet the global need for a more intelligent electric power grid. To meet these opportunities, we also face some important challenges in upgrading the foundations of our field, and really living up to its full promise. At the same time, talk about the "smart grid" has stimulated many people in all fields-computational intelligence, control theory, and many other fields-to try to get rich quick, without really thinking about where we are trying to get to with the power grid or with intelligent systems in general. Many of us have a special responsibility to understand the larger target, first, before deciding what to propose or what to fund from the huge menu of possibilities. This paper will review the larger needs and strategic situation, introducing specifics in the context of the needs that they serve.

1.2.2 Roadmap for the Intelligent Grid-needs and Opportunities

The fourth generation intelligent grid as illustrated in Figure 1.1 has an important role to play in the larger urgent task of humanity to achieve a sustainable global energy system. In the long term, we would want all the decisions made in the power grid-from switching of low level relays and generator controls to global decisions made by Regional Transmission Organizations (RTOs), from millisecond-to-millisecond decisions to control unwanted harmonics through to multiyear planning decisions-to be the best possible set of decisions in some sense, with enough sensor inputs to make it possible to compute the best decisions and enough actuators or control authority to get close to the full potential efficiency of the system. In other words, the total collection of algorithms used all across the system should somehow implement a true intelligent optimal control of the system as a whole, with foresight and adaptation and resilience in coping both with random disturbances and systematic threats from terrorists. It should be a true intelligent system.

This vision is not the same as the usual concept of multiagent systems, in which each individual component of the system is some kind of intelligent agent. From game theory and economics, we know that systems of multiple optimizing agents will converge, at best, to something called a Nash equilibrium. In the general case, a Nash equilibrium will typically be far inferior to any of the best possible outcomes (Pareto optima), unless there is a special effort to design the larger market system or coordination to achieve some kind of collective optimality. That special effort is one of the key defining elements of the fourth generation vision.

Of course, this vision will not become real overnight. The need for these full capabilities was not fully appreciated until recently. Therefore, I will first review earlier visions of the smart grid, and some tangible near-term needs, which can serve as steppingstones to the real thing.

1.2.3 The First Generation Vision

Prior to 1998, the IEEE Power Engineering Society (PES, recently expanded to become the Power and Energy Society) pushed hard for greater national attention to the needs of the power grid. Many engineers argued that the grid is aging rapidly, even as we try to place ever more challenging new demands upon it, such as new nonlinear loads (like computers) which generate harmonics and make it difficult to maintain power quality. They argued that the people needed to maintain that grid are aging even more rapidly-creating a serious workforce crisis, which is more serious now than ever. Bob Thomas of Cornell and of PES argued that we need to revisit the theory of Large-Scale Nonlinear Systems (LSNS), to help us do a better job of minimizing blackouts and cascading outages. In 1986, NSF invited Thomas to come to set up a program in LSNS and in power engineering, starting in

1987, which evolved to become part of the expanded program, EPAS, in existence today.

Many power engineers recognized even then that computational intelligence had a lot to offer in coping with these well-established challenges. For example, Dejan Sobajic of the Electric Power Research Institute (EPRI) recognized the importance of Time-Lagged Recurrent Networks (TLRN) and Back propagation Through Time (BTT) as a tool for prediction and diagnosis in power systems, and did important substantial work in collaboration with Bernie Widrow. Bob Marks and Mohammed El-Sharkawi from PES helped set up the IEEE Neural Networks Council, and applied neural networks to areas like security assessment. An important annual conference was created, Intelligent Systems Applied to Power (ISAP), which included applications of neural networks, fuzzy logic, evolutionary computing and other methods to power engineering. Mo Yuen-Chow and Kwang Lee developed new, breakthrough systems for real-time diagnostics of electric motors and power systems, using TLRN. Curt Lefebvre, who had earlier developed the NeuroDimensions software package with unique capabilities in handling TLRNs, then used these tools in real-world applications to generator control, which gave him a position as founder and for many years president of a new company, NeuCo. He now estimates that these tools are used in 20% of US coal-fired generators.

New third generation and fourth generation concepts aim for a truly intelligent power grid, addressing new requirements for a sustainable global energy system, making full use of new methods for optimization across time, pluggable electric vehicles, renewable energy, storage, distributed intelligence and new neural networks for handling complexity and stochastic challenges.

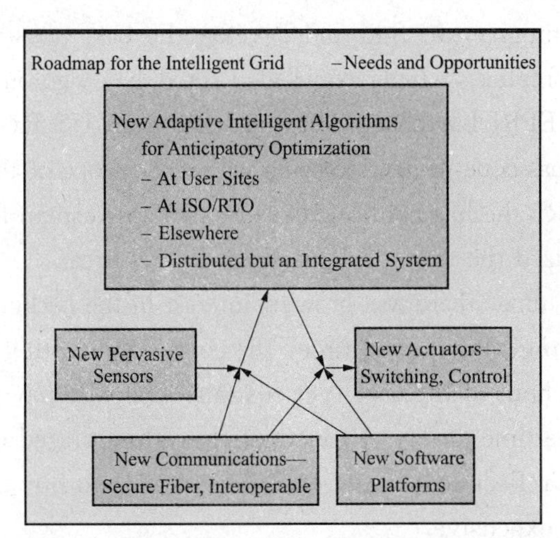

Figure 1.6 Vision of the fourth generation intelligent grid.

At the same time, the power industry itself also recognized many more mundane needs for upgrades, such as the use of automated electricity meters (which did not require human meter readers to go out and measure people's electricity use) and simple sensors and communications to let them know what was actually going on in a complex, diverse grid, with many components dating from before television, let alone the internet.

In essence, the first generation vision of the smart grid is to put lots of money into all of these things—to modernize the grid, to support the kind of research reported at PES and ISAP at a higher level, to build new wires, to install new meters and install some new sensors and communications, to make the existing grid better able to cope with blackouts.

Certainly we need new meters, new wires, new sensors and communications before we can really implement the full fourth generation vision. At the same time, it's important that we not get locked into standards or legacy investments which make it more difficult to move ahead to the fourth generation; for example, overreliance on wireless communications in some parts of the grid may make it actually harder to achieve the security requirements which will become ever more important, as potential adversaries develop access to intelligent systems and the ability to drive by key power installations.

1.2.4 The Second Generation Vision

The next chapter in the story begins, strangely, at a workshop organized by NASA Ames, close to EPRI. A university researcher named Massoud Amin presented an excellent paper on time-lagged recurrent networks, which caught the eye of Dejan Sobajic and others at EPRI. They brought Amin to EPRI, where from 1998 to 2001 he pioneered a new effort on smart grids, and created what I would view as the second generation vision of the smart grid.

The key element here was essentially an extension of the LSNS idea-an effort to go to the control theory community, to find and develop the best possible methods to achieve stable control of this complex system. Amin also acted as a vigorous spokesman for EPRI in this field. He presented EPRI bar charts showing that total US R&D into the power grid proper, of the funding has done to developments within the scope of the first generation smart grid. After leaving EPRI, he has continued to act as a spokesman for this vision.

Amin also popularized the term "self-healing" in this area.

At the same general time, there was growing interest in the background for a key concept called *time of day pricing*. For a long time, the cost of generating electricity has varied a great deal from hour to hour of the day. Yet customers usually pay the same price for electricity, regardless of the time of day. Economists have long urged us to change this situation, so that prices can reflect costs, and encourage people to buy more of their electricity at times when it is less expensive.

This is called "load shifting". Before 2000, most policy analysts felt that Public Utility Commissions (PUCs) would never allow time-of-day pricing, no matter how much it could add to efficiency and to reducing congestion, but this slowly changed.

Another key development in this period was the acceleration of the "deregulation" of the electric power industry, which many called "regulation" or "new market rules". Because electric power is a kind of natural monopoly, it is not realistic to talk about just getting rid of all regulations; however, major efforts began to try to build as much of a market-based decision system as possible. The key idea was to make the generation of electricity a truly open, competitive market, while creating new Independent Systems Operators (ISOs) and Regional Transmission Organizations (RTOs) to make the higher-level decisions controlling the use of transmission systems, subject to continued regulation by PUCs of transmission companies (transcos) and local distribution companies (discos) which take electricity to "the last mile" to small-scale users. Even now, much of the research and policy thinking on power grids does not properly account for the new central realities of the ISOs and RTOs, described very clearly and mathematically in a series of workshops held in 2010 by the Federal Energy Regulatory Commission. Marija Ilic, who ran electric power at NSF in that period, worked hard to mobilize the PES community to perform R&D to help in these transitions. She also contributed heavily to the more recent workshops at FERC, which focused on the most important near-term research challenges related to ISOs and RTOs.

Prior to deregulation, electric utilities were generally allowed to send 1% of their revenues to support research at EPRI, as part of the rate base. Because this was generally lost during deregulation, EPRI went through substantial struggles in later years, especially with regards to its ability to support university-based basic research.

1.2.5 The Third Generation Vision

For part of 2001, I was asked to manage the electric power area at NSF, in addition to the long-standing program in computational intelligence, after Marija's return to university.

At that time, the newspapers were full of headlines about California losing many billions of dollars, and many jobs, due to an electric power crisis. The headlines and the high-level decision makers were full of conventional wisdom saying that nothing could be done to reduce the physical cost of electricity to California in less than three years——but I was skeptical. At a recent ISAP conference in Brazil, people in Cepel ("the EPRI of Brazil") had demonstrated new transmission technology which could have saved California many billions, so far as I could tell. Because Brazil does have a kind of European level of higher education, was investing more in grid R&D than the US, and faced major challenges in the electric power area, they had started to deploy technologies for digital control of power flow (Flexible AC Transmission Systems, FACTS, and also magnetic control technology borrowed from Russia) and developed a new technology "SIL" for pumping more electricity safely on existing lines. To help California with its urgent crisis, I contacted EPRI and proposed that we hold a joint workshop in California, inviting advanced system for that task even today.

This led to a fateful meeting in a hotel in Washington between Fritz Kalhammer (a vice-president of EPRI), Massoud Amin, James Momoh (then at Howard but already selected to take over electric power at NSF as a new rotator), myself, and Edris and one other person from EPRI. When I asked Kalhammer to cosponsor the workshop on transmission, he said yes, on one condition. The condition was that we should also cosponsor a workshop on what he viewed as the biggest unmet need for R & D in that industry——the need for global dynamic optimization of the entire power grid as one system. In today's grid, he said, the engineering is all too narrow and stove-piped. People design independent little pieces based on how they would behave in isolation or under some kind of hypothetical model, but there is nothing to guarantee that the pieces all work smoothly together as a larger system. The pieces should be designed to contribute as much as possible to the larger system——as pieces of the larger system. There should be one larger optimization. "I expect you will tell me that this is a pipe dream, and impossible, just like all the other power engineers I have spoken to about this, but this is what we really need more and more for the system as a whole." He was rather surprised at first that Momoh and I agreed so quickly. But in fact, Momoh had spent years developing the world's most advanced Optimal Power Flow (OPF) system, marketed by EPRI, which is probably the most advanced system for that task even today. OPF performs a truly global optimization of the grid already, integrating all kinds of local decisions-but on a single time-slice, as a kind of static optimization. And I myself had spent many years pioneering the new area of adaptive, approximate dynamic programming (ADP), to address the general problem of optimization across multiple time periods, with foresight and learning, in the face of nonlinearity, random disturbance, and complexity, such as what mammal brains must be able to cope with.

So we reached agreement very quickly.

The first workshop, held in October 2001 (one month after 9/11), was an unusual experience. It was chaired by ChenChing Liu, well-known in ISAP for his work applying fuzzy logic to power systems. The Brazilians presented extremely impressive detailed plans for how to insert their technology into the Western power grid, which could have stopped the bleeding as soon as six weeks after start of work. The SIL technology would have allowed a quick upgrade of existing power lines between California and the Rockies, allowing underutililized coal plants to sell California low-cost electricity without raising costs or supply in other areas. But high-level political appointees in other agencies of the federal government would not come, because they felt insulted about the whole idea that the US could learn something from Brazil. Speakers for the major ISOs did come, and agreed that the plans from Brazil would be workable immediately-but asserted that the market rules then in place made it impossible for anyone in the US to pay for the upgrades. First, they said, we need the research to change the market rules to empower someone to do the work. EPRI representatives made convincing presentations that the crisis was costing California on the order of $20 billion per year, not really ending "after the crisis". They also quoted major corporations who said they would start outsourcing many, many jobs from Silicon Valley if electricity supply were not fixed up; it wasn't, and they did. The costs to the US economy, and to the funding of California state education, may be substantial to this day.

The second pair of workshops, held back-to-back in April of 2002 in Playa Del Carmen in Mexico (with cosponsorship from Conacyt of Mexico), were far more encouraging. The first workshop focused on global dynamic optimization of the grid, drawing mainly on people active in PES or ISAP. The second workshop focused on algorithms for global dynamic optimization in general. It was a great warning that some people still asked: "Global dynamic optimization of the grid, and algorithms for global dynamic optimization-what relation could those very different areas possibly have to each other?"

These workshops led to the third generation vision for an intelligent grid, described in the chapters by Momoh and myself. The key idea was not to simply get rid of the existing OPF methods, but to augment OPF by training and adding a value function or critic network. Advanced neural networks would be used to approximate the value function, because of the superior function approximation abilities of traditional multilayer perceptrons (MLPS) and because we know that neural networks in brain can handle much greater spatial complexity than MLPs. OPF already inputs a measure of present utility as part of the optimization; in dynamic stochastic OPF (DSOPF), it would input the same utility measure plus a value measure representing the future. That neural network critic network could be initialized as something already meaningful to the power grid, such as ElSharkawi's trained neural network representing the degree of network security. In addition to upgrading OPF, the measures of value λ, output by OPF would be transmitted as price signals throughout the grid. They would be used to provide a kind of dynamic measure of price, superior to the static measures of locational marginal price in common use today.

Even now, much of the research and policy thinking on power grids does not properly account for the new central realities of the ISOs and RTOs, described very clearly and mathematically in a series of workshops held in 2010 by the Federal Energy Regulatory Commission.

At the first of these workshops, Venayagamoorthy also presented new empirical laboratory results on the new controller for turbogenerators he had developed with Wunsch and Harley, based on Dual Heuristic Programming (DHP). He showed how it could withstand disturbances three times as large as those which would end up shutting down turbogenerators controlled even by the most capable alternative controllers in use today. This began a collaboration with the "EPRI of Mexico" in Cuernavaca, which may be able to deploy this kind of new technology more easily than the more conservative systems in the US. Note the control of individual turbogenerators requires training a value function of only a dozen variables or so, which can be done relatively easily by online learning or particle swarm optimization or the like with MLPs; more complex systems, like wide-area control, require moving up to more powerful neural networks like Object Networks, and returning to more biologically plausible local learning rules like modulated backpropagation.

Though ADP has finally become more popular in mainstream control theory and operations research lately, some of the recent work in control theory has neglected the stochastic case. These applications require returning to the general stochastic case and to model-based approaches (even if the models themselves are a hybrid of neural networks and first principles models) to cope with the complexity of the systems. Of course, the same is required to fully understand similar capabilities in the mammal brain.

At one NSF workshop, a speaker working in classical linear robust control once said: "If all you do is maximize value added or profits, you will lose the reliability of the system. That is hard enough by itself to achieve. You are too optimistic. We prefer that the system should be based on solid theorems proving absolute, unconditional stability. We will not use anything else in the real world." The power company executive who had been funding his work got up and said: "If you have no room for value added, we have no room for you." My reply was perhaps more moderate: "It is you who is overly optimistic. If you think you can offer 100% ironclad guarantees that blackouts will never occur in the power system, you are not really addressing the real world. You are relying too much on imperfect models. With the uncertainties and nonlinearities that we face in the real world, the best that we could do is to minimize the probability of a blackout, or minimize the expected value of the damage due to blackouts. That is an optimization problem, which ADP addresses head-on, as much as we possibly can. I would call this resilient control, as opposed to robust control. But in fact, as economists would tell us, the proper utility function should also include a term to account for value added. There is some value to quality of service or insurance against blackouts, but in order to achieve a Pareto optimum between these concerns and concerns about cost or value added we need to formulate utility functions which account for both of them."

There is a still a role for simple common-sense reliability rules, of course, but the third generation grid would converge to an intelligent balance between the competing goals here. Research innonlinear robust control theory tells us that the most robust controller in the general case is a controller "which solves the Hamilton-Jacobi-Bellman equation" that is exactly what ADP does, as accurately as we know how to do. There is room for more research to improve our arsenal of general- purpose ADP algorithms, but ADP is exactly that family of methods which exploits adaptation, learning and approximation to "solve the Hamilton-Jacobi Bellman" equation as effectively as possible.

1.2.6 The Fourth Generation Vision

The fourth generation vision has crystallized out from many discussions of global energy needs and real-world markets since 2001. Though I presented an early version of that vision in 2009, this paper itself is perhaps the most complete and reliable statement of that vision to date.

The main objective in the fourth generation grid is to help humanity as much as we can, to overcome energy problems which threaten its very existence.

A. The Vision for Cars and the Grid

The first and most urgent of these problems is the growing dependence on fossil oil, a resource in finite supply which creates severe near-term risks of unmanageable conflict and economic shocks. The technology already exists which would let us become totally independent of the need to use fossil oil: the technology of GEM-fuel flexible plug-in hybrid electric vehicle cars (PHEV). ["GEM" refers to gasoline/ethanol/ methanol-more precisely, the ability to use gasoline, blended (E85) ethanol, blended (M85) methanol, or any combination of the three as a fuel, without requiring the driver to flip any kind of switch when switching between them.]

One key goal of the fourth generation grid is to maximize our ability to use and afford these kinds of GEM-PHEVs, especially in case of a sudden oil shock. Computational intelligence can play a crucial role here, not only at the grid level, but also within the cars themselves——if it is used in partnership with other key technologies.

For example, in 2008, Danil Prokhorov of Toyota showed how use of neural network control could improve the mileage of the Prius hybrid by 15% without increasing the cost of the vehicle at all. This is a huge increase, by automotive industry standards. This made heavy use of control by TLRNs, which is also seen as an important core technology in work from Ford, Siemens, and other participants in the IEEE CIS task force on alternative energy. It is very unfortunate that Bernie Widrow's classic example of the truck backer-upper (a TLRN cleverly trained with BTT) has not been so widely used in university courses as it deserves to be.

More recently, there has been a major scare claiming that a scarcity of rare earth materials, required in the permanent magnet motors of hybrid cars, would limit our ability to shift to PHEVs. IEEE has recently coined the more general term "PEV" (pluggable electric vehicles) to include PHEVs, pure electric vehicles (EV) and fuel cell cars which can be plugged in. The scarcity of rare earths would threaten all of these PEVs, and conventional hybrid cars as well. But it turns out that two alternative types of electric motors——induction motors IM and switched reluctance motors SRM——do not require rare earths, and actually allow greater average efficiency across the entire driving cycle, if a resilient enough controller can be found for this very challenging nonlinear control problem. (Harley——one of the few really front-line experts in such motors in the US——has reported that SRMs and IMs are about equally good here.) Toyota has recently reported that, thanks to breakthroughs in control, they at least will no longer need the rare earths in the main motor (the traction motor) of their cars. Intelligent nonlinear control can make this more widely available.

When I asked Kalhammer to cosponsor the workshop on transmission, he said yes, on one condition. The condition was that we should also cosponsor a workshop on what he viewed as the biggest unmet need for R&D in that industry——the need for global dynamic optimization of the entire power grid as one system.

Major automobile companies already have control chips able to handle GEM flexibility at fairly low cost——but full use of optimal adaptive methods could allow cars to really optimize performance on the fly as fuel mixes vary, and to use "virtual sensing" to reduce sensor hardware cost. TLRNs and ADP are the two key technologies needed here, in partnership with domain experts. Work by Sarangapani demonstrating efficiency improvements by use of ADP on Otto and diesel engines may be a useful first step towards this goal.

Even more important to PHEVs is the goal of reducing the cost of batteries and power electronics, which ends up requiring longer battery lifetimes and flexibility in getting full use of new types of power electronics. Battery manufacturers have told me that lack general, flexible battery management systems good enough for use in cars is the main obstacle to companies like GM getting full use of the lowest-cost effective batteries now available——let alone starting the use of new types of batteries. Ordinary ADP and TLRNs should do well enough, with the right data input and variables, for overall control of batteries——but to control individual battery cells better, and handle new switching degrees of freedom within batteries (as explored, for example, by Song Ci), might require use of Object Nets, because of the complexity and network properties of the task. For power electronics, new NSF-funded work by Khaligh and Emadi has demonstrated that new power chips and high-frequency designs (http:// hybrid. iit. edu) could halve the cost of power electronics for advanced PHEVs like the Chevrolet Volt, while adding a fast recharge capability able to plug into standard 480 volts AC "level three" recharging as defined in the National Electric Code. This opens up new opportunities for control, and also makes a radical change in what the future grid might look like, since 480 volts AC is much easier for the grid to provide than the fast DC recharge stations now being deployed in the US and Japan.

Many believe that the primary obstacle to greater use of PHEVs occurs at the local level of the power grid. There, too, the technology of high frequency power conversion may play a crucial role. The demonstrations of Khaligh and Emadi basically tell us that we could use the new chip-based technology at the local distribution level as well, but we do not yet know whether costs would be reduced enough to justify a massie replacement of old transformers with new (higher rated) chip-based power converters. The benefit of the new type of converters may depend on our ability to actually exploit the new switching capabilities they require, which requires that they be paired with more intelligent control and a smarter distribution grid. In short, greater intelligvence may be a crucial part of the essential transformation here. This new switchability would also allow better defense against serious near-term threats like massive solar storms.

The results in Germany show that the potential contribution of demand response to load shifting is far greater than traditional field or elasticity studies reveal, because new technologies and automation allow demand to respond more intelligently to price in the future than it can at present.

B. The Vision for Renewables and Peak Shaving

The most crucial advantage of DSOPF over traditional OPF and static optimization methods concerns foresight——the ability to juggle supply and demand across time, in the face of uncertainty.

The balancing of supply and demand across time is already a major economic issue for power grids. For example, a large part of the cost of transmission systems is the cost of wires designed to handle the rare times of peak load, such as air conditioning at noon in the hottest few days of the summer. Even today, a better balancing across time could lead to substantial savings, and less need to build new wires (or less risk of blackouts with a given level of buildout).

One key feature of the fourth generation grid will be many new ways to do time-shifting of the traditional sources and generators: (1) better demand response——ability of loads to adapt as prices change over the day; (2) better "ramping" ability to change generation levels efficiently, as in the talk by Alstom at the FERC workshops; and (3) more storage in the grid, in part perhaps because of PEVs hooked up to the grid (as in "vehicle to grid" technology, V2G) but also because of greater use of new batteries——both central and distributed——as well as compressed air storage and more pumped hydro. (Note that I do not include hydrogen in this context, even though it is a very efficient way to consume taxpayer dollars, and a wonderful fuel for reusable rocketplanes, which, if done right, would add energy from space to the menu of affordable renewable energy options.) To take full advantage of all these new degrees of freedom, we need to be able to perform better optimization across time. To account for uncertainties and unpredictable events as part of this optimization, there is essentially no alternative to the development of more powerful ADP systems, at all levels of the grid.

Among the key uncertainties, of course, is uncertainty about when the wind will blow, as well as uncertainties about clouds floating over solar farms, rooftop PVs and uncertainties in load.

Very sophisticated optimization methods are already in use at ISOs and RTOs, for all the different traditional time scales used in electric power, from regulation to planning. Thus it is already apparent that better capabilities and use for ADP, on the scale of complexity encountered in power grids, could be of great assistance in what ISOs and RTOs already do. This is already under discussion in the FERC community, but benefits exist with large logistics systems where similar methods may be used.

Many US researchers have argued that household demand for electricity can be shifted a few seconds or minutes, but not enough to make major changes in our ability to use renewable energy sources like wind, which require that load shifting from one time of day to another or more. Studies prove that this is true, for more conventional types of load shifting or demand response. However, a combination of simulation studies and field studies in Germany using a new software platform "OGEMA" have demonstrated that massive load-shifting can be achieved in a system which allows intelligent agents to be inserted both at the grid level and at the household level. The true fourth generation grid would insert true intelligent systems (based on ADP) as services at both levels, designed so that the combination of these "services" itself is an implementation of a larger virtual ADP decision-making system including both levels. The results in Germany show that the potential contribution of demand response to load shifting is far greater than traditional field or elasticity studies reveal, because new technologies and automation allow demand to respond more intelligently to price in the future than it can at present.

Of course, the optimization to be performed here would actually be a multicriterion optimization, respecting the right of individual household to specify the parameters of their respective parts of the greater utility function. Even today's OPF systems do this kind of thing implicitly. The development of Pareto optimality theorems and stability results for multiplayer ADP systems is one of the useful potential areas for future research here, probably requiring collaboration of economists and engineers, as with other aspects of market design. This should be a very viable line of research, since Dynamic Stochastic General Equilibrium theory and DSOPF have many common assumptions, and the "lambda" vectors common to all three of them are essentially the same vectors.

Another aspect of this challenge is to expand the intelligent optimization to also include the many new degrees of freedom offered by recent breakthroughs in the technology for controlling flows of electricity in the grid as well as the new sensor information resulting from NSF-funded research on phasors and other relevant types of sensors. It has been estimated that effective use of such new technology could cut the cost of transmission lines for renewable energy in half. These kinds of cost reductions, combined with other essential breakthroughs and R&D in energy technology (also within the scope of EPAS funding), could make it possible for humanity as a whole to transition to renewable energy without nuclear proliferation risks, and without paying more for electricity than we do today.

Greater use of civilian nuclear fission power in developing nations could result in a massive increase in the availability of sensitive materials and technologies, even to substate actors; the renewable path would be far safer, if we can also make it more affordable.

McElroy and his collaborators have estimated that the onshore wind resources of the US (and other nations) are many times larger than their entire electricity demand, and should cost only 6.7 cents per kwh for generation as such; if we can make deep reductions in the additional costs due to the difficulty of using 80% wind on the power grid, this would already allow us to make a massive transition at an affordable cost. Many of us believe that solar farms, especially with new energy conversion systems for solar thermal power, have the potential to be even more reliable and less expensive. Energy from space, such as space solar power or laser-induced deuterium-deuterium fusion in space, also looks likely to be a lowcost energy source, if necessary work on lower cost reusable access to space could be initiated. These three sources, between them, are very likely to be able to meet all the needs of the earth at an affordable cost, sooner than we expect——if we develop a power grid fully able to use them.

1.2.7 Summary and Conclusions

Enormous investments are now being made to upgrade electric power grids, and to implement the first and second generation visions of the smart grid. This paper has provided a roadmap for reaching a fourth-generation power grid, which would build on those kinds of investments, and would use intelligent system-wide optimization to allow up to 80% of electricity to come from renewable sources and 80% of cars to be pluggable electric vehicles (PEV) without compromising reliability, and at minimum cost to the world economy. It is one of the crucial elements of a global strategy to address urgent issues of sustainability in economic growth and progress which appear to be a matter of life or death for humanity as a whole.

1.3 A New Implementation Method of Wavelet-Packet-Transform Differential Protection for Power Transformers

1.3.1 Introduction

There are many types of disturbances that may be experienced in different elements of a power system. The electromagnetic-energy storage nature of such elements may produce oscillatory disturbances with complex characteristics, such as nonperiodic, nonstationary, short-duration, fast-decaying, impulse-superimposed, and/or high-frequency components. Magnetizing inrush current and various internal-fault currents in power transformers are among the oscillatory disturbances that initiate high currents with the aforementioned features. There have been many proposed techniques to classify the various currents flowing through any power transformer in order to develop an accurate, efficient, and reliable transformer protection. Accurate classification of currents in a power transformer is considered the key requirement for preventing maloperation of the protective equipment under different nonfault conditions, including magnetizing inrush current, through-fault current, saturation current of current transformers (CTs), tap changers, ratio mismatch, currents in the grounding impedance, etc. This requirement for power-transformer protection has become a main challenge due to the need for accurate, fast, and reliable differentiation between internal-fault and magnetizing inrush currents so that a proper action can be carried out.

Recently, wavelet-based techniques have been used for the differential protection of power transformers. A wavelet packet transform (WPT)-based disturbance detector and classifier was developed and tested on two different three-phase power transformers for different disturbances under different loading conditions and grounding arrangements. This type of digital relays is designed using a two-level wavelet multiresolution analysis to extract the second high-frequency subband coefficients of the differential current.

The WPT-based differential relays have demonstrated accurate, reliable, and fast responses to fault currents without depending on transformer parameters, loading conditions, grounding arrangements, or core type. The WPT-based differential relay is implemented using a digital signal processing (DSP) board for carrying out the needed discrete filtering functions. In most cases, digital differential relays may not have such a signal-processing circuitry. Moreover, the employed DSP boards, along with their auxiliary circuits, require a voltage supply that may be connected to the system and can be affected if the system voltage suffers from any voltage disruptions. Furthermore, the DSP and digital circuitry required for the digital filters (DFs) may be sensitive to the magnetic fields created by power transmission lines and/or the protected power transformer. In addition, in some microgrid power systems, digital relays may be considered as expensive protective devices. This paper aims to develop and test a WPT-based differential relay for protecting power transformers using Butterworth passive (BP) filters. The BP filters are designed to extract the second-level details consisting of high-frequency components of the three-phase differential current in order to detect and diagnose fault currents. The main reason for selecting BP filters is their inherent capabilities to provide monotonic and ripple-free magnitude responses and their ability to provide an accurate approximation of the WPT-associated DFs.

1.3.2 WPT-Based Disturbance Detection and Classification

A. WPT

WPT is one type of wavelet-based signal processing that offers a detailed localized frequency-time analysis of discretetime (DT) signals. This analysis is obtained as a result of successive time localization of frequency subbands generated by a tree of low-pass and high-pass filtering operations. As a result, the frequency resolution becomes higher, while the time resolution is reduced with increase in the number of used filter banks. Starting with the DT signal $x_d[n]$ of length N, the first level $j=1$ decomposition produces two subband DT signals

$$a^1[n]=\sum_{k=0}^{N-1}g[k]x_d[n-k] \tag{1.12}$$

$$d^1[n]=\sum_{k=0}^{N-1}h[k]x_d[n-k] \tag{1.13}$$

where $a^1[n]$ is the first-level approximations, $d^1[n]$ is the first-level details, k is an integer, and $g[n]$ and $h[n]$ are the low-pass filter (LPF) and the high-pass filter (HPF) associated with the used wavelet function, respectively. In order to increase the frequency resolution and ensure the time localization of each frequency subband, the outputs of both the LPF and HPF are downsampled by two at the end of each filtering stage. The second-level $j = 2$ decomposition produces four subbands

$$aa^2[n] = \sum_{k=0}^{\frac{N}{2}-1} g[k]a^1\left[\frac{N}{2}-k\right] \tag{1.14}$$

$$ad^2[n] = \sum_{k=0}^{\frac{N}{2}-1} h[k]a^1\left[\frac{N}{2}-k\right] \tag{1.15}$$

$$da^2[n] = \sum_{k=0}^{\frac{N}{2}-1} h[k]d^1\left[\frac{N}{2}-k\right] \tag{1.16}$$

$$dd^2[n] = \sum_{k=0}^{\frac{N}{2}-1} h[k]d^1\left[\frac{N}{2}-k\right] \tag{1.17}$$

where $dd^2[n]$ represents the highest frequency subband of the second-level of the WPT decomposition. Figure 1.7 shows the successive LPF and HPF stages that implement the WPT decomposition. It is to be noted that $G(\omega)$ and $H(\omega)$ are the discrete Fourier transforms (DFTs) of $g[n]$ and $h[n]$ as

$$G(\omega)\ \underleftrightarrow{DFT} g[n]$$

$$H(\omega)\ \underleftrightarrow{DFT} h[n]$$

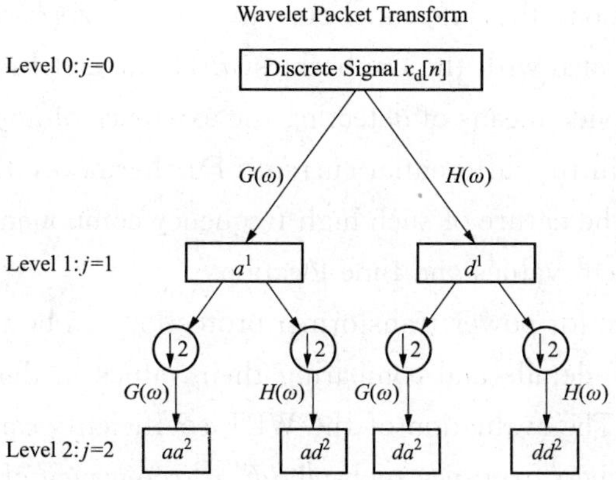

Figure 1.7 Decomposing a discrete signal $x_d[n]$ using a two-level WPT.

The coefficient of $g[n]$ and $h[n]$ are related to each other by the following relation:

$$h[k] = (-1)^k g[L-k],\ 0 \geqslant k \geqslant L-1 \tag{1.18}$$

where L is the length of $g[n]$. In general, extracting frequency subbands of any signal using stages of cascaded filter banks can be implemented using two possible structures that can be stated as follows:

1) The filters of each stage have different cutoff frequencies and bandwidths, while the processed signal is kept unchanged.

2) The filters of each stage have the same cutoff frequencies and bandwidths, while the processed signal is downsampled at the end of each stage to increase the frequency resolution.

The first filter-bank structure includes different filters, which can be difficult to implement and to realize practical applications. On the other hand, the second filter-bank structure can be realized using digital quadrature mirror filters (QMFs). The QMF banks are associated with wavelet functions and are mostly constructed using half-band DFs. These DFs have coefficients determined by solving refinement equations

$$\phi(t) = \sum_{k=0}^{N-1} g_{\phi}[k]\phi(2t-k) \tag{1.19}$$

$$\psi(t) = \sum_{k=0}^{N-1} g_{\phi}[k]\phi(2t-k) \tag{1.20}$$

where $\phi(t)$ and $\psi(t)$ are the scaling and wavelet functions, respectively.

B. WPT-Based Transformer Differential Protection

The WPT-based disturbance detection and classification are realized through evaluating the frequency-subband coefficients. The existence and the time location of the WPT coefficients can provide the needed signatures of the analyzed signal, and such signatures can be used to identify the time-frequency structure of that signal. As the number of levels of resolution increases, the number of frequency subbands increases, offering better, accurate, and more detailed representation of the extracted frequency subbands. In the case of transformer differential protection, the differential currents are decomposed using a two-level WPT with HPF and LPF associated with the Daubechies(*db*4) wavelet function. The first level of resolution is able to provide means of detecting the existence of any transient high-frequency components present in the differential current. Furthermore, the second level can provide information about the nature of such high-frequency components according to their frequency-subband coefficient values and time location.

The WPT algorithm for power-transformer protection can be realized by evaluating the coefficients of the WPT details and comparing their values in the second-level highest frequency subband to zero. The evaluation of the WPT coefficients can be achieved by extracting the second-level highest frequency subband $dd^2[n]$ component of the transformer differential currents. The required digital HPFs for extracting the second-level highest frequency subband $dd^2[n]$ component are separated by a down sampling stage to increase the frequency resolution. Figure 1.8 shows the magnitude of the frequency responses of the two digital HPFs associated with the *db*4 wavelet function which are used for the WPT-based transformer differential protection. It is to be noted that the phase response of $H_{\varphi}(\omega)$ has almost no impact on the performance of the WPT-based transformer differential protection due to the fact that only the magnitude of the $dd^2[n]$ component is required to classify the type of the analyzed differential current.

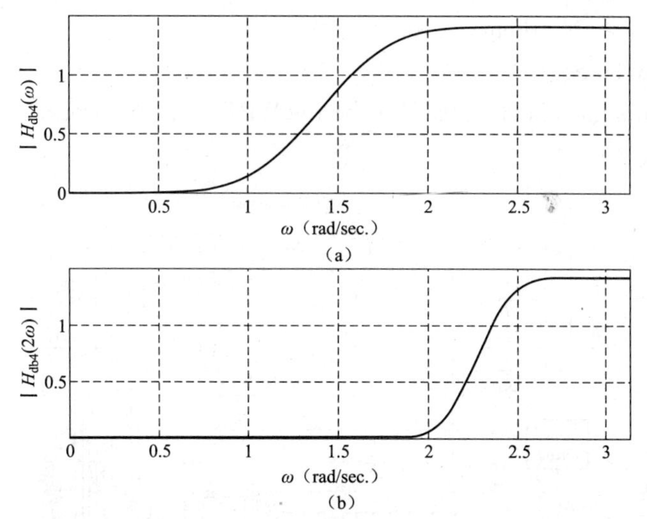

Figure 1.8 Magnitude responses of the cascaded digital HPFs associated with the db4 wavelet function: (a) $|H_{db4}(\omega)|$ of the first-level ($j=1$); (b) $|H_{db4}(2\omega)|$ of the second-level ($j=2$).

The WPT-based transformer differential protection can be implemented using the following steps.

- *Step* 1: Initialize the samples' counter $n=0$.
- *Step* 2: Read one sample from each of the three-phase differential currents I_a, I_b, and I_c and store these samples in three vectors of length N as $I_{da}[N]$, $I_{db}[N]$, and $I_{dc}[N]$.
- *Step* 3: Perform the first-level high-pass filtering for $I_{da}[N]$, $I_{db}[N]$, and $I_{dc}[N]$ (disturbance detection) as

$$d'_a[n]=\sum_{k=0}^{N-1}h[k]I_{da}(N-k) \tag{1.21}$$

$$d'_b[n]=\sum_{k=0}^{N-1}h[k]I_{db}(N-k) \tag{1.22}$$

$$d'_c[n]=\sum_{k=0}^{N-1}h[k]I_{dc}(N-k) \tag{1.23}$$

- *Step* 4: Downsample $d'_a[n]$, $d'_b[n]$, and $d'_c[n]$ by two.
- *Step* 5: Perform the second-level high-pass filtering for the down sampled by two versions of $d'_a[n]$, $d'_b[n]$ and $d'_c[n]$ (disturbance classification) to determine dd^2 using (1.17).
- *Step* 6: Declare FAULT and initiate a trip signal.

if

$$\left|\sum_{k=0}^{\frac{N}{2}-1}dd_a^2[k]\right|>0 or \left|\sum_{k=0}^{\frac{N}{2}-1}dd_b^2[k]\right|>0 or \left|\sum_{k=0}^{\frac{N}{2}-1}dd_c^2[k]\right|>0$$

else

$$n = n + 1.$$

- *Step* 7: If $n \geqslant N$, then $n = 0$.
- *Step* 8: Go To *Step* 2.

Figure 1.9 shows the schematic diagram for the WPT-based transformer differential protective relay.

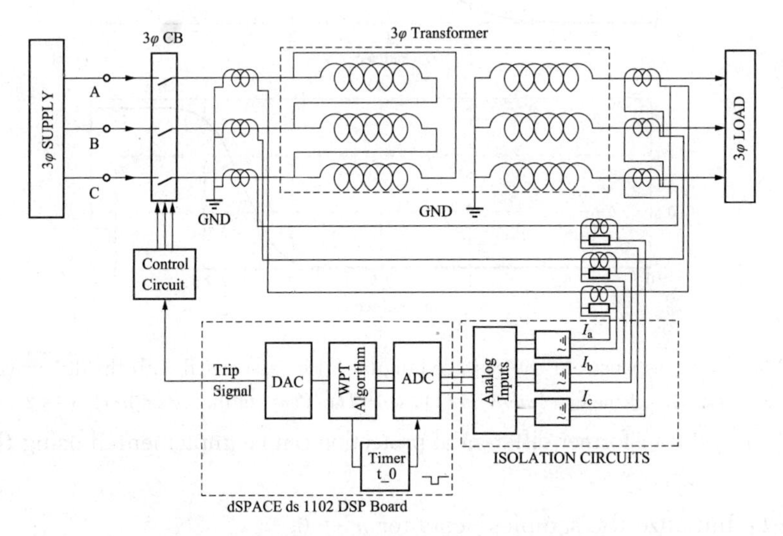

Figure 1.9 Schematic diagram for the WPT-based transformer differential protective relay.

The WPT-based differential relay is supplied by three-phase differential currents in order to detect and classify disturbances that may result from load changes, inrush currents, through faults, and internal faults. These detection and classification functions are carried through the successive high-pass filtering of the sampled versions of the three-phase differential currents. These DFs require digital circuitry for their implementation, which is supplied from the power-system line voltage. As a result, disturbances that affect the line voltage can cause severe impacts on the overall functionality of the WPT-based transformer protective relays. As an alternative, wavelet DF functions can be approximated using passive *LC* elements. Such approximate implementation can increase the reliability and security of the WPT-based transformer protective relays. Moreover, it can extend the applications of such transformer protective relays to various remote locations of the power systems, as well as to microgrid and renewable-energy supplies.

1.3.3 Design of BP HPFs to Implement the WPT

The main function of the WPT-based transformer differential protection is to detect and distinguish between fault and nonfault currents flowing in the protected transformer. This main function is achieved by extracting the coefficients of the second level high frequency subband $dd^2[n]$ present in the three-phase differential currents using two stages of digital

HPFs separated by a downsampling step. The used digital HPFs have the same cutoff frequency and bandwidths, where the downsampling step increases the frequency resolution to facilitate the time localization of the extracted frequency subbands. This digital filtering process can be implemented using BP HPFs under the following constraints.

1) The cutoff frequency of the first filtering stage HPF is set to the system frequency. This selection is made to ensure that only high-frequency components of the differential currents will be passed to the second filtering stage. Furthermore, normal currents (even with high magnitudes) will not be passed to the second filtering stage.

2) The cutoff frequency of the second filtering stage HPF is set to half the cutoff frequency of the first stage HPF. This selection is made to account for the downsampling by-two stage. It is to be noted that downsampling by two increases the frequency resolution by two, which can be realized by narrowing the passband of the second stage HPF to half.

The passive HPFs can be designed using the Butterworth filters. The selection of the Butterworth method is mainly due to the monotonic magnitude response with no ripples in both the passband and the stopband. Also, Butterworth filters have a maximally flat passband magnitude response and contain moderate group delays. The design of a Butterworth HPF can be carried out by specifying the cutoff frequency (−3 dB frequency) Ω_c. The transfer function of a Butterworth filter is

$$A(s) = \frac{1}{1+\left(\frac{-s^2}{\Omega_c^2}\right)^p} \tag{1.24}$$

where p is the order of the filter. In order to simplify the implementation of the WPT-based transformer differential protection, the order of the desired Butterworth HPFs is designed with $p = 3$. The locations of the poles of a Butterworth filter can be determined using the following relation

$$s_k = \Omega_c e^{j\frac{\pi}{2}} e^{j\frac{2k+1}{2p}\pi},\ k = 0,1,2,\cdots,p-1. \tag{1.25}$$

For $p = 3$ and $\Omega_c = 120\pi$ rad/s, the first-stage BP HPF filter has the following poles

$$s_0 = -188.5 + \mathrm{j}\ 326.48$$

$$s_1 = -377$$

$$s_2 = -188.5 - \mathrm{j}\ 326.48$$

For $p = 3$ and $\Omega_c = 60\pi$ rad/s, the second-stage BP HPF filter has its poles as

$$s_0 = -94.25 + \mathrm{j}\ 163.24$$

$$s_1 = -188.5$$

$$s_2 = -94.25 - \mathrm{j}\ 163.24$$

Figure 1.10 shows the magnitude responses of the designed BP HPFs. The designed BP HPFs can be realized using inductors and capacitors. The prototype values (values for a cutoff frequency of Ω_c for $p = 1$) are determined using the following relations

$$C_j = 2\sin\left(\frac{2j-1}{2p}\pi\right), j = 1,3,\cdots,p \tag{1.26}$$

$$L_j = 2\sin\left(\frac{2j-1}{2p}\pi\right), j = 2,4,\cdots,p \tag{1.27}$$

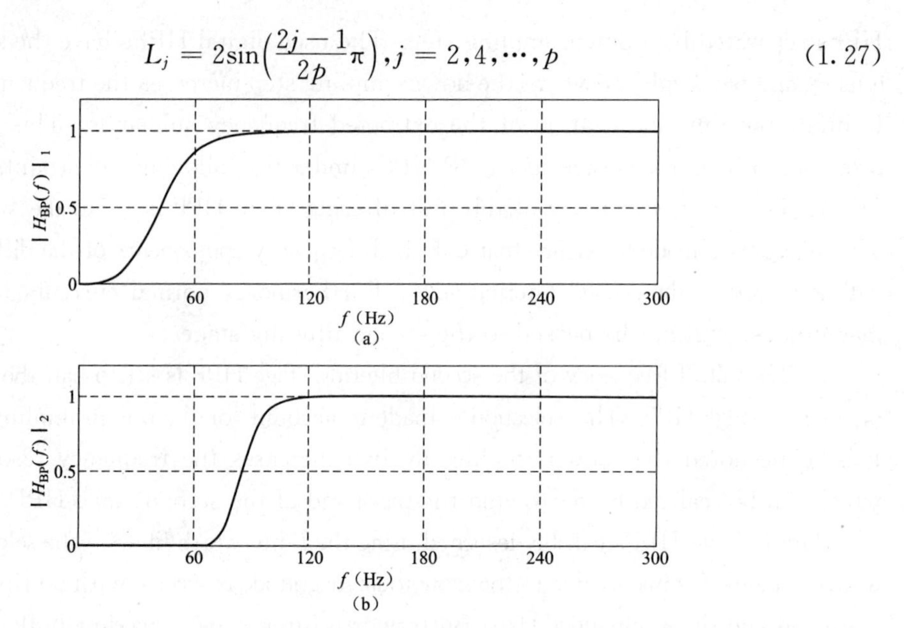

Figure 1.10 Magnitude responses of the designed BP HPFs for two cascaded stages of HPFs: (a) $|H_{BP}(\Omega)|_1$ with a cutoff frequency of 30 Hz; (b) $|H_{BP}(\Omega)|_2$ with a cutoff frequency of 60 Hz.

For $p = 3$, the prototype values for the required elements are calculated as $C_1 = 1$, $C_3 = 1$, and $L_2 = 1.414$. The actual values of the filter elements are determined by frequency and impedance scaling of the prototype values as

$$C = \frac{C_j}{\Omega_c R_L} \tag{1.28}$$

$$L = \frac{L_j R_L}{\Omega_c} \tag{1.29}$$

where R_L is load resistance that is 40 Ω. The actual values of the first-stage HPF ($\Omega_c = 120\pi$ rad/s) elements are $C_1 = C_3 = 66.32$ μF and $L_2 = 150$ mH. Also, the actual values of the second-stage HPF ($\Omega_c = 60\pi$ rad/s) elements are $C_1 = C_3 = 132.36$ μF and $L_2 = 300$ mH. Figure 1.11 shows the circuit diagrams of the designed BP filters.

The designed stages of BP HPFs can approximate the responses of the WPT-associated DFs in extracting the secondlevel high-frequency subband $dd^2[n]$ present in the differential current in order to distinguish fault currents. This approximation is evident from the cutoff frequencies of both types of filters, as well as the ripple-free magnitude responses, as shown in Figure 1.8 and Figure 1.10.

1.3.4 Offline Simulation of BP Filter WPT-Based Transformer Protection

The proposed BP filters for implementing the WPT-based differential protection is to be simulated using data collected from a three-phase(3φ) 5kVA 230/550-575-600 V 60Hz core-type Δ-Y laboratory prototype power transformer. The collection of several differential currents is carried out using the procedure detailed. Many currents are investigated for the

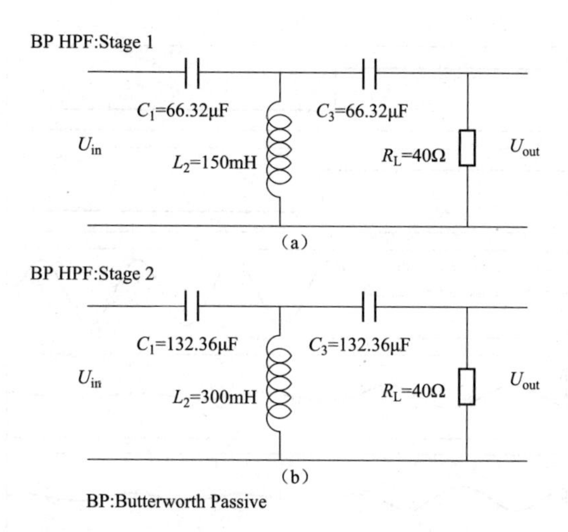

Figure 1. 11 Circuit diagrams of the designed BP filters: (a) The Stage 1 HPF; (b) The stage 2 HPFs.

purposes of offline testing the performance of the proposed differential protection. It is to be noted that in order to demonstrate the accuracy of the BP-filter-based implementation of the WPT, the collected data are tested using the DF-implemented WPT. The following cases are presented in this paper:

1) Fault currents: line-to-ground fault, loaded.

2) Magnetizing inrush current, unloaded.

It is noted that in some of the investigated currents, the transformer is not loaded, which provides the worst conditions of operation. The investigated cases are presented in the following sections.

Case 1—Line-to-Ground Fault: The line-to-ground fault (phase A-to-ground) has occurred on the primary side of the 3φ 5-kVA power transformer while supplying a load. Figure 1. 12 shows the three-phase differential currents I_a, I_b, and I_c and the responses (trip signals) of the DF-based WPT and proposed BP-filter-based WPT. The BP-filter-based WPT trip signal changed its status from high to low, indicating that an internal fault had been detected. Also, the DF-based WPT trip signal changed its status from high to low, indicating that an internal fault had been detected. Both trip signals had been initiated with small delay between them, as shown in Figure 1. 12 (a) and (b). These trip signals, initiated as a result of the line-to-ground fault, showed that the BP-filter-based WPT was capable of initiating a trip response close to the DF-WPT for detecting the internal fault.

Case 2—Unloaded Magnetizing Inrush Current: For the case of magnetizing inrush current, the protected 3φ 5kVA power transformer was without a load and with a rated primary voltage of 230 V. Figure. 1. 13 shows the responses of the DF-based WPT and the proposed BP-filter-based WPT, along with the three-phase differential currents. Both the

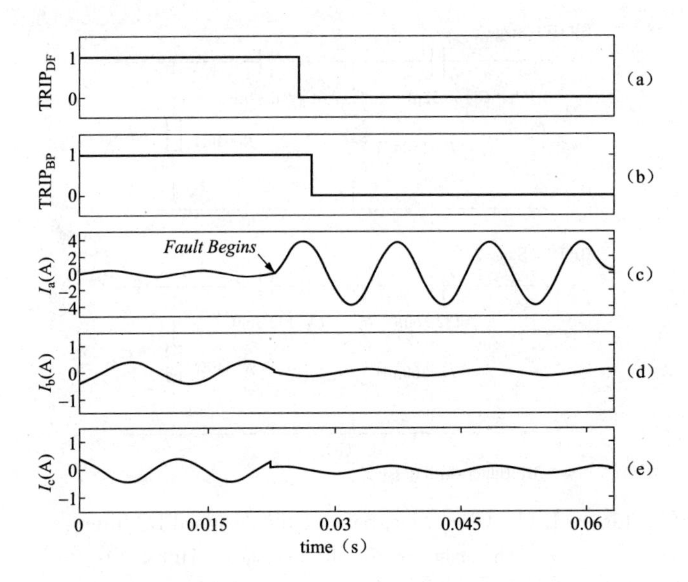

Figure 1.12　Line-to-ground fault occurring on the primary side: The responses of the DF-based WPT ($TRIP_{DF}$) and BP-filter-based WPT ($TRIP_{BP}$) and the three-phase differential currents.

BP-filter-based WPT and DF-based WPT trip signals remained high, indicating a nonfault condition, as shown in Figure 1.13 (a) and (b). These trip signals indicated that the proposed BP-filter-based WPT was capable of accurately identifying the magnetizing inrush current as a nonfault current.

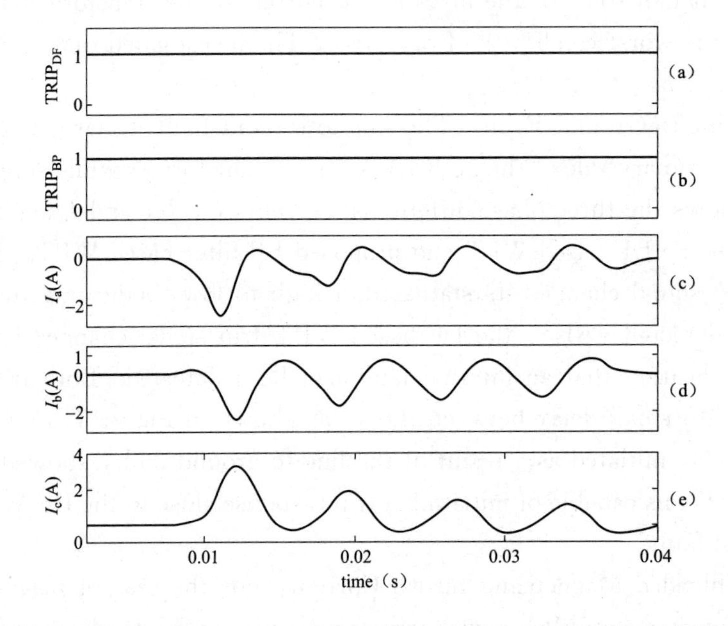

Figure 1.13　Unloaded magnetizing inrush current: The responses of the DF-based WPT ($TRIP_{DF}$) and the BP-filter-based WPT ($TRIP_{BP}$) and the three-phase differential currents.

The offline simulation results of the WPT-based transformer differential protection showed fast, reliable, and accurate responses. Moreover, these response characteristics were observed when the WPT algorithm was implemented using both the DFs and the third-order passive BP filters. In addition, the successful classification of the magnetizing inrush current as a nonfault disturbance demonstrated the accuracy of both implementations of the WPT-based transformer differential protection.

1.3.5 Experimental Testing of the BP-Filter-Based WPT for Transformer Differential Protection

In order to establish the validity and applicability of the BP-filter-based WPT transformer differential protection, its performance was experimentally investigated. The proposed differential protection was tested on a 3φ 5kVA 230/550 - 575 - 600 V 60Hz core-type Δ - Y laboratory prototype power transformer. Moreover, the accuracy of the proposed protection technique was verified through comparisons with the DF-based WPT transformer differential protection under similar operating conditions.

A. The Experimental Setup

The setup for the experimental testing of the proposed BP-filter- based transformer differential protection was composed of the following components.

1) Three identical CTs were connected in Y on the primary side (Δ - connected windings), and another three identical CTs were connected in Δ on the secondary side (Y-connected windings) of the power transformer. It is worth mentioning that the configuration of the CTs on both sides of the transformer was set to eliminate any phase shifts due to the Δ - Y configuration of the transformer windings.

2) Three identical BP filter banks (each filter bank was composed of two cascaded stages for processing one phase current).

3) The outputs of the BP filter banks were fed into an inverting-summing operational amplifier to provide the response (trip signal).

4) A control circuit that was employed to trigger three identical triac switches, which were used to provide a short contact between the supply and the tested transformer.

The experimental setup is shown in Figure 1.14. The values of the BP filter parameters were selected as follows: $C_{11} = C_{13} = 66.32\mu F$, $L_{12} = 150$ mH, $R_{L1} = R_{L2} = 40\ \Omega$, $C_{21} = C_{23} = 132.36\mu F$, and $L_{22} = 300$ mH. The triggering circuit was designed to activate the triac switches by supplying their gates with a dc voltage of 10 V. For all nonfault conditions (the trip signal was high), the triggering circuit maintained the 10V dc voltage on the triac switches' gates, while it disconnected the dc voltage for any fault conditions (the trip signal was low). The triggering circuit details and structure are given in Appendix. The trip signal, as well as the three-phase differential currents for all tests, was collected using a *Tektronics* 1002 *B* storage digital oscilloscope.

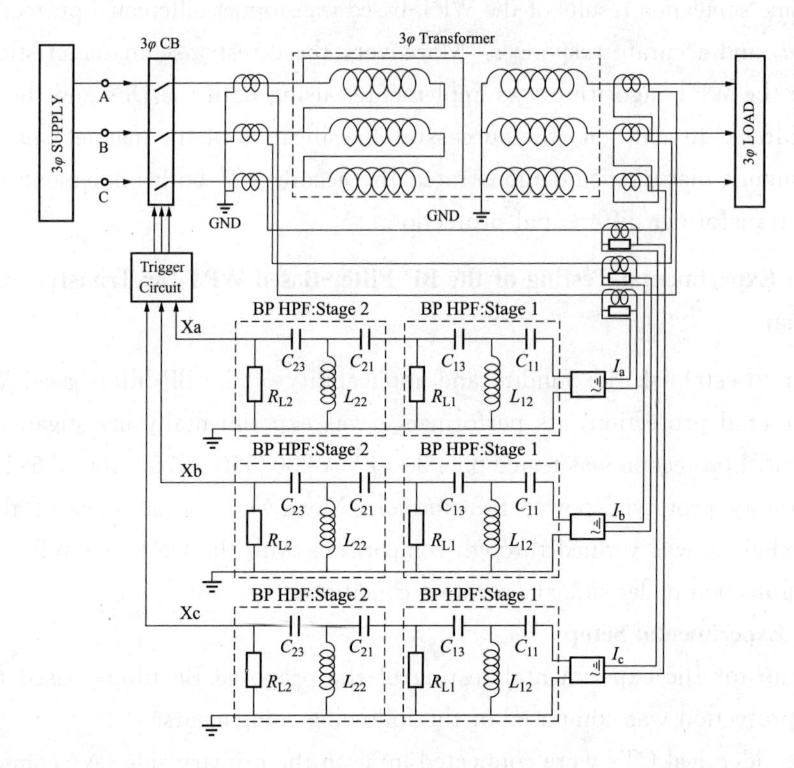

Figure 1. 14 Schematic diagram for the experimental setup of BP-filter-based WPT transformer differential protective relay.

B. Experimental Test Results

Many currents were investigated for experimental testing of the proposed BP-filter-based WPT transformer differential protective relay. The following experimental test results are presented in this paper:

1) normal load current (through current);

2) magnetizing inrush current, unloaded;

3) magnetizing inrush current, loaded;

4) fault currents: line-to-line fault, unloaded;

5) fault currents: line-to-ground fault, loaded;

6) fault currents: 3φ-to-ground fault, loaded.

Case 1—*Normal Load Current*: The test for this current was carried out after the transformer was energized. The transformer secondary was connected to a balanced 3φ Y-connected *R-L* load of impedance of $Z = 17.14 + j34.23\ \Omega$/phase. Figure 1.15 shows the BP-filter-based WPT response (trip signal) and the three-phase differential currents.

It was evident from the status of the trip signal (remaining high) that the normal current was distinguished as a nonfault condition.

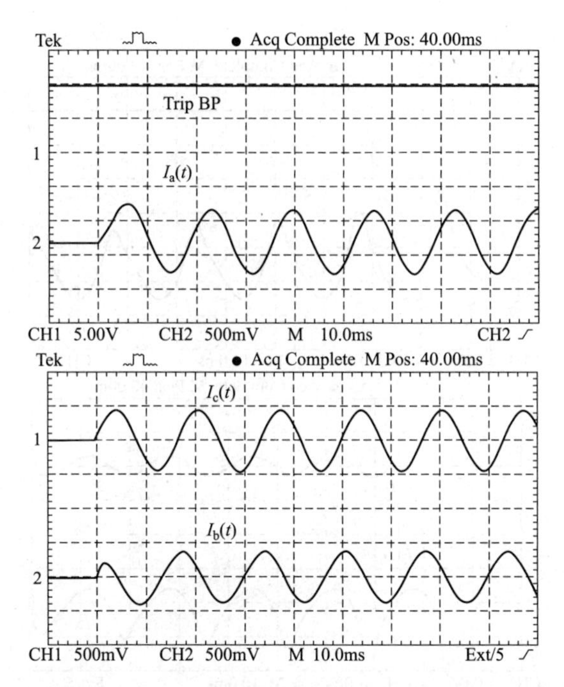

Figure 1.15 Normal current for a balanced 3φ Y-connected R-L load of an impedance $Z = 17.14+\mathrm{j}34.23\ \Omega$/phase: The response of the BP-filter-based WPT transformer differential protective relay and the three-phase differential currents. Current scale is 2 A/div and time scale is 10 ms/div.

Case 2—*Unloaded Magnetizing Inrush Current*: Many tests were carried out for investigating the magnetizing inrush current. For this tested power transformer, in most cases, the magnetizing inrush current had a high magnitude with longer time to decay. The BP-filter-based WPT trip signal remained high, indicating that the detected magnetizing inrush current was classified as nonfault condition. As a result, there was no change in the status of the trip signal. The experimental test results for the case of magnetizing inrush current without a load are shown in Figure 1.16. The status of the trip signal (remaining high) provided an indication that the unloaded magnetizing inrush current was detected and distinguished as a nonfault current.

Case 3—*Loaded Magnetizing Inrush Current*: This test was carried out for investigating the magnetizing inrush current, when a load was connected on the secondary side. In this test, a balanced 3φ Y-connected R-L load of impedance of $Z = 17.14 + \mathrm{j}34.23\ \Omega$/phase was connected on the tested 5-kVA core-type power transformer. The BP-filter-based WPT trip signal remained high, indicating that the detected magnetizing inrush current was classified as nonfault condition. Furthermore, The trip signal remained unchanged even for a high magnitude inrush currents. As a result, there was no change in the status of the trip signal. The experimental test results for the magnetizing inrush current with a load on the secondary side is shown in Figure 1.17. It was evident from the status of

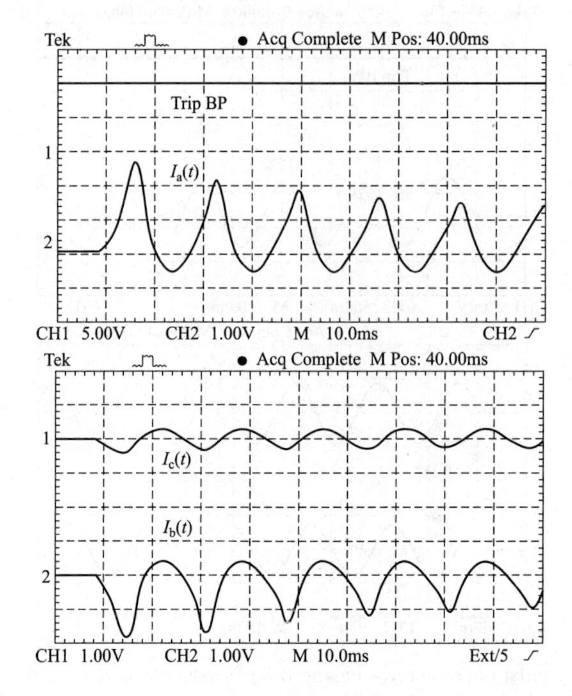

Figure 1. 16 Unloaded magnetizing inrush current: The response of the BP-filter- based WPT transformer differential protective relay and the three-phase differential currents. Current scale is 2 A/div and time scale is 10 ms/div.

the trip signal (remaining high) that the magnetizing inrush current with a load connected to the secondary side was detected and distinguished as a nonfault condition.

Case 4—*Line-to-Line Fault*: The test for this fault was carried out by connecting phase A to phase B on the primary side of the test 3φ 5-kVA power transformer. The primary line-to-line fault test was carried out to demonstrate the ability of the proposed BP-filter-based WPT differential protective relay to respond to internal-fault currents. Figure 1. 18 shows the BP-filter-based WPT trip signal and the three-phase differential currents for this fault. The BP-filter-based WPT differential protective relay identified the internal-fault current and initiated a trip signal in less than half a cycle.

Case 5—*Line-to-Ground Fault*: The test for this fault was carried out through connecting phase C to the ground, while a balanced 3φ Y-connected R-L load of impedance of $Z = 17.14 + j34.23\ \Omega$/phase was connected to the secondary side of the test 3φ 5-kVA power transformer. The primary line-to-ground fault test was carried out to demonstrate the ability of the proposed BP-filter-based WPT differential protective relay to respond to internal faults occurring during the transformer operation. The BP-filter-based WPT trip signal and the three-phase differential currents for this fault are shown in Figure 1. 19. The BP-filter-based WPT identified the internal-fault current and initiated a trip signal in less than half a cycle. Moreover, Figure 1. 19 showed that the proposed BP-filter-based WPT differential

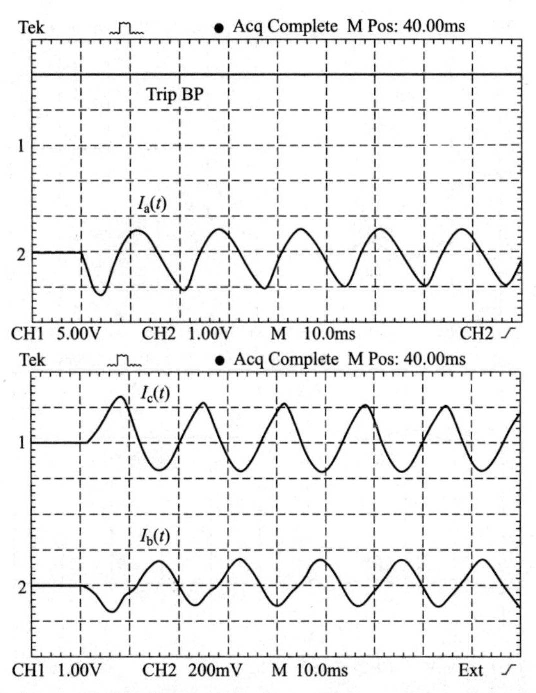

Figure 1. 17 Loaded magnetizing inrush current: The response of the BP-filter-based WPT transformer differential protective relay and the three-phase differential currents. Current scale for $I_a(t)$ and $I_c(t)$: 2 A/div; for $I_b(t)$: 0. 4 A/div. The time scale is 10 ms/div.

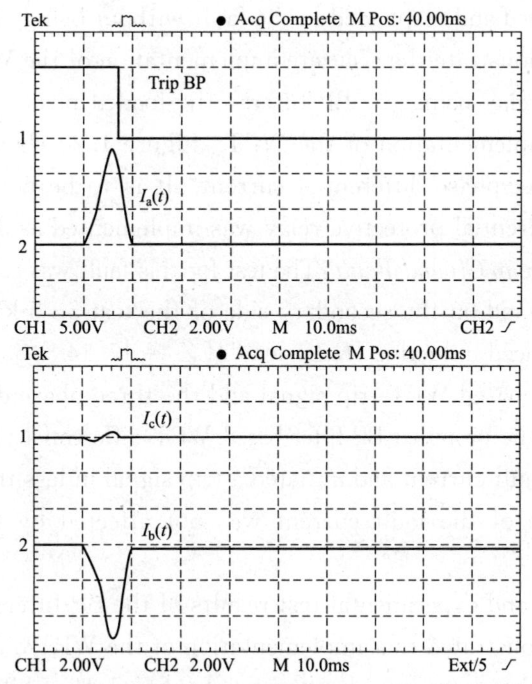

Figure 1. 18 Unloaded phase-A-to-phase-B fault occurring on the primary side: The response of the BP-filter-based WPT transformer differential protective relay and the three-phase differential currents. Current scale: 4 A/div. Times cale: 10ms/div.

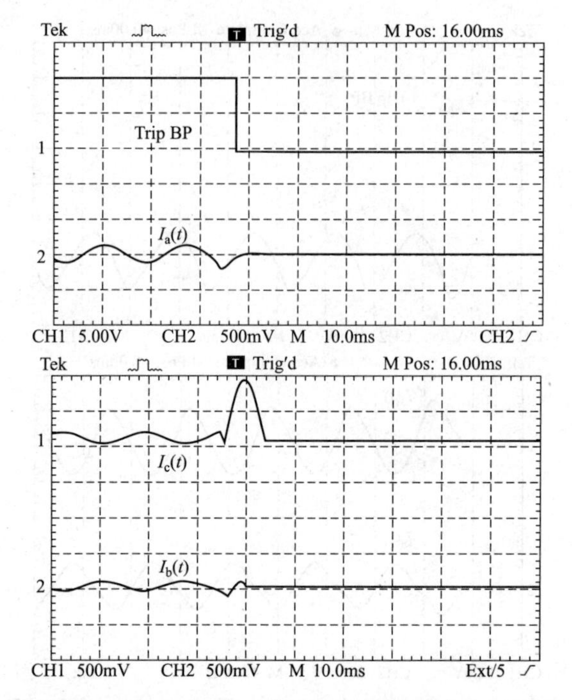

Figure 1. 19　Loaded phase-C-to-ground fault occurring on the primary side: The response of the BP-filter-based WPT transformer differential protective relay and the three-phase differential currents. Current scale: 4 A/div. Time scale: 10ms/div.

protective relay identified and responded to the fault without being affected by the CT saturation. In order to demonstrate the accurate implementation of the WPT-based transformer differential protective relay using the BP filters, the loaded phase- C-to-ground fault was tested using the DF implementation of the WPT. Figure 1. 20 shows the DF WPT-based trip signal and the three-phase differential current. It is to be noted that the DF WPT-based transformer differential protective relay was implemented as detailed.

Case 6: *Three-Phase-to-Ground Fault*: The test for this fault was carried out by connecting the three phases to ground on the secondary side of the test 3φ 5-kVA power transformer while supplying a balanced inductive (R-L) load of $Z = 17.14 + j34.23\ \Omega$/phase. In Figure 1. 21, the BP-filter-based WPT trip signal and the three-phase differential currents for this fault are shown. The proposed BP-filter-based WPT differential protective relay detected and distinguished the fault current and initiated a trip signal in less than half a cycle. Moreover, the identification of the fault current was not affected by the CT saturation, as shown in Figure 1. 21.

Offline simulation and experimental test results of the BP-filter-based WPT differential protective relay demonstrated that implementation of the WPT differential protective relay was accurate. This accuracy was demonstrated through the ability of the BP filters to detect, identify, and respond to faults in almost the same time as the DF WPT-based differential relay. These results provided evidence for the applicability of the proposed BP-

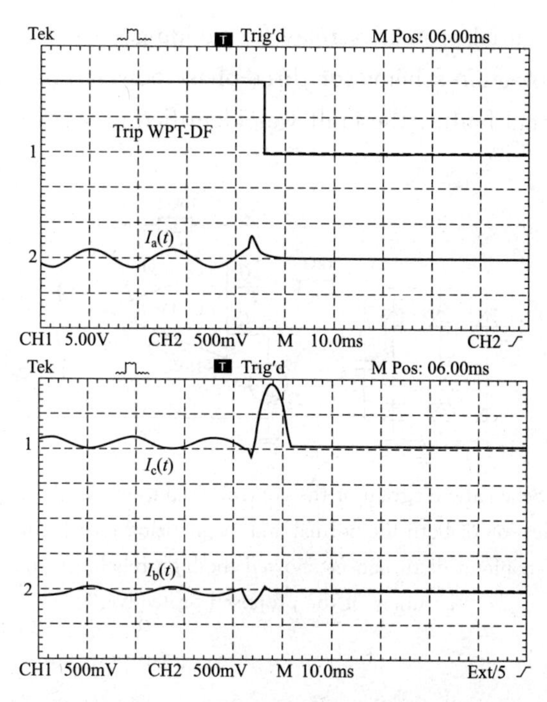

Figure 1. 20 Loaded phase-C-to-ground fault occurring on the primary side: The response of the DFWPT-based transformer differential protective relay and the three-phase differential currents. Current scale: 4 A/div. Time scale: 10 ms/div.

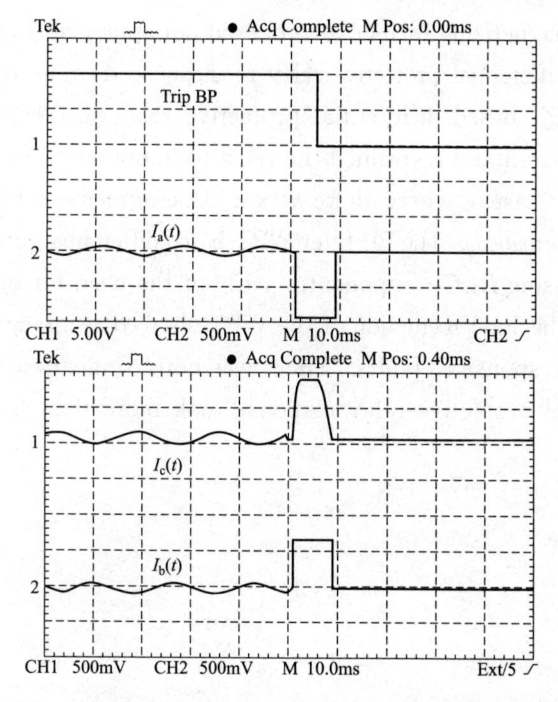

Figure 1. 21 Loaded 3φ-to-ground fault occurring on the secondary side: The response of the BP-filter-based WPT transformer differential protective relay and the three-phase differential currents. Current scale: 5 A/div. Time scale: 10ms/div.

filter-based WPT differential protective relay for building a reliable, accurate, fast, and economic differential protection scheme for three-phase power transformers. In addition, in all cases of tested internal faults, the fault was identified in less than half a cycle based on a 60Hz supply.

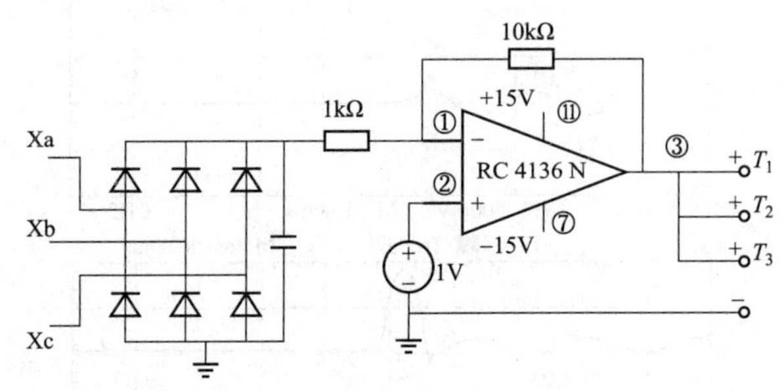

Figure 1.22　Schematic diagram of the control circuit. relay. Also, the successful classification of both the normal and magnetizing inrush currents as nonfault disturbances showed the high reliability and accuracy of the proposed protective.

1.3.6　Conclusion

The WPT is a new approach for power-transformer differential protections. A new method for implementing the WPT-based differential protective relay using BP filters has been successfully tested for both offline and online performances. The different and unpredictable characteristics of magnetizing inrush currents did not appreciably affect the ability of the BP-filter WPT-based differential protective relay to diagnose them as nonfault currents. The proposed BP-filter WPT-based differential protective relay successfully distinguished the magnetizing inrush current and restrained the relay from operation for all investigated magnetizing inrush current cases, where there was a close agreement between the simulation and experimental performances. The BP-filter WPT-based differential protective relay was quite simple to implement using *RLC* components, without the need for digital circuits and DSP boards. Furthermore, this implementation of the WPT showed accuracy and reliability for initiating trip signals in response to faults. There was not a single case in which the BP-filter WPT-based differential protective relay response took more than half a cycle based on a 60Hz system (4-7ms).

2

专业英语语法(1)

2.1 冠词的基本用法

科技英语写作大量使用冠词，经常会出现误用或漏用的场合。为此，下面介绍冠词的基本用法。

1. 名词前使用冠词的常见情况

（1）在单数可数普通名词前一定要用不定冠词，泛指时也多用不定冠词。例如：

1）A node is usually indicated by a dot in a circuit. 电路中的节点通常用一个点来表示。

2）A computer system consists of a computer and some peripherals. 计算机系统是由计算机和一些外部设备构成的。

3）These chemical properties can be observed when a substance undergoes a chemical reaction. 当物质经受化学变化时，就可以观察到这些化学性质。

4）An active element is capable of generating energy while a passive element is not. 有源元件能够产生能量，而无源元件则不能。

需要注意的是，用 a 还是 an 取决于不定冠词后紧跟名词的第一个音素（而非第一个字母）是否为元音，若为元音，则必须用 an，例如：

5）This is a unit.（第一个字母 u 为元音字母）这是一个单位。

6）This is an RS flip-flop. 这是一个 RS 触发器。

7）This is an n-valued function. 这是一个 n 值的函数。

8）A magnet has an S pole and an N pole 磁铁具有一个 S 极和一个 N 极。

9）An 8-volt battery should be used here. 这里应该使用 8V 的电池。

（2）名词前使用定冠词的一些情况。

1）文中前面曾提及过的事物（名词）；

2）同一句中第二次提及的事物（名词）；

3）心中特指的事物（名词）；

4）带有后置修饰语的事物（名词）。例如：

a）The collector（带有后置修饰语）of the transistor（文中前面曾提及过）must be connected to the positive terminal（带有后置修饰语）of the power supply（文中前面曾提及过）. 该晶体管的集电极必须连接到电源的正极。

b）The resistance（带有后置修饰语）of a given section of an electric circuit is equal to the ratio（带有后置修饰语）of its voltage to the current（第二次提及）through this section of the circuit（第二次提及）. 电路某一段的电阻等于其电压与流过该段电路的电流之比。

c）If a voltage is applied across the terminals（带有后置修饰语）of a closed circuit, an electric current will flow in the circuit（第二次提及）. 如果把电压加在闭合电路的两端，电路中就会有电流流动。

d）The reader（心中特指的事物）should be aware that the binary data code（第二次提及）is converted to base 10 for human consumption. 读者应该明白，将二进制数据码转换成十进制是为了方便人类的使用。

需要注意的是：

1）很多情况下，若不强调特指，则带有后置定语的复数名词前可以不用定冠词。例如：

Thermometers（未用定冠词）used in chemistry are marked in degrees Celsius. 化学中所使用的温度计是用摄氏度来标记的。

In this case, all we need to do is to measure the potential difference across two terminals（未用定冠词）of the circuit. 在这种情况下，我们只需要测出该电路两端的电位差。

Elements（未用定冠词）that are similar chemically fall directly beneath one another in the Periodic Table. 在周期表中，化学性质类同的元素是一个接一个直接挨在一起的。

2）有些情况下，在带有后置定语而不强调特指的可数名词单数前，也可以使用不定冠词，例如：

A transformer（使用不定冠词 ）having 2400 turns on the primary side and 48 turns on secondary side is used as an impedance-matching device. 一台一次侧 2400 匝、二次侧 48 匝的变压器用作阻抗匹配器。

A temperature scale（使用不定冠词 ）in common use in the United States today is based on the work of Daniel Fahrenheit. 现今美国常用的一种温度标是基于丹尼尔·华伦海特的工作。

A compound is a pure substance（使用不定冠词 ）that can be broken down into two or more elements. 化合物是能够分解成两种或多种元素的纯净物质。

A temperature of 45° corresponds to a mercury level 45%（使用不定冠词 ）of the way from the 0° to the 100° mark. 45℃相当于从 0°标记到 100°标记之间水银柱高度为 45%的位置时的温度。

2. 名词前不用冠词或省略冠词的常见情况

（1）泛指的物质名词与不可数名词前不加冠词；表示一类的复数名词前不用冠词。例如：

1）Information（不加冠词）is knowledge. 信息就是知识。

2）Electricity（不加冠词）is widely used in industry（不加冠词）, agriculture（不加冠词）and our daily life. 电广泛地应用于工农业和人们的日常生活。

3）Electrical energy（不加冠词）can be changed by electric motors（不加冠词）into mechanical energy（不加冠词）. 电能可由电动机转换成机械能。

4）Machines（不加冠词）are run by electricity. 机器靠电力运转。

（2）论文、书籍的各级标题前的冠词可以省略，例如：

（An）Analysis of RC Resonant Circuits　RC 谐振电路的分析

（A）Comparison of Radio Waves with Sound Waves　无线电波与声波的比较

（A）Study of Wavelets　小波研究

（The）Research on Image Processing　图像处理的研究

（An）Introduction to Electric Circuits　电路入门（书名）

（The）Fundamentals of Electric Circuits　电路基础（书名）

18.7　（The）Interconnection of Networks　18.7　网络连接

需要注意的是，若书籍章节的标题为可数名词，则常用复数形式表示，也可用单数名词或定冠词加单数名词表示。例如：

Chapter 1　Basic Concepts　第一章　基本概念

Chapter 9　Sinusoids and Phasors　第九章　正弦量和相量

Chapter 11　AC Power Analysis　第十一章　交流功率分析

5.4　Inverting Amplifier　5.4　反向放大器

Chapter 7　The Memory Element　第七章　存储元件

8.4 The Source-Free Parallel RLC Circuit　8.4　零输入并联 RLC 电路

（3）专有名词前冠词的使用。在科技文写作中，专有名词主要涉及人名、地名、单位名称、机构名称和国家名称，下面分别加以说明。

1）人名前不加冠词，例如：Maxwell（麦克斯韦）、Faraday（法拉第）、Einstein（爱因斯坦）。

2）地名单独使用时通常不加冠词，例如：New York（纽约）、Xi' an（西安）、Shaanxi（陕西）。

注意：由于英语中没有四声，无法区分“陕西”和“山西”，因此“陕西”的英语为 Shaanxi，即其中多了一个字母 a。

3）由若干词构成的国名、组织机构名，即由普通名词构成的专有名词前通常加定冠词，例如：the State Department（美国国务院）、the China Travel Service（中国旅行社）。

（4）图示说明文字中一般可省略冠词，例如：Figure 2.1 Generation of sine wave by vertical component of rotating vector.

图 2.1 用旋转矢量的垂直分量来产生正弦波。

需要注意的是，generation 前省略了 the；sine wave 前省略了 a；vertical 前省略了 the；rotating 前省略了 a。另外，科技英语中不论图示的说明是否是一个句子，末尾均要加句号。例如：

1）Figure 1.1（The）Block diagram of（a）typical digital computer. 图 1.1 典型数字计算机的方框图。

2）Figure 25.6（The）Electric field due to（a）ring of charge. 图25－6电荷环产生的电场。

3）Figure 2.5（The）Effect of（the）ammeter resistance on（the）current in（a）circuit. 图2－5安培表电阻对电路中电流的影响。

4）Figure 7.78 Circuit for an automobile ignition system. 图7.78汽车点火系统电路。

使用冠词的例句如下：

5）Figure 2.57 The potentiometer controlling potential levels. 图2.57控制电位的电位仪。

6）Figure 7.77 A relay circuit 图7.77继电器电路。

（5）可数名词单数形式泛指时可省略冠词，特别是在between A and B、from A to B、the variation of A with B等表达中，A和B之前可以省略定冠词，例如：

1）A transistor consists of three parts：emitter，base and collector. 晶体管是由发射极、基极和集电极三部分构成的。

2）Ohm first discovered the relationship between current，voltage，and resistance. 欧姆首先发现了电流、电压、电阻之间的关系。

3）Figure 1.5 shows the variation of output with input. 图1.5画出了输出随输入的变化情况。

4）The resulting pressure difference between top and bottom of the ball causes a "lift" force. 该球的顶部与底部之间所产生的压差产生了一个"提升"力。

5）The distance from earth to sun is about 1.5×10^{11} m. 地球到太阳的距离大约为1.5×10^{11}米（m）。

6）This sine wave travels along the tube from left to right. 此正弦波沿着管子从左边传播到右边。

（6）人名的所有格之前不用冠词，例如：

1）Kirchhoff's laws and Ohm's law are very useful to the analysis of networks. 基尔霍夫定律和欧姆定律对分析网络很有帮助。

2）Maxwell's equations are very useful to the analysis of Electromagnetic fields. 麦克斯韦方程对于分析电磁场问题非常有用。

3）The voltage induced in the primary winding is proportional to the primary inductance according to Faraday's law. 根据法拉第定律，在一次绕组中感应出来的电压与一次电感成正比。

4）Shannon's sampling theorem is very important in information theory. 香农取样定理在信息论中非常重要。

5）Thevenin's theorem is very important in circuit analysis. 戴维南定理是电路分析中的重要定理。

这时，用人名的所有格形式表示某人发现的定律、定理、原理、现象、效应、疾病等，再例如：Euler's equation（欧拉方程）、Ohm's law（欧姆定律）、Archimede's principle（阿基米德原理）、Boltzmann's constant（波尔茨曼常数）、Charles's law（查尔斯定理）、Hooke's Law of elasticity（胡可弹性定律）。

但是，如果人名直接修饰普通名词，则其前一般要使用定冠词，例如：

1）We first find the Thevenin equivalent circuit looking into the base at AA's，as shown in Figure 4. 27. 如图 4. 27 所示，我们首先求出从 AA's 处向基极看进去的电路的戴维南等效电路。

2）This model is called the Bohr model. 这个模型被称为波尔模型。

3）The Reynolds number is a dimensionless quantity. 雷诺数是一个没有量纲的量。

4）These are the Maxwell expressions. 这些是麦克斯韦表达式。

5）This equation can also be obtained from the Karnaugh map shown in Figure 1－3. 这个式子也可以从图 1－3 所示的卡诺图中获得。

6）The Wien bridge is also useful as a frequency-selective network. 维恩电桥也可用作频率选择网络。

7）The Meacham bridge can not be used here. 这里不能使用米契阿姆电桥。

这时表示的是属于由某人发明的东西，不能用所有格，而采用复合名词形式，再例如：the Kelvin scale（开尔文温标）；the Thevenin resistance（戴维南电阻）；the Laplace transform（拉普拉斯变换）；a Bursen burner（本生灯）；a Diesel engine（迪塞尔内燃机）；a Wheatstone bridge（惠斯顿电桥）；由两人或多人发明或发现的东西也要用复合名词形式，例如：the Stefan-Boltazmann law（斯提范伯尔曼定律）、the Joule-Tompson effect（焦耳汤普森定律）。

若属于由某人发明的方法、技术或化学反应等，则上述两种表达形式即人名的所有格形式和非所有格形式均可采用。

（7）chapter、section、page、figure、table、equation 等后紧跟数字表示“第……”时，其前面不得加冠词，例如：

1）This point will be discussed in Chapter 2. 这一点将在第二章加以讨论。

2）This phenomenon was described in Section 3. 5. 这现象在第 3. 5 节介绍过了。

3）The key to Exercise 2 appears on Page 121. 练习 2 的答案在第 121 页。

4）Figure 3. 1 shows the block diagram of a control system. 图 3. 1 画出了一个控制系统的方框图。

5）The range of resistivities of several metals at room temperatures is shown in Table 1－1. 表 1－1 给出了几种金属在室温下的电阻率的范围。

6）Equations（1）and（2）can be cast in matrix form. 方程（1）和（2）可以表示为矩阵形式。

7）For $t<0$，the switch is in position A，but for $t>0$，the switch is in position B. $t<0$ 时，开关位于位置 A，$t>0$ 时，开关位于位置 B。

（8）在解释方程、公式、表达式等里面的参数时，一般不用冠词。例如：

Hence，the natural response is

$$u_{on}=e^{-t}(A\cos 2t+B\sin 2t)\mathrm{V},t\geqslant 0$$

where A and B are unknown constants to be determined. 因此，自然响应为 $u_{on}=e^{-t}(A\cos 2t+B\sin 2t)\mathrm{V},t\geqslant 0$，其中 A 和 B 为待定常数。

用冠词的例子如下：

The amount of charge stored, represented by q, is directly proportional to the applied voltage u so that

$$q = Cu$$

where C, the constant of proportionality, is known as the capacitance of the capacitor.

电容储存的电荷量用 q 表示，它与施加的电压 u 成正比，即有 $q = Cu$，式中 C 为比例常数，称为电容（量）。

(9) 在表示下定义的句型“by A is meant B”中，A 之前不用冠词。例如：

1) By linear operation is meant the ability of an amplifier to amplify signals with little or no distortion. 所谓线性工作指的是放大器以很小的失真或毫无失真地放大信号的能力。

2) By frequency is meant the number of times something repeats itself per second. 所谓频率指的是某物每秒钟重复的次数。

(10) 在学科名称前不用冠词，例如：

Mechanics belongs to a branch of physics. 力学属于物理学的一个分支。

(11) 表示“在某一方面”时不用冠词，例如：

1) Aiken's machine was limited in speed by its use of relays rather than electronic devices. 爱肯机的速度受限是由于它使用的是继电器而不是电子器件的缘故。

2) Machines differ greatly in size. 机器的体积各不相同。

3) This computer is very good in performance. 这台计算机的性能很好。

(12) 在“每（per）”之后不用冠词，例如：

1) Its frequency stability is six parts per million (=six parts in a million). 其频率稳定度为百万分之六。

2) Radio waves travel 300,000 kilometers per second. 无线电波每秒钟传播 30 万 km。

(13) 不用冠词的其他情况，例如：

1) To use Cramer's rules, we put Eqs. (1) and (2) in matrix form as

$$\begin{bmatrix} 3 & -2 \\ -1 & 2 \end{bmatrix}\begin{bmatrix} i_1 \\ i_2 \end{bmatrix} = \begin{bmatrix} 1 \\ 1 \end{bmatrix}$$

为了利用克莱姆法则，将方程式（1）和（2）表示成矩阵形式，即

$$\begin{bmatrix} 3 & -2 \\ -1 & 2 \end{bmatrix}\begin{bmatrix} i_1 \\ i_2 \end{bmatrix} = \begin{bmatrix} 1 \\ 1 \end{bmatrix}$$

2) These are data in polar form. 这些是极坐标形式的数据。

3) This integral can be evaluated in closed form. 这个积分可以按闭合形式求出来。

4) They take place in reversed order. 它们逆序发生。

但是，在用“domain”时，却要在其前面用定冠词；在用“fashion”时可用不定冠词。例如：

5) After transformation, the signal in the time domain is changed into the signal in the frequency domain. 变换后，时域信号变为频域信号。

6) Doing so will modify the signal in a deterministic fashion。这样处理会以确定的方式修正信号。

（14）表示职位的名词或者表示独一无二的人之前一般不用冠词，例如：

Dr. Samuel Rankin，head of the mathematical sciences department. 数学系主任塞缪尔·兰金博士。

This is called a hertz in honor of Heinrich Hertz，discoverer of radio waves. 这被称为1赫兹（Hz），以纪念无线电波的发现者海因里希·赫兹。

The unit of power is a joule per second，which is called a watt（W），in honor of James Watt，developer of the steam engine. 功率的单位是每秒1焦耳（J），称为1瓦特（W），以纪念蒸汽机的研发者詹姆斯·瓦特。

若非“独一无二”，则必须加上冠词。例如：

“Scientists see this as the last industrial moment”，said Frank Y Fradin，a physicist at the Argonne National Laboratory near Chicago. “（a physicist，不定冠词a）科学家们视此为最后的工业契机”，位于芝加哥附近的阿贡国家实验室的物理学家弗朗克·Y·弗拉金说道。

Georg Simon Ohm，a German physicist，is credited with finding the relationship between current and voltage for a resistor.（a German physicist，不定冠词a）电阻的电流和电压关系是德国物理学家乔治·西蒙·欧姆发现的。

3. 名词前冠词可加可不加的特殊情况

（1）用“系表”结构定义某个参数，被定义的名词参数前加或不加不定冠词均可。例如：

1）A scalar/Scalar quantity is one that possesses magnitude only. 标量是只具有大小的一个量。

2）A velocity/Velocity is a vector quantity involving both magnitude and direction. 速度是既涉及大小又涉及方向的一个矢量。

3）An absolute/Absolute error is the actual difference between the measured value and the accepted value. 绝对误差是测得值与认可值之间的实际差别。

4）A gauge/Gauge pressure is the difference between true pressure and atmospheric pressure. 表压是真正的压力与大气压之差。

5）A heat/Heat engine is any device that converts heat energy into mechanical energy. 热机是能够把热能转换成机械能的设备。

（2）被称谓的名词前加或不加定冠词均可，例如：

1）This ratio is called the magnetic intensity. 这个比值被称为磁强度。

2）This is measured by a quantity called the regulation. 这是由一个被称为调整率的量来度量的。

3）This quantity is called the diffusivity. 这个量被称为扩散率。

4）The energy liberated by nuclear fusion is called thermonuclear energy. 由核聚变释放出来的能量被称为热核能。

5）The ability of some elements to give out radiations is called radioactivity. 某些元素产生辐射的能量被称为放射性。

4. 名词前加定冠词的特殊情况

（1）表示某个参数的单位时一般要用定冠词，例如：

1）The unit of potential difference is the volt. 电位差的单位为伏特。

2）The unit of capacitance is the farad. 电容的单位为法拉。

3）The unit of resistance is labeled the ohm，after George Ohm，who first discovered the relationship between current，voltage，and resistance. 电阻的单位以乔治·欧姆的名字命名为欧姆，他首先发现了电流、电压、电阻之间的关系。

（2）带有同位语的参数等名词前一般使用定冠词。例如：

the current i，电流 i；the initial voltage u（0），初始电压 u（0）；

the subscript g，下标 g；the coefficient μ，系数 μ；the variable x，变量 x。

（3）当 any、some、most 等不定代词作以下用法时，其中名词前必须用定冠词，即 any（all，both，none，neither，either，some，most，one，each，much，many，the rest）of the＋名词例如：

1）Once all of the IED hardware，LAN technologies，and IED and LAN protocol issues have been resolved，the next question is how to display all of this integrated information to the substation operator in an economical fashion. 一旦所有的 IED 硬件、LAN 技术以及 IED 和 LAN 通信规约条款问题得到解决，下一步就是如何以最经济的方式将所有这些采集到的信息显示给变电站的值班人员。

2）每个系统均由一台发电机向四个用户供电。

3）One of the transformer windings is connected to a source of ac electric power，and the second（and perhaps third）transformer winding supplies electric power to loads. 变压器的第一个绕组与交流电源连接，第二个绕组（也许还有第三个绕组）为负载提供电功率。

4）Much of the apparatus used on a power system has small physical dimensions when compared to the length of general transmission line circuits. 用于电力系统的大多数电气设备与一般输电线路的长度相比，实际尺寸都比较小。

5）Both of the above system of protection leave the neutral ends of the stator windings unprotected against earth faults（typically the bottom 5%～10% of the windings）. 上述两种保护系统对发电机定子绕组靠近中性点部分都有接地故障保护死区（一般在底部，约为定子绕组的 5%～10%）。

6）None of the texts available mention this problem. 现有的教科书均没有提到这个问题。

7）If neither of the foregoing conditions is satisfied，then Eq.（4）is not readily solved. 如果前面两个条件都不满足，则式（4）难以求解。

8）Some of the connections have a black dot indicating a connection. 有些连接是用一个黑点来表示的。

5. 冠词共用的情况

由一个并列连词连接或者由一个并列连词连接并同一个介词界定的若干名词并列时，

一般只需在第一个名词前加冠词，该冠词为后续各名词所共用。例如：

1）Variations in the load have little effect on the frequency or amplitude of oscillations. 负载的变化对振荡的频率或振幅影响很小。（the frequency or amplitude of oscillations）

2）These convenient sources of electrical signals are useful in the test，maintenance，or operation of a wide variety of electrical apparatus. 这些方便的电信号源对各种电气设备的测试、维护或运行很有用。（the test，maintenance，or operation of a wide variety of electrical apparatus）

3）To apply Ohm's law as stated in Eq.（2.3），we must pay attention to the current direction and voltage polarity. 要应用好如式（2.3）所示的欧姆定律，必须注意电流方向和电压极性。（the current direction and voltage polarity）

4）Capacitance depends on the size，shape，and separation between any two conductors. 电容的大小取决于任意两个导体的尺寸、形状及其间距。（the size，shape，and separation between any two conductors）

5）It is convenient to include the functions of a voltmeter，ammeter，and ohmmeter within one instrument. 把伏特表、安培表和欧姆表的功能包括在一只仪表内是很容易做到的。（a voltmeter，ammeter，and ohmmeter within one instrument）

6）After the hot and cold water have been mixed，we will have 2 kg of water at a temperature of 50℃. 将冷热水混合后，我们会得到两 kg 温度为 50℃的水。（the hot and cold water）

7）The technology applications，and scope of high voltage engineering are discussed in this book. 本书讨论了高电压工程的技术、应用及范围。（The technology applications，and scope of high voltage engineering）

8）To find the Thevenin resistance R_{th} and Thevenin voltage U_{th} at the terminals $a-b$ in the circuit in Figure. 4.31，we first use Schematics to draw the circuit as shown in Figure. 4.53. 为了求得端子 $a-b$ 的戴维南电阻 R_{th} 和戴维南电压 U_{th}，首先用 Schematics 画出该电路，如图 4.53 所示。（the Thevenin resistance R_{th} and Thevenin voltage U_{th} at the terminals $a-b$）

9）Microwave engineering deals with the transmission，control，detection，and generation of radio waves whose wavelength is short compared to the physical dimensions of the system. 微波工程讨论的内容为无线电波的发送、控制、检测和产生，而且这种无线电波的波长比系统的物理尺寸短。（the transmission，control，detection，and generation of radio waves）

6. 抽象名词前不定冠词使用情况

表示“作一分析”“计算一下”“作一讨论”“比较一下”“了解一下”“考察一下”“作一比较”“作一研究”等的抽象名词前一般使用不定冠词。例如：

（1）A close observation of Figure 7.35 reveals that $v(t)$ is a multiplication of two functions：a ramp function and a gate function. 仔细观察图 7.35 可知，$v(t)$ 是两个函数

即斜坡函数和门函数之积。

(2) The prerequisite is a good knowledge of electric circuit fundamentals. 首先要很好地了解电路基础内容。

(3) Let us make a rough estimate of the depth there. 让我们粗略地估算一下那里的深度。

(4) The manner in which the Wheatstone bridge is used maybe understood from an analysis of the circuit. 可以通过对该电路的分析来理解惠斯登电桥的使用方法。

(5) A quantitative analysis of this experiment is rather involved. 对该实验作定量分析是相当复杂的。

(6) An examination of the two experiments shows that a definite relationship exists between current, voltage and resistance. 考察一下这两个实验就可以看出在电流、电压、电阻之间存在一种确定的关系。

(7) A more detailed description of the operation of transistor in saturation will be given in section 3. 在第 3 节中对晶体管处于饱和状态下的工作情况会做更为详尽的描述。

(8) A short calculation will convince you that this conclusion is indeed true. 略微计算一下你就会相信这个结论的确是正确的。

(9) The scope of this book does not permit a detailed discussion of all of these Applications. 本书范围有限，无法对所有这些应用都作详细的讨论。

(10) A general knowledge of the characteristics of electrical transmission is essential if the reader is to gain an understanding of data communications. 读者若想要对数据通信有所了解，就必须对电传输的特性有个大致的了解。

注意：从句 if …be to do，主句 essential，must…，表示：若要…，就要（就必须），例如：

If there is to be revolution, there must be a revolutionary party. 若要革命，就要有一个革命党。

(11) There is a growing awareness that this procedure is of value. 现在人们越来越认识到该方法是很有价值的。

7. 序数词前冠词使用情况

当序数词并不强调次序而表示“另一个”“又一个”“再一个”的含义时，其前用不定冠词而不用定冠词，也可仍译成“第一”“第二”“第三”等。

(1) As the first example, consider the circuit in Figure 13.7. As a second example, consider the circuit in Figure 13.8. 举第一个例子，来看一下如图 Figure 13.7 所示的电路。再举一个例子，来看一下如图 Figure 13.8 所示的电路。

类似的句式：First, ……As a second example, ……As a third example, ……首先，……，再看（举）一个例子，……，再看（举）一个例子……。

(2) A second approach is as follows. 另一种方法如下。

The advent of electronics is reckoned from the discovery that the current in a vacuum diode can be controlled by introducing a third electrode. 电子学的诞生是从发现真空二极管中的

电流可通过引入第三个电极来加以控制算起的。

（3）By adding a third signal v_3，we get v（t）. 再用第三个信号 v_3 就得到 v（t）。

2.2 定义句式

科技写作通常需要对某种物质、实物或过程等下定义。定义不是例子，因此不能通过举例来下定义。例子可以跟在定义之后，却不能代替定义本身。此外，定义的第一部分应该总体描述，细节应留到后面再具体描述。也就是说，被定义的事物首先应该用它的总类词（包含被定义的对象所在类别的词语。几个具体词与总类词的对应关系举例如下：orange—fruit；copper—metal；biology—science）来描述以说明其类别，然后再用该事物的特有属性、用途或起源等来描述。英语的定义句有好几种，下面分别加以叙述。

2.2.1 简单定义句

这种定义方式只用一句话，因而是最为简单的。按照其定义的对象是一般的总类或特定的一类又可以分为两种：第一种称为总类定义句；第二种称为特定类定义句。它们均有自己的句式。

1. 总类定义句

这种定义句一般可以表示为

$$T=G+d\ (=d_1+d_2+\cdots+d_n)$$

其中，T 为待定义的事物（thing），G 为总类词，d（$=d_1+d_2+\cdots+d_n$）为待定义事物与总类的其余成员在特性上的差别（difference），“=”表示“是”。例如：

A catalyst（T） is a substance（G） which alters the rate at which a chemical reaction occurs（d_1），but is itself unchanged at the end of the reaction（d_2）.

催化剂（T）是改变化学反应的速度（d_1），但在反应结束时本身并不改变的（d_2）物质（G）。（$T=G+d_1+d_2$）

这种定义句的具体句式为：

An x/y is a 总类词＋
- which is＋动词＋ed ……
- 动词＋ed……
- for＋动词＋ing……
- 关系代词＋动词＋s……
- 动词＋ing……
- 介词＋关系代词或关系副词 ……
- with＋名词短语……
- with the property of 动词＋ing……

其中，x 代表可数名词，y 代表不可数名词。例如：

which is used for stepping up or down voltages.

A transformer is a device which voltages.

- used to step up or down voltages.
- for stepping up or down voltages.
- steps up or down.
- stepping up or down voltages.
- by means of which voltages can be stepped up or down.

变压器是一种升压或降压的设备。

下面为“with＋名词短语”结构的定义句：

A triangle is a plane figure with three sides.（＝A triangle is a plane figure which has three sides.）

三角形是有三个边的平面图形。

An electron is a particle with a mass of 9.107×10^{-28} grams.（＝An electron is a particle which has a mass of 9.107×10^{-28} grams.）

电子是一种质量为 9.107×10^{-28} g 的粒子。

Tungsten is a metal with the property of retaining hardness at red-heat.

钨是一种金属，其性质是在赤热状态下仍能保持坚硬。

上句的意思非常接近于 Tungsten is a metal which retains hardness at red-heat.

钨是在赤热状态下仍能保持坚硬的一种金属。

从上面的叙述可以看出：

（1）很多定义使用第三人称单数一般现在时主动语态的定语从句，即“关系代词＋动词＋s”。对于这种主动语态的定语从句，可以用“动词＋ing”来简化，例如：

An engine is a device converting one form of energy into another.（＝ An engine is a device which converts one form of energy into another.）

发动机是将能量的一种形式转换为另一种形式的装置。

（2）定义也常常使用第三人称单数一般现在时被动语态的定语从句，即“which is＋动词＋ed”。对于这种被动语态的定语从句，可以用“动词＋ed”来简化，例如“used ＋for＋动词＋ing”以及“for＋动词＋ing”就是这种简化的定语从句，但它们只能用于定义工具以及仪器仪表等这一类词，例如：device、apparatus、instrument、machine、tool、microphone、ruler、voltmeter、knife、crane。下面为一被动结构的简化定语从句：

Aluminium is a metal produced from bauxite.（＝Aluminium is a metal which is produced from bauxite.）

铝是一种从铝土矿提炼出的金属。

这两种情况可以表述为，在总类词后面跟一个关系代词引导的定语从句或简化的定语从句（现在分词短语或过去分词短语）。但是，关系代词前面有时也可能出现介词，即有“介词＋关系代词”。当定义所包含的两个句子的主语不相同时就会出现这种情况，例如：

Acoustics is a branch of physics. The properties of sounds are studied in it.

可以改写为：

Acoustics is a branch of physics in which the properties of sounds are studied.

声学是研究声音性质的物理学分支。

下面给出“介词+关系代词”或“关系副词”引导的定语从句：

A two-winding autotransformer is a transformer in which both the primary and the secondary are in a single winding.

二绕组自耦变压器是一种一次绕组和二次绕组为同一绕组的变压器。（或译为：二绕组自耦变压器的一次和二次为同一绕组。）

A steam condenser is a vessel where exhaust steam is condensed by contact with cooling water.

凝汽器是一种容器，排出的蒸汽在其中通过与冷却水接触进行冷凝。

下面给出几个简单定义句的例子：

1）An electric circuit is an interconnection of electrical elements.

电路是电路元件的互联结构。

2）An ideal dependent (or controlled) source is an active element in which the source quantity is controlled by another voltage or current.

理想的非独立源（受控源）是其电源量受另一电压或电流控制的有源元件。

A phasor is a complex number that represents the amplitude and phase of a sinusoid.

相量是一个表示正弦量幅值和相角的复数。

A sinusoid is a signal that has the form of the sine or cosine function.

正弦量是一个函数波形为正弦或者余弦函数的信号。

2. 特定类定义句

总类定义句所定义的对象几乎全是一般的，因此所用的是一个孤立的名词，没有另外的名词或形容词限定它。实际上也经常需要对某种特定的对象下定义，例如：

A step-up substation is a place where the stepping-up of voltages takes place.

升压变电站是将电压升高的地方。

这就是一个特定类定义句或者说是特指定义，即它是给某类特定的变电站（专为升压的那一类）下定义，而不是给普遍的变电站下定义。这种定义句的句式一般可以表示为

$$A+T=\begin{Bmatrix}T\\G\end{Bmatrix}+d(=d_1+d_2+\cdots+d_n)$$

其中，A 为修饰语。例如：

A short circuit is a circuit element with resistance approaching zero and an open circuit is a circuit element with resistance approaching infinity.

短路是一个电路中电阻接近于零的电路，开路则是一个电路中电阻接近无穷大的电路。

An ideal transformer is a unity-coupled, lossless transformer in which the primary and secondary coils have infinite self-inductances.

理想变压器是一种全耦合、无损、一次和二次绕组自感均为无穷大的变压器。

A step-down transformer is one whose secondary voltage is less than its primary voltage.

降压变压器是一种二次电压低于一次电压的变压器。

A two-port network is an electrical network with two separate ports for input and output.

二端口网络是一个输入和输出端口分离的网络。

An equilateral triangle is a triangle with all three sides equal in length.

等边三角形是三边长度相等的三角形。

An equilateral triangle is a plane figure with all sides equal in length.

等边三角形是三边长度相等的平面图形。

上面这个例句中，plane figure 为总类词。实际上，有时非使用总类词不可，试比较：

A suspension bridge is a bridge with a long central span suspended from cables.

吊桥是中心跨距大、吊挂在索缆下的桥。

A wheatstone bridge is an apparatus for measuring the resistance of an electric circuit.

惠斯通电桥是测量电路电阻用的一种装置。

显然，惠斯通电桥不是一个真正的桥，所以下定义的时候不使用总类词 apparatus 是不对的，例如：

A wheatstone bridge is a bridge for measuring the resistance of an electric circuit.

由上面的例句可以看出，在特定类定义句中，is 之后的名词常常带某种修饰语，例如：a plane figure 中的 plane。

2.2.2　扩展定义句

当定义所要描述的内容比较多时，这样的句子就会比较长，因此为了使叙述更为清楚，可以将一个定义句所要表达的内容拆分为两个句子或多个句子，即首先对某一对象下定义，然后再用一两个句子（或短语）加以补充说明，这种定义句称作扩展定义句。下面对常见的扩展定义句形式分别加以说明。

1. 定义＋例子

这种定义句式用于对某一对象下了明确的定义以后，为了明确起见，还需要再举例说明的情况。其英语的表述式通常是

Definition sentence＋
- A common example is……
- Common examples are *a*，*b*，*c* and *d*.
- Typical examples are *a*，*b*，*c* and *d*.
- For example/for instance，*a*，*b*，*c* and *d*.
- Main types are *a*，*b*，*c* and *d*.
- Such as/e. g. *a*，*b*，*c* or *d*.
- Therefore，it is used ……
- As a result，one of its main uses is ……
- It consists of … main parts：……
- Its main components are ……

该表述式中，定义句式即为简单定义句式或其中的主要部分。例如：

A transformer is a device by means of which electrical energy can be transferred from one alternating-current circuit to another.

变压器是一种把电能从一交流电路转移到另一交流电路的设备。

Common examples are core type and shell type.
Typical examples are core type and shell type.

常见的例子是铁芯式变压器和外壳型变压器。

典型的例子是铁芯式变压器和外壳型变压器。

又例如：

A transformer is a device by means of which electrical energy can be transferred from one alternating-current circuit to another, [for example, / for instance, / such as / e. g.] core type and shell type.

再例如：

A time-varying current is a current that varies with time. A common example is alternating current.

时变电流是一种随时间变化的电流，一个常见的例子就是交流电。

A branch represents a single element such as a voltage source or a resistor.

支路是一个单独的元件，例如电压源或者电阻器。

显然，“定义＋例子”的句式只用于某一类而非某一具体的事物，例如，变压器（transformer）、绝缘体（insulator）、半导体（semiconductor）、车辆（vehicle）、燃料（fuel）、合金（alloy）、气体（gas）等，因为只有在这种情况下才能对这种表示总类的名词进行举例说明。

2. 定义＋用途

当所定义的对象不是一个总类名词（如金属），而是具体名词（如铜）时，就不能用“定义＋例子”的句式，而要用“定义＋用途”的句式。这种句式可以表示为

Definition ＋ {Therefore, / Consequently, / As a result,} {it is used … / one of its main uses is …}

例如：

Copper is a metal which is a good conductor of electricity [Therefore, / Consequently, / As a result,] it is used for making wires.

铜是一种良导体，因此它可以用来制造导线。

Copper is a metal which is a good conductor of electricity. Therefore, one of its main uses is in the electrical industry.

铜是一种良导体，因此其主要用于电力工业。

An op amp is an active circuit element designed to perform mathematical operations of addition，subtraction，multiplication，division，differentiation，and integration.

运算放大器是可以进行加、减、乘、除和微积分这些数学运算的有源电路元件。

一般而言，凡具体名词［cement（水泥）、hydrogen（氢）、mercury（水银）、喷气机（a jet plane)］都可用于这种定义式。

3. 定义＋主要组成部分

当对所定义的对象除了给出其定义外，还想同时指出其组成部分即内部结构时，就可以用这种句式，一般可以表示为

Definition＋{ It consists of ……
It consists of n main parts：……
Its main/chief components are …… }

例如：

A transformer is a device by means of which electrical energy can be transferred from one alternating current circuit to another. [It consists of two main parts： / Its main/chief components are] windings and magnetic core kept in a closed tank.

变压器是一种把电能从一交流电路转移到另一电路的设备，它由两部分组成：封闭在箱中的绕组和磁芯。

A transformer is generally a four-terminal device comprising two magnetically coupled coils.

变压器通常是由两个磁耦合线圈组成的四端元件。

类似地，还可以对断路器（breaker)、避雷器（arrester)、电视机（television receiver）等作这种定义说明。

最后，给出扩展式定义句式的小结如下

Definition＋{ Common examples are a，b，c，and d.
Typical examples are a，b，c，and d.
Main types are a，b，c，and d.
Such as a，b，c，and d.
Therefore，it is used ……
As a result，one of its main uses is ……
It consists of … main parts ……
Its main components are …… }

2.3 双重名词

实际上，有些英语的可数和不可数名词，很难明确区分。有很多名词，特别是科技英语里的不可数名词向可数名词转化后既可以用作可数名词，又可以用作不可数名词，这里

称之为双重名词。当双重名词用作可数名词时，既有复数形式，又可加冠词 a（an）。双重名词的可数形式和不可数形式意义有时相差很远，有时却无甚差别。

下面列举不可数名词向可数名词转化的几种类别：

（1）物质名词或抽象名词或改变原来的含义表示某种可数事物或直接表示某种可数事物时，即转化为可数名词。例如：

不可数	单数	可数复数
grass	a grass	grasses
草（一般含义）	一种牧草（具体一种草）	牧草（各种牧草）
stone	a stone	stones
石头（物质）	石头（一种石头或一块石头）	石头（各种石头）
wire	a wire	wires
金属丝（一般含义）	金属丝（实在的一段金属丝）	金属丝（各种金属丝）
wood	a wood	woods
木料（物质）	树木（一段树木）	树木（各种树木）
diamond	a diamond	diamonds
金刚石（坚硬物体）	钻石（宝石）	钻石（各种宝石）
fertilizer	a fertilizer	fertilizers
肥料（一般含义）	肥料（具体例子）	肥料（各种肥料）
food	a food	foods
食物（一般含义）	食物（某种具体食物）	食物（各种食物）
fire	a fire	fires
火（一般概念）	炉火（某一种火）	火（各种火或火灾）
football	a football	footballs
足球（运动）	足球（比赛时用的球）	足球（两个以上的足球）
glass	a glass	glasses
玻璃（物质）	玻璃杯（某种玻璃容器）	玻璃（各种玻璃）；眼镜
ice	an ice	ices
冰（冻结的水）	冰激凌（某种冻结的水）	冰（各种各样的冰）

iron 铁（Fe）	an iron 熨斗（某种铁制品）	irons 铁制品（各种铁制品）
light 光（一般概念）	a light 灯（某种光源）	lights 光（各种各样的光）
paper 纸（物质）	a paper 纸件（报纸，论文）	papers 论文（各种论文和报纸）
rope 绳子（一般含义）	a rope 绳子（某一根绳子）	ropes 绳子（很多根绳子）
man 人（一般概念，包括女人和小孩）	a man 男人（某一个人）	men 男人（两个以上的男人）
rubber 橡胶（一般含义）	a rubber 橡皮擦（某种橡胶制品，用以擦除错字等）	rubbers 橡胶制品（各种橡胶制品）
wheat（一般含义） 小麦	a wheat 小麦（某一种小麦）	wheats 小麦（各种小麦）
soil（一般含义） 土壤	a soil 土壤（某一种土壤）	soils 土壤（各种土壤）
composition 写作，作曲（指动作）	a composition 作文，曲子（指作品）	compositions 各种作文、曲子
pleasure 快乐（一般含义）	a pleasure 乐趣（具体含义）	pleasures 各种乐趣

再如：

1）Light travels faster than sound. 光线比声音传播得快。

2）There were bright lights and harsh sounds everywhere. 到处是明亮的灯光和刺耳的声响。

（2）不可数名词的复数用来表示程度的不等、种类的不同等含有分门别类意味的概念时就转化为可数名词。试比较：

Have you put the fruit on the table? 你可曾把水果放在桌子上了？

China has some wonderful fruits. 中国出产好几种出色的水果。

some wine 一些酒 several wines 几种酒

（3）物质名词或抽象名词的复数形式用来表示强度之高、数量之大、范围之广时，可转化为可数名词。试比较：

Water is changed into steam by heat and into ice by cold. 水受热变成蒸汽，受冷结成冰。

This is where the waters of theChangjiang River and the Jialing River meet. 这就是长江和嘉陵江的汇合处。

再如：

wood	woods
木材（物质）	树林
sand	sands
沙	茫茫黄沙
dew	the dews of heaven
露水	天上的露水
hope	hopes and fears
希望	希望和恐惧

（4）物质名词若用来表示某一具体的类别或型号，抽象名词若用来表示"一种""一个方面""这种""那种"等具体概念时，它们转化为可数名词。例如：

1）A petrol（=A type of petrol）I like very much is Brand X. 我非常喜欢用的一种汽油是爱克斯牌的。

2）Mathematics is a science. 数学是一门科学。[science：科学（一般含义）；a science：学科（特定的科学）]

（5）物质名词作为度量的单位时，作可数名词用。例如：

rope	a rope
绳（一般含义）	（实在的）一根、一段绳子
wire	a wire
金属丝（一般含义）	（实在的）一根、一段金属丝
sugar	Two sugars（= two lumps of sugar）
糖（一般含义）	两块糖
coffee	three coffees（= three cups of coffee）
咖啡（一般含义）	三杯咖啡

（6）物质名词如表示"一场""一阵""一个具体情况"等概念时，就转化为可数名词，且和不定冠词连用。例如：

a heavy rain 一场大雨　　a fire 一堆火

（7）抽象名词若表示某种具体的、特殊情况或场合，就转化为可数名词，可以有复数形式，也可以与不定冠词连用。例如：

analysis　　an analysis　　analyses

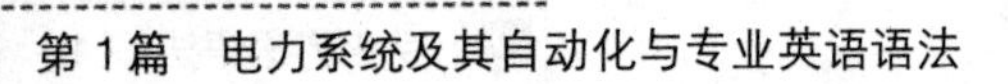

分析，分解（一般含义）	分析（具体例子：一次或一番分析）	分析（多次分析）
calculation	a calculation	calculations
计算（一般含义）	计算（具体的例子：一次计算）	计算（多次计算）
study	a study	studies
研究，观察（一般含义）	研究，观察（具体的例子：一次研究，观察）	研究，观察（多次研究，观察）
insulation	an insulation	insulations
绝缘（一般含义）	绝缘(具体的例子)	绝缘（多种绝缘）

再如某一文章标题：

A Calculation of Uncertainties in Virtual Instrument

虚拟仪器中不确定性的计算

但是，在科技论文标题中，如果以可数名词单数开头，可以加不定冠词 a（an），也可以将之省略，例如：A Study of …或 Study of …，A Comparison of …或 Comparison of …，然而，这种省略只能用于论文标题，对文摘或论文正文则不能使用。这种省略虽未见于语法著作的介绍，但是在英美科技刊物中却大量使用。

某些具有一般性概念的词，在某些情况下是可数名词，例如对单词 temperature（温度、气温、体温）有下列四个例句：

A thermometer measures temperature. 温度计测量温度。

Temperature is usually expressed in degrees. 温度通常用度来表示。

A temperature of over 50° Centigrade was recorded. 温度已记录下来，其值超过 50℃。

The child had a high temperature. 这小孩发高烧。

由此可以看出，表示一般性概念的名词可以是可数名词的情况主要有：

1）在动词 have 之后（have a water：有一种水＝have a temperature 发烧）。此外，还有表示尺寸、重量和性质的名词，例如：pressure 压力，moisture 湿度，strength 强度，voltage 电压，mass 质量，velocity 速度，acidity 酸度，weight 重量，population 人口（群体）等。

2）在 of 短语之前，例如 a temperature of 250°，a voltage of 220kV 等。

3）带形容词时，例如 a high temperature。能这样用的词还有：pressure 压力，velocity 速度，force 力，voltage 电压，density 密度，growth 增长，gravity 重力，strength 强度，water 水，moisture 湿度，current 电流，weight 重量，mass 质量，light 光度，population 人口（群体），velocity，speed（速度），acidity 酸度，resistance 电阻，capacitance 电容，inductance 电感，life 寿命、性命，coefficient 系数，capacity 能力、容量，range 范围，freezing（boiling）point 冰（沸）点等。

最后需要指出的是，有些单词通常在普通英语里是不可数名词，在很专门的科技写作中有时会变成可数名词。例如，对于 steel 和 wheat 可以有：

Cheaper mild steels are now being produced.

较为价廉的软钢正在生产之中。

Improved wheats will be introduced on a wide scale this year.

改良的小麦今年将大规模引种。

出现这种现象可能有两个原因：其一是为了简练，显然，improved varieties of wheat 比 improved wheats 占篇幅；其二是，对于农业和冶金不是很深知的人，可能看不出在各种小麦之间以及在各种钢材之间存在着什么差别，可是从专家来看，这种差别则是很明显的，因此，干脆就使用 steels 和 wheats。

2.4 独指形容词作前置定语与定冠词连用

独指形容词（unique adjectives）主要有 following、same、wrong、right、last、next、present、usual、only、sole、ultimate，它们与形容词最高级或最高级形容词一样，也用于指明独一无二的人或事物，故这类形容词之前一般要用定冠词。这种用法并非是一条恒定不变的规则，却是一种很强的趋势。

（1）following（第二，下一；下述的，下面的）和 same（同样的，相同的；同一个）毫无例外地全部前面加“the”。

1）The following morning，he began experiment. 第二天早上，他开始做实验。

但是需要注意的是，following 作后置定语时其前不加“the”，如 on the day following（第二天）。

2）They even used the same transformers there. 在那里他们甚至使用了相同的变压器。

（2）wrong 和 right 除了前用物主代词或名词所有格以及数词外，几乎均加定冠词。

1）Three wrong answers and you fail the test. 答错三个题考试就不及格。

2）If he gives the wrong answer，the machine stops. 他若是回答有误，机器就会停止运转。

3）We are all in the wrong business. 我们都进入了不该进入的行业。

以上例句表明错误答案或行业可能不止一个。但是，个别情况下加不定冠词。例如：

4）That would create a wrong impression. 那会造成错误的印象。

5）He had given me a wrong address.（对比：the wrong number 号码弄错）他给我的地址不对。

6）She gave me a wrong instruction. 她给我的指令不对。

7）We’ve taken a wrong turn. 我们拐错弯了。（对比：We are late because we took the wrong road. 我们来迟了，因为走错路了。）

8）Give him the right advice. 不要给他错误的建议。

9）Pour the right amount into each bottle. 将合适的分量倒入每个瓶子里。

（3）last 的用法。

1）意为“最后”时，除了前用物主代词或名词所有格外均加“the”。

This is our last measurement. 这是我们最后一次测量的结果。

The last simulation results showed that the proposed method is valid. 最后的仿真结果表明，所提出的方法是正确的。

I began the last section of the operation manual. 我开始阅读该操作手册的最后一部分。

但是，个别情况下与不定冠词连用：

He paid a last visit to the substation. 他最后一次到访该变电站。

2）和表示时间的名词连用，意为“上一个，去（年），昨（晚），过去，（最近）几个时”，除了前用物主代词或名词所有格外，有时加“the”，有时不加。

Our last experiment was conducted in June last year. 我们上一次实验是去年六月做的。

He's been researching the faults on power systems for the last few years. 在过去几年里，他一直在研究电力系统故障问题。

3）意为“最不可能的”或“最新的”和“决定性的”时，均前面加“the”。

This is the last experiment result. 这是最不可能的实验结果。

This is the last thing in energy-saving devices. 这是最新的节能装置。

（4）next 的用法。

1）意为“下一个”时，若是指时间，常不加冠词，而指其他东西时，多前加“the”。

We'll discuss the causes of overvoltages next time（next week，next Friday）. 下次（下周，下星期五）我们将讨论过电压产生的原因。

The next step of the experiment is to measure the current through the capacitor. 该实验的下一步是测量流过电容器的电流。

The next accident took place four hours later. 后面一次事故发生在 4h 之后。

2）意为“第二（天、年等），以后（几天、几周等）”时，多前面加“the”，有时则不加。

He calculated it again the next morning（next morning）. 第二天上午，他又计算了一遍。

3）意为“隔壁的，离……最近的”时，均前面加“the”。

The next room to ours is High Voltage Lab. 我们隔壁的房间是高电压实验室。

（5）present（现在的）除了前用物主代词或名词所有格外均加“the”。

What's your present plan? 你目前的计划是什么？

I'm not satisfied with the present experiment results. 我对当前的实验结果不满意。

（6）usual（平常的，通常的）除了前用物主代词或名词所有格外均加“the”。

1）Reading Electric Power Engineering Handbook is his usual hobby. 通常，阅读电力工程手册是他的业余爱好。

2）The usual phenomena are evident in this company's case. 对于这家公司而言，惯常发生的现象十分明显。

（7）only 除了在“an only child（独生子女）；an only daughter（独生女）；an only son（独生子）；I am an only child（我是个独生孩子）”以及前加物主代词或名词所有格外，均前面加“the”。

1）He was almost her only acquaintance. 他是她唯一的熟识。

2）This is Smith's only exercise. 这是史密斯唯一的运动。

3）This is the only course I have learned about electricity. 这是我学过的唯一一门关于电气的课程。

（8）sole 同 only 一样，但不加不定冠词，这一点不同于 only。

1）His sole end was to prevent the accident. 当时他唯一的目的就是阻止事故发生。

2）The company has the sole right of selling the high-voltage electricity installation. 这是一家唯一有权销售这种高压电气设备的公司。

但是，sole 也有前加零冠词的用法，这时其词义为“独有的”“无人分担的”之意，并非“唯一”之意。例如：

3）He has sole responsibility for the accident. 他独自一人承担这次事故的责任。

4）The company bought sole rights to two kinds of installations. 这家公司购买了两种设备的独家销售权。

（9）ultimate 除了前用物主代词或名词所有格外均加“the”。例如：

1）This is our ultimate (final) proposal, and no other changes will be considered. 这是我们的最后建议，不会考虑再作修改。

2）Scientists are searching for the ultimate truths. 科学家在探求终极真理。

3）The hypothesis regarding the ultimate structure of matter remains to be proved. 关于物质最终结构的假设还有待证明。

但是，在作“最大的”“最根本的”之意解时，除了前加“the”外，有时也前用零冠词。例如：

4）The sun is the ultimate store of power. 太阳是最大的能源。

5）Ultimate authority is exercised by the king. 国王行使最大的权力。

6）Ultimate principles are used here. 这里使用了最根本的原理。

2.5 to be＋动词不定式

连系动词 to be 后跟动词不定式，即 be＋to do 或 be＋to be done，在句中构成复合谓语，这种结构在科技英语中所表示的意义有下列五种。

（1）表示一种按计划、安排、规定的要求将要发生的动作、行为或出现的状态，一般译为“要……”“打算……”“应该……”“将……”等。例如：

1）We were to produce this kind of insulator. 我们准备生产这种绝缘子。

2）The newly established power station is to be put into service next year. 新建的发电厂定于明年开始发电。

3）Where a voltage greater than the rated voltage of the glow-discharge tube or the Zener diode is to be regulated, two or more of these devices may be connected in series across the load. 在待调节的电压比辉光放电管或稳压二极管的额定电压要大的场合，可以在负载两端串联两个或更多的这类器件。

4) Hardness is one of the most important characteristics required of any material if it is to be used successfully for a metal-cutting operation. 任何材料如果要成功地用于金属切削，则硬度是最为重要的性能之一。

5) The fishing ground is to be closed due to over-catching. 由于捕捞过度，这个渔场将被关闭。

6) Hybrid rice is to be planted to 1 million hectares in the province next year. 明年，这个省将种植一百万公顷杂交稻。

7) Hence secondary voltage is produced only as the input voltage approaches or passes through the point where the magnetic flux is to change its direction. 从而，仅仅当输入电压接近磁通量改变方向的点时，二次电压才会产生。

8) We will indicate how the results are to be extended to the general case. 我们将说明如何将这些结果推广到一般情况。

9) The sixth chapter is to discuss the structure of atoms.

(按预定计划) 第六章将讨论原子结构。

(2) 表示一种预期的结果或目的。to be＋动词不定式的主动式或被动式结构用于陈述语气的条件从句（if 从句或 when 从句）时表示一种预期的结果或目的，近似于 want to、intend to，意为“想要”“想”“要”。主句会配有 must、have to 或 essential 等表示必须的词语，主从句译为：如果（假如）想（要）……，就必须（要）……。例如：

1) In the N-P-N amplifier the input voltage must be positive if it is to aid the forward-bias of the emitter-base junction. 在 NPN 放大器中，如果输入电压要使发射极基极结正向偏压，那么输入电压必须是正的。

2) We must continue to carry out experiments, if we are to get further results. 如果我们还要得出进一步的结果，就必须继续做实验。

3) If a power plant is to be built, a suitable site must first be chosen. 要建造一个发电厂，就得先选择合适的厂址。

4) If you are to finish the testing at 7：00 p. m., you have to use that equipment. 假如你想在下午 7 点之前完成测试，就必须使用那台设备。

(3) to be＋动词不定式的被动式等于情态动词。

1) 当 to be 后跟肯定的不定式被动形式即“is (are) to be＋过去分词”的结构时，有时具有“可以，能够”“应该，必须”(can be, must be, ought to be) 的含义。例如：

a. Such books are to be found in any library. 这种书哪个图书馆都可找到。(are to be found＝can be found)

b. The calculations are to be rechecked. 这些计算须再核对。(are to be rechecked＝must be rechecked)

c. This substance is rarely to be found free in nature. 这种物质很少能发现其在自然

界以游离状态而存在。(is rarely to be found=can rarely be found)

d. The proposal states that only materials such as polymer electrolyte membranes are to be used in low temperature fuel cells. 申请书规定，只有类似聚合物电解质膜这种材料才可以用于低温燃料电池。(are to be used=can be used)

e. The problem is to be solved in two ways. 此问题可用两种方法解决。(is to be solved=can be solved)

f. No voltage was to be measured. 测不到电压。(was to be measured=can be measured)

g. Not a sound was to be heard. 一点声音也听不见。(was to be heard=can be heard)

h. These instructions (The lab regulations) are to be strictly observed. 这些说明（实验室规则）必须严格遵守。(are to be observed=must be observed)

i. They are to be congratulated on their brilliant discovery. 应当祝贺他们了不起的发明。(are to be congratulated = ought to be congratulated)

2）当 to be 后跟否定的不定式被动形式即 is (are) not to be+过去分词时，它不同于它的肯定主动式。这种句型表示的意思是“禁止”“不许”“不准”等。例如：

a. These reference books are not to be taken out of the library. 这些参考书不准携进（图书）馆外。

b. The devices are not to be taken out of the lab. 仪器不准携带出实验室。

c. This liquid is not to be exposed to air, for it will soon evaporate. 这种液体不可暴露在空气中，因为它会很快蒸发掉。

d. The transformers are not to be used without being tested. 这些变压器不测试就不准使用。(are not to be used = must not be used)

(4) to be+动词不定式的主动式等于情态动词 should，must。表示根据规章，客观规律或情况准会发生或必须进行的动作。例如：

1）You are to stop (=should stop) the machine carefully. 你们要小心地停下这台机器。

2）We are to measure (=must measure) the diameter of this small axle by means of a micrometer. 我们必须用千分尺来测量这根小轴的直径。

有时也后接动词不定式的被动式，例如：

3）Whenever occasion arises intensive observations are to be done to get the information on the fine structure of the atmospheric event. 每当时机出现时，都应进行精确观测以便获得这一大气事件的精细结构的资料。

需要注意的是还有一类“be+动词不定式”结构的句子，其意义是不同的，例如：

1）One of the chief uses of zinc is to protect the surface of iron against rust. 锌的主要用途之一是防止铁的表面生锈。

2）The main objective of this book is to present circuit analysis in a manner that is clearer, more interesting, and easier to understand than earlier texts. 本书的主要目的是比以往教材更加清晰、有趣、易懂地讲述电路分析的内容。

3）Our main task now is to repair these instruments. 我们当前的主要任务是修理这些

仪器。

4）Our wish is to serve the people. 我们的志愿是为人民服务。

这两类句子的区别在于，从逻辑上看，后者句中动词不定式短语即为主语所指的具体内容，而前者句中动词不定式短语则并非是主语所指的具体内容；从语法上来说，后者句中的“to be”是系词，意思是“是”，其后的动词不定式是表语，但前者句中“to be”不是系词，其后的动词不定式也非表语。

2.6 单词作定语

英语中绝大多数单词作定语时都是前置的，但是也有少数是可以后置的。下面介绍前置定语的一些特殊含义以及后置定语的情况。

2.6.1 单词作前置定语

单词定语一般多置于其所修饰的中心词之前，例如 our lab（我们的实验室）、renewable energy（可再生能源）、sliding mode control（滑模控制）、automated factory（自动化工厂）等。但即便是单词定语，倘若是表示具体事物的名词，而中心词是表示抽象性状的名词时，则必须按照汉语习惯变动原文词序，将原文定语与中心词的偏正关系在汉文中予以对调。例如，life expectancy 应译为“预期寿命”，smudge resistance 应译为“抗污性”，cost difference 应译为“差价”，诸如此类。此外，有些形容词作定语，修饰的是中心词的词根，或是复合名词的第一部分，因此翻译时必须正确理解，切不可看待为一般的作为前置定语的形容词。对于前者，例如 painless dentist 中的 painless 并不是修饰 dentist，而只修饰 dentist 的词根 dent，故 painless dentist 应译为“无痛牙科的牙医”，而不是“无痛的牙科医生”。同理，physical chemist 应译为“物理化学家”，nuclear physicist 应译为“核物理学家”，artificial florist 应译为“纸花商”。对于后者，例如 an old bookstall 中的 old 所修饰的乃是 bookstall 这个复合名词的第一部分 book，而不是 bookstall，应译为“旧书摊”。同理，a public schoolboy 应译为“公立学校的男生”，a historical playwright 应译为“历史剧作家”。像这种与中心词具有特殊关系的形容词为数不少，汉译时都必须正确理解才不致造成误译。一类是接近于名词化的形容词，例如 nuclear deployment 中的 nuclear，它与 deployment 之间的关系，严格言之，并非定语与中心词的关系，而是组成复合词两个部分之间的关系。Nuclear deployment 可分析理解为 deployment of nuclear arms，汉译应为“核武器部署”。同理，a mental hospital 汉译应为“精神病医院”（不是“精神医院”），a deaf and dumb teacher 汉译应为“聋哑人（的）教师”（不是“聋哑教师”）。另一类是斯威特（H. Sweet）所称的压缩形容词，是一种省略了真正所修饰的中心词的形容词，例如 a scientific writer 中的 scientific 所修饰的中心词并非 writer 而是省略了的 topics 或 articles 之类的词。Scientific writer 应理解为 a writer who writes on scientific topics（articles），汉译应为“科学作品（文章）作家”（不是“科学的作家”）。同理，a long credit 应译为“长期信贷”，不能译为“长的信贷”。

2.6.2 单词作后置定语

有时单词定语可能置于中心词后面，有下面几种值得注意的情况：

（1）某些副词充当定语，例如 the note below（下面的注解）、the equation above（上面的方程）等；再如：

Things far off look as if they are smaller than they really are. 远处的物体看起来比他们原来的尺寸小。

（2）有些以-able、-ible 结尾的描述性形容词，这些词往往放在被修饰的名词之后，如 every means conceivable（一切可能设想的办法）；再如：

1）They have solved the problem in spite of difficulties unspeakable. 尽管经受了种种难以言传的困难，他们还是解决了这个问题。

2）Engine revolution should not exceed the maximum permissible. 发动机的转数不应超过所允许的最大数值。

3）The working stress allowable is based not only upon the material but also upon the type of loading. 许可工作应力不仅根据材料来定，还要根据负荷的类型来决定。

当然，这类形容词大多也可以前置，其基本含义不变，但是略有细微差异：作前置定语时，表示在一般正常条件下可能发生的现象；作后置定语时，表示在某特定的条件下发生的一时的现象即具有暂时和条件的含义。试比较：

The sun is one of the visible stars to the naked eye. 太阳是肉眼看得见的恒星之一。（一般现象）

This is one of the stars visible to us through a radio telescope. 这是只有借助射电望远镜才能被我们看见的恒星之一。［在这种特定条件（或时间）下的可见星星：特殊现象］

再例如：

This is a river navigable. 这是一条（暂时）可通航的河流。

This is an instrument suitable. 这是一台适合（某种用途）的仪器。

（3）有些用单个的分词或形容词所表示的定语，通常后置。例如：

the parties concerned	有关各方面
the victory won	所取得的胜利
the mistakes made	所犯的错误
a position unique	独一无二的地位
the day following	下一天
the ranges given	所规定的范围
the methods involved	所涉及（或用到）的方法

还有的形容词，如 next、last，同 following 一样一般位于名词之前，也可以后置，如 Monday last、Tuesday next 等。concerned 作前置定语表示“担心”，例如，“He has a very concerned look（他表情忧虑）”，involved 作前置定语表示“复杂”，例如，the involved methods（复杂的方法）。

再例如：

Carbon occupies a position unique in the science of chemistry. 碳在化学中具有独一无二的地位。

Variations outside the range given are not unlikely. 超出规定范围的变化，并非不可能。

（4）保留有较强的动词意义的过去分词起强调作用。例如，the pressure required（所需要的压力）、the force exerted（所施加的压力）、the subject discussed（所讨论的题目）。再例如：

1）The heat energy produced is equal to the electrical energy utilized. 所产生的热量和所使用的电能相等。

2）The work obtainable equals that expended. 可得到的功和所消耗的功相等。

3）All told，the chemical elements known have several hundred isotopes. 总的算来，已知的化学元素有几百种同位素。

4）As the rated voltage increases，the dimensions required become so large that this form of bushing is not a practical proposition. 随着额定电压的增高，所需要的体积也增大，使得这种形式的套管不再实用。

必须强调指出，这种分词有时也可置于中心词之前，但与后置时的含义大不相同，汉译时要注意加以区别。例如：

This is heretofore the best process known. 到现在为止，这是所知道的最好方法。

This is heretofore the best known process. 到现在为止，这是最有名的方法。

（5）具有表语作用的形容词。

1）以 a-开头的形容词，通常只能作后置定语。例如：

He is a well-known scholar of the 1940's alive. 他是仍然在世的 20 世纪 40 年代的著名学者。

2）necessary，available 和 present 等。例如：the engineers present 在场的工程师、the number present 出现的数目。再如：

The amount necessary is about 50 c. c. 所需要的用量约 50 mL。

The voltage available is 50V. 可用电压是 50V。

Even at low pressures there are still larger number of molecules present. 即使在低压时，仍然有大量分子出现。

This is the only information available. 这是唯一可得到的资料。

3）possible 所修饰的名词前已有 all、only、every 之类的词或形容词最高级。例如：all the ways possible（一切可行的方法），the best way possible（可能的最佳途径）。再如：

We must find the faults in every way possible. 我们必须用各种可能的方法找出故障。

This is the only solution possible. 这是唯一可能的解释。possible 有时也可前置，例如：the best possible way（极可行的方法）、every possible way（每种可行的方法）。

4）imaginable 所修饰的名词前已有 all、only、every 之类的词或最高级形容词。例如：

This is one of the most difficult problems imaginable. 这是可以想象得出的最困难的问题之一。

This is the only reason imaginable. 这是想得到的唯一理由。

They have overcome this difficulty by all (every) means imaginable. 他们用各种想得到的方法克服了这一困难。

综上表明，含有被动意味的形容词作定语，而被修饰词前面常常另有最高级形容词或all、only、every 之类的词作定语时，形容词后置，但有时也可以前置。例如：

This is the only data available. 这是仅有的数据。

It is a matter of the available resources and technology in China. 这是一个中国现有的资源和技术问题。

(6) 其他一些单词受外来语的影响，或者由于习惯用法，作定语时也常后置，但翻译为中文时仍然置于中心词前面。例如：sum total（总额）、payment due（到期应付之款）、their experience overseas（他们在海外的经历）等。再例如：

The sum total of innumerable relative truths constitutes absolute truth. 无数相对真理之总和，就是绝对真理。

2.7 短语定语的位置

英语中的不定式短语、分词短语、介词短语、形容词短语等作定语时，通常在所修饰的词之后。例如：

1) The force to change the motion of a body is proportional to the mass of body. 改变物体运动的力与该物体的质量成正比。

2) The work done is the product of the force acting on the body and the distance traveled by the body. 所做的功等于作用在物体上的力与该物体通过的距离的乘积。

3) Transistors designed for low-noise applications operate at considerably lower current densities than those producing the highest power gain. 设计用于低噪声用途的晶体管，与给出最大功率增益的晶体管相比，其工作电流密度相当小。

4) Air layers near to the earth have greater density than those at higher altitude. 靠近地球的空气层密度大于高度较高的空气层。

5) Transistors suitable for power amplification must deliver power efficiently with sufficient gain at the frequency range of interest. 适合于功率放大的晶体管，应能高效地供给功率，而且在有关频带内又有够高的增益。

但是，英语中的分词短语较短时，放在所修饰的名词之前的现象也并不少见。这种短语通常以“表示方式或程度的副词＋现在分词或过去分词”的形式出现。例如：

1) If a negatively charged rod is brought near a suspended ball, we observe the ball to be attracted at first. 如果把一根充有负电荷的棒放到一只悬着的球附近，我们首先会观察到球被吸引住。

2) The most extensively used indicator on weather radars is the plane-position indica-

tor. 气象雷达最广泛使用的显示器是平面位置显示器。

3) Fluids transmit pressure with litter loss because they consist of freely moving molecules. 流体传送压力时损耗很小，因为流体由自由移动的分子所组成。

2.8 科技英语中常见的复合结构

复合结构是英语中的一种特殊结构，它由存在着逻辑主谓关系的前后两部分组成，两者在句法上构成一个整体。科技英语中常见的复合结构有复合宾语、不定式复合结构、分词复合结构以及动名词复合结构，它们的共同特征是具有逻辑意义上的主谓关系。汉译时必须将这种主谓意义表示出来。

这里着重介绍科技英语中常见的一些用法特殊或容易译错的复合结构。

2.8.1 分词的复合结构

1. have 或 get＋宾语＋现在分词（宾语补足语）

这种结构表示主动意义，即使某人或某事做某事。翻译时常在宾语前加“使”“把”“让”“促使”、“要使”“要求”“允许”等词。例如：

(1) I shall have the engine going by the time you get back. 等你回来的时候我就把汽车引擎发动起来了。

(2) Can you get the old generator working again? 你能使这台旧发电机重新运转吗？

(3) Once we get the automobile rolling, it would tend to roll a uniform rate in a straight line. 一旦我们把汽车开动行驶，它往往就会沿直线作匀速运动。

2. have 或 get＋宾语＋过去分词（宾语补足语）

这种结构一般含有两种完全不同的意思：一种是有意识的，即以主语的意志为转移的，或者说主语的意志致使某事发生或被做到，简称为“致使”；另一种是无意识的，即该动作不是以主语的意志为转移的，或者说与主语的主观意志无关，意为被动。过去分词所表示的动作往往是别人完成的，有时该动作也由主语完成。若是有意识的，则翻译时可在分词之前加“叫人”“找人”“请人”“让人”“使人家”等；但根据具体情况，也可不加任何词，尤其是在表示主语自己完成某件事时。若是无意识的，则不需加词，直接译出。

(1) 表示主语的意志（“致使”），例如：

1) The engineer had the results of the experiments checked and rechecked. 这个工程师叫人核对再核对了他的实验结果。(别人完成)

2) You should have the two steel pipes welded together. 你应该把这两根钢管焊接起来。(主语自己完成)

3) We often have our test-tubes broken. 我们经常把试管打破。(主语自己完成)

4) Electric heaters have electric energy turned into heat. 电热器使电能转变为热。(不能译成：电热器具有转变为热的电能。)(主语自己完成)

5) Jigs and fixtures have the work accurately machined. 钻模和夹具使工件加工准确。(主语自己完成)

6）If you have an object oiled，you will find friction reduced. 假如你给一物体上了润滑油，你会发现摩擦减少了。（主语自己完成）

7）We must get our products examined. 我们得请人把产品检验一下。（别人完成）

8）I must get the lab work done tomorrow. 明天我一定把这项实验做完。（主语自己完成）

（2）与主语的意志无关（被动），例如：

1）I had my leg broken in the accident. 我在事故中摔断了腿。

2）She got her purse snatched away on the 6th Avenue. 她的钱包在第六大道被人抢走了。

3. 视听感觉动词

see、feel、hear、notice、watch 等动词或 consider、declare、find、imagine、keepleave、observe、set、smell、start、want 等＋宾语＋现在分词或过去分词（宾语补足语）。

如果这些动词为主动语态，分词作宾语补足语；如果为被动语态，分词则作主语补足语。通常将这种结构中的分词译成其原动词之意，例如：

（1）With the help of a special instrument，we can see the electric waves traveling along. 借助一种专门的仪器，我们能看到电波向前传播。（宾语补足语）

（2）To keep a 100,000 ib. airplane flying takes only about 1,000ib. of propeller pull. 要让一架 100000 磅重的飞机继续飞行，只需大约 1000 磅的螺旋桨推力。（宾语补足语）

（3）To start a body moving requires a greater force than to keep it moving. 要让一物体开始运动，所需的力要比维持它继续运动的力大得多。（宾语补足语）

（4）They were seen making an experiment on high voltage insulation. 曾有人看见他们在做高压绝缘实验。（主语补足语）

（5）Sulfur is also found combined with many metals. 发现硫也同许多金属化合在一起。（主语补足语）

2.8.2 动名词的复合结构：名词（代词）＋动词-ing

动名词前面可以带有自己的逻辑主语，两者一体合称为动名词的复合结构。其逻辑主语可以是物主代词、名词的所有格或名词的普通形式（名词通格）等。依据动名词的含义可知，动名词的复合结构在句中的主要作用与普通动名词相同，在科技英语中主要作介词宾语，还可以作主语和表语（极少有）。下面仅介绍动名词的复合结构作介词宾语时容易发生误解的两种情况。

1. 介词＋its（their）＋动名词

在这一结构中，物主代词 its（their）与动名词组成动名词复合结构，常作介词的宾语。这里的 its（their）是动名词的逻辑主语，所以不要译成定语即“它的（它们的）”，而译成“它（它们）”或译成所代的名词，而将“its（their）＋动名词”译成汉语的主谓结构。例如：

（1）The transformer cannot be called a machine because of its having no moving parts. 变压器不能称为机器，因为它没有可动部件。

(2) Robots gain an ever increasing application due to their being constantly improved. 由于机器人经常加以改进，所以其适用范围不断扩大。

(3) Electronic computers cannot replace man because of their lacking the ability to think. 电子计算机不可能代替人，因为它缺少思维能力。

(4) The earth, in its travelling around the sun, meets many of meteor fragments. 在地球绕太阳运行时，它碰到许多流星碎片。

2. of＋名词＋动名词（短语）

现代英语中，动名词的逻辑主语多用名词来表示。由名词和动名词组成的动名词复合结构，常作主语或介词宾语，在科技英语中尤为普遍。特别是“名词＋动名词（短语）”常作介词 of 的宾语，在形式上与“of＋名词＋现在分词（短语）”相同，所以容易把动名词（短语）看作是作后置定语的现在分词（短语）。当动名词是主动语态时，将“名词＋动名词（短语）”译成汉语的主谓结构；若动名词是被动语态，则还可译成汉语的动宾结构，也就是说把被动语态的动名词译成动词，而把逻辑主语的名词译成宾语。例如：

(1) We know of the earth behaving as a large magnet. 我们知道，地球相当于一个大磁体。（“behaving…”并非是现在分词短语作“the earth”的后置定语，“the earth behaving as a large magnet”整体作为一个动名词短语起名词作用，作介词 of 的宾语）

(2) The process of heat travelling from one end of a wire to the other is known as conduction. 热从导线一端传到另一端的过程称为传导。（译为主谓结构）（不要将“travelling…”视为现在分词短语，作 heat 的后置定语而译成“从导线一端传到另一端的热的过程”）

(3) Most commonly, fire is the result of oxygen combining with carbon. 大多数情况下，火是氧与碳化合的结果。（不可译成：大多数情况下，火是与碳化合的氧的结果。）

(4) The process of one substance mixing with another because of molecular motion is called diffusion. 由于分子运动，一种物质与另一种物质混合的过程称为扩散。（译为主谓结构）

(5) Man has had the idea of the earth revolving about the sun only since the 18th century. 从 18 世纪以来，人们才有了地球绕太阳转的思想。（译为主谓结构）

(6) Here is a clear case of magnetism being converted into electric current. 这里就有一个磁转化为电流的明显例子。（译为主谓结构）

(7) The possibility of sea water being used to generate electricity has been realized. 用海水发电的可能性已经认识到了。（译为动宾结构）（不可译成：正在用来发电的海水的可能性已经认识到了。）

(8) There is every possibility of plastic material being used instead of steel and cast iron here. 完全有可能用塑料来代替钢和铸铁。（译为动宾结构）

(9) The idea of the atom being indivisible is untrue. 原子是不可分裂的概念是不正确的。（译为主谓结构）

(10) The sophisticated design reduces the risk of foreign matter entering the device

under the water. 这种尖端的设计方法减少了该仪器在水下工作时，外来物质进入其中的危险。(译为主谓结构)

一般而言，当动名词的复合结构作主语时，大多用名词的属格（所有格）和代词的所有格作逻辑主语；在这种复合结构作宾语时，则多半用名词的普通形式（名词通格）或代词宾格作逻辑主语，尤其在口语中更是如此。理由之一是很多长句便于上口。例如：

We are discussing the possibility of the house（不用 house's）being converted into a substation. 我们在讨论把这栋房子改为变电站的可能性。(译为动宾结构)

理由之二是为了避免误解。例如：

He didn't like the idea of his daughter going to work at a substation at night. 他不喜欢他女儿晚上在变电站上班的念头。（若用 daughter's，在口语中和 daughters 的发音一样，听的人以为他不止一个女儿。）

尽管如此，有些名词所有格还是可作为动名词的逻辑主语。例如：

(11) The metal's being workable is one of the reasons why it is so widely used in industry. 金属的可加工性是它在工业上得以广泛使用的原因之一。

名词＋动名词（短语）作其他介词的宾语比较好辨认，例如：

They objected to the local government building a nuclear power plant there. 他们反对地方政府在那里建一座核电站。

2.8.3 动词不定式的复合结构

动词不定式也可以构成复合结构，下面讨论在科技英语中易发生误解的几种不定式复合结构。

1. too＋形容词或副词＋for＋逻辑主语（名词或代词）＋不定式

这种结构中，for 引导不定式的逻辑主语，从而构成不定式的复合结构，它是"too＋形容词（副词）＋不定式"结构的变形，即为不定式添加了一个逻辑主语。不定式复合结构"for＋逻辑主语＋不定式"作结果状语说明副词 too，表示"太……以致不能……"的意思，翻译时多采用状语分译法，把作结果状语的不定式复合结构断开，译成一个否定句。例如：

(1) The signal is too large for the transistor to handle. 这个信号太大了，以至于该晶体管无法处理。

(2) The speed of light is too great for us to measure in simple units. 光的速度太快，我们不能用普通的单位来计量。

(3) Modern high-strength alloys are often too hard for ordinary machine tools to cut. 现代高强度合金太硬，普通机床往往不能切削。

(4) Thermal expansion of solids is too small for our eyes to detect. 固体的热膨胀很小，以至于我们的眼睛觉察不到。

2. 形容词＋enough＋for＋逻辑主语（名词或代词）＋不定式

在这种表述方式中，不定式复合结构作副词 enough 的状语，两者组成副词短语，作形容词的程度状语，一般可译成"……得足以使……"。例如：

(1) The voltage is great enough for an ion to acquire between collisions sufficient energy. 电压大得足以使离子在两次碰撞之间获得充分的能量。

(2) The temperature is high enough for metal to melt. 温度高得足以使金属熔化。

(3) The attracting force of the earth is just large enough for the moon to travel in its orbit. 地球的引力恰好大得足以使月球在自己的轨道上运行。

3. there be 句型的动词不定式复合结构

There be 句型的动词不定式复合结构虽然并不常见，但却是一个难点。它是一种书面语，在科技文体中，主要有以下两种情况。

(1) 动词不定式复合结构形式：for there to be＋名词。这是由 there be 句型构成的带逻辑主语的不定式复合结构，介词 for 负责引出不定式的逻辑主语，但其引出的逻辑主语要放在逻辑谓语 to be 之后。这种形式的不定式复合结构在句中可作为不定式可作的各种非谓语成分，表示非主谓关系的"有（存在）"之意。翻译时可译成"有（存在）……"等。例如：

1) In such cases it is impossible for there to be an escape of oil from the transformer. 在这些情况下不存在油从变压器溢出的可能。（"for there to be an escape of oil from the transformer"在句中作真正的主语，it 为形式主语）

2) It is possible for there to be a current in the circuit under such a condition. 在这种条件下，电路中有可能存在电流。（"for there to be a current in the circuit under such a condition"在句中作真正的主语，it 为形式主语）

3) For there to be life there must be air and water. 要有生命存在，就必须有空气和水。或译为：为了使生命能够存在，必须要有空气和水。（"for there to be life"在句中作目的状语）

4) For there to be no mistake, you must recheck the results got from the experiment. (＝In order that there may be no mistake, you must recheck the results got from the experiment.) 为了不出差错，你必须再次核对实验结果。（" For there to be no mistake"在句中作目的状语）

(2) 复合主语或复合宾语形式：there … to be …。例如：

1) When one of the charged bodies has a polarity opposite to the other, there is said to be a difference of potential between them. 当两个带电体极性相反时，我们就说，在它们之间存在一个电位差。（"there is said to be a difference of potential between them"构成复合主语，there 作主语，而 to be …为主语补足语）

2) Now we assume there to be just two variables in the function. 现在我们假设，在这个函数中只存在两个变量。（"there to be just two variables in the function"构成复合宾语，there 作宾语，而 to be …为宾语补足语）

2.8.4　let＋there＋be＋名词

动词 let 所要求复合宾语为"名词（代词）＋不带 to 的不定式"，如 Let the iron touch the magnet（让铁与磁铁相接触）。"let＋there＋be＋名词"结构是"let＋名词＋不带

to 的不定式”结构的变形，这里的“there＋be＋名词”是由 there be 句型变化而来的复合宾语，作宾语的名词要放在补足语 be 之后。翻译时可按此结构表示“让……有（存在）……”的意义来灵活处理。例如：

（1）Let there be no mistake about the program. 这个程序不要搞错。

（2）Let there be an isolated point charge situated at this point. 把孤立的点电荷放在这里。

2.8.5 With 的复合结构

With 的复合结构的一般形式为

With＋名词或代词＋
- 分词（短语）
- 介词短语
- 形容词（短语）
- 副词
- 不定式短语
- 名词（短语）

其否定形式用 without 或 with＋no/neither/none 来表示。这种结构在科技英语写作中使用广泛，这一短语在结构上有它自己的特点，也就是在逻辑上，短语中有其自身的逻辑主语和谓语，即介词 with（或without）后的名词或代词是整个结构的逻辑主语，紧随其后的其他词就是该短语中的逻辑谓语（或表语）。with 在此失去原有词义，仅起语法功能作用，翻译时省略不译。但 without 除了具有语法功能外仍具有否定的词义，必须译出否定之义，常加“……的”，置于其所修饰的名词之前，修饰该词。这种由 with（或 without）引导的在逻辑上具有主、谓结构的介词短语在句中作状语或定语，具体讨论如下：

（1）with 结构通常在句中作状语：处于句首表示条件、时间和原因等，处于句尾表示附加说明、方式、条件、让步、伴随情况等。这时可视具体情况译成状语从句或并列分句，译为主谓结构或动宾结构。

1）位于句首：

a. With its base grounded，Q4 is a very high impedance.（with 短语表示条件）在其基极接地的情况下，Q4 的阻抗值很高。

b. With its base voltage 0，transistor Q1 will be cut off.（with 短语表示条件）若基极（电压）为零，晶体管 Q1 就会截止。

c. With friction present，a part of power has been lost as heat.（with 短语表示原因）由于存在摩擦，所以一部分功率转化为热能而损耗掉了。

d. With Q2 on，the voltage fed to the base of Q3 rises.（with 短语表示时间）Q2 导通时，加给 Q3 基极的电压就上升。

e. With these definitions in hand，Eq.（6）can be transformed term by term.（with 短语作条件状语）有了这些定义后，就可对式（6）逐项进行变换了。

f. With the volume of gas being held constant，the pressure it exerts in its container depends upon its temperature.（with 短语表示时间）体积保持不变时，气体对容器的压力

取决于气体的温度。

g. With the atomic theory of matter placed on an experimental basis, chemistry developed as a separate science.（with 短语作原因状语）由于把物质原子论放在实验基础上，化学发展成为一门独立的科学。

h. Without the temperature or pressure changed matter can never change one state into another.（with 短语表示条件）如果温度或压力不变，物质永远不能从一种状态变成另一种状态。

i. Let us test the temperature of a basin of water with both our hands placed together in the water.（with 短语表示方式）让我们把两只手都放在水中来试一盆水的温度。

j. With the voltage $u(t)$ in Eq.（3），we can find the current $i_R(t)$. 利用式（3）已经求得的 $u(t)$，就可以求得 $i_R(t)$。

2）位于句末：

a. Each planet revolves around the sun in an elliptical orbit，with the sun at one focus of the ellipse.（with 短语作伴随情况状语）每颗行星在椭圆轨道上绕太阳运行，而太阳则处于椭圆的一个焦点上。

b. A slide projector is to be used with its lens 29 ft from a screen.（with 短语作伴随情况状语）应该使用一台透镜离屏幕 29 英尺的幻灯机。

c. Standard screws are all right-handed，with left-hand ones employed only for special purposes.（with 短语作伴随情况状语）标准螺钉都是右旋的，左旋螺钉只用于特殊目的。

d. In gases，the particles are far apart，with empty space between.（with 短语作伴随情况状语）在气体中，这些微粒相距很远，其间都是真空。

e. The condition of resonance can be achieved with L and C either in series or in parallel.（with 短语作条件状语）将 L 和 C 串联或并联均可获得谐振状态。

f. This parameter shall be measured with E grounded. 这个参数应该在 E 接地的情况下加以测量。

g. The density of air varies directly as pressure，with temperature being constant.（with 短语作条件状语）温度不变，空气的密度与压力成正比。

h. The article deals with microwaves，with particular attention being paid to radio location.（with 短语作伴随情况状语）这篇文章论述微波，特别注意无线电定位。

i. An object may be hot without the motion in being visible.（without 短语作让步状语）一个物体，即使其内部运动不可见，仍可能是热的。

j. The satellite is circling the earth，with its solar batteries being charged by the sun. 卫星正围绕地球运转，其太阳能电池由太阳充电。（with 短语表示伴随情况）

k. The article deals with high voltage engineering and testing，with particular attention being paid to basic testing techniques.（without 短语表示伴随情况）本文专论高电压工程与测试，尤其注意基本测试技术。

l. The piece of iron may move towards the magnet without the magnet itself moving.（without 短语表示附加说明）这块铁可向磁铁移动，而磁铁本身不动。

m. In a transformer, energy is transferred by magnetic coupling, without electrical connection between the primary circuit and second circuit.（without 短语表示附加说明）变压器中的能量传输是通过磁耦合进行的，其一次电路和二次电路之间并无电气上的联系。

（2）with 结构有时也可作后置定语，修饰其前面的另一名词。例如：

1）Atoms with the outer layer filled with electrons do not form compounds. 外层布满了电子的原子不能形成化合物。

2）Equations with radicals in them are normally solved by squaring both sides of the equation. 含有根式的方程通常是通过对方程两边进行平方来求解的。

3）The device with buttons on it is a keyboard. 上面带有按键的装置就是键盘。

4）This is an inequality with zero on the right side. 这是一个右边为零的不等式。

5）For this purpose we construct a new set of axes with their origin at the Q point. 为此，我们建立原点处于 Q 点的一组新轴。

6）A body with a constant force acting moves at constant speed. 作用在物体上的力不变，该物体即以恒速运动。

7）The body with a spring connected to it is ready to work. 接上弹簧的物体，可以随时做功。

8）Aluminum, with a weight one-third that of steel, can be given a strength approaching that of steel when it is alloyed with small quantities of copper, manganese and magnesium, and subjected to heat treatment processes.（with 短语作后置定语，说明 aluminium）铝重量是钢的 1/3，如果和少量的铜、锰和镁熔合并加以热处理，能获得接近于钢的强度。

9）The iron core with the coils wound on it is called an armature. 缠绕着线圈的铁芯叫做电枢。(译为动宾结构)

10）Modern computers with all elements miniaturized can be made much smaller. 所有元件都已小型化的现代计算机能够制造得更加小巧。(不要译为：带有已小型化的所有元件的现代计算机能够制造得更加小巧。)

11）A body with a constant force acting on it moves at a constant speed. 受恒力作用的物体作等速运动。

12）Pig iron is alloy of iron and carbon with carbon content more than 2 percent. 生铁是铁碳合金，其含碳量在 2%以上。

13）There are a number of materials known as semiconductors with properties midway between those of conductors and insulators. 还有许多叫作半导体的材料，它们的特性介于导体和绝缘体之间。

14）When two loops with or without contacts between them affect each other through the magnetic field generated by one of them, they are said to be magnetically coupled. 若两个彼此之间有或无接触的回路通过其中一个回路产生的磁场而相互影响，则称为磁耦合。

需要注意的是：

1）结构“without＋名词＋分词＋…”一般不作定语，作状语。

2）with＋名词＋逻辑表语（形容词、介词短语或副词）结构中，分词是 being，常常省略，而留下作逻辑表语的形容词、介词短语或副词，其作用与译法与“with＋名词＋分词＋…”结构完全相同。在翻译时可把名词和它后面的形容词、介词短语或副词作为主语和表语的关系来处理。例如：

a. The product is very pure with little impurities.（with 短语表示伴随情况）这种产品很纯，内含杂质很少。

b. In general, all the metals are good conductors, with silver the best and copper the second. 一般来说，金属都是良导体，其中以银为最好，铜次之。（with 短语表示伴随情况）

c. An uncharged object contains a large number of atoms, each of which normally contains an equal number of electrons and protons, but with some electrons temporarily free from atoms.（with 短语表示让步）不带电的物体含有大量原子，每个原子通常含有等量的电子和质子，但是有一些电子暂时脱离了原子的束缚。

d. The electronic current flows through the circuit with the switch on. 开关接通时，电流便在线路中流动。（with 短语表示时间）

2.9 科技英语中的情态动词

除了一般现在时之外，科技英语中最常见的动词形式是含有情态动词的形式。而常用的情态动词是：

第一组：	can	may	might	could
第二组：	will			
第三组：	should	must	have to	

情态动词与动词的原形连用会赋予句子额外的意义。在英语口语中，很难非常准确地指出这些额外的意义，但在科技英语中则比较容易做到这一点。下面分别分析上述三组情态动词的含义。

1. 第一组情态动词（can、may、might、could）常用来表达可能性和或然性（也许、或许）

1）The glass bottle breaks when dropped. 这玻璃瓶子掉下时会破碎。（一般现在时：每当这种瓶子从这样的高度掉到这样的地面上时，它就破碎。破碎的概率约为98%～100%。）

2）The bottle can break when dropped. 这瓶子掉下来时很可能破碎。（can：破碎的可能性相当大，概率约为 40%～70%。）

3）The bottle may break when dropped. 这瓶子掉下来时有可能破碎。（may：破碎有一定的可能性，概率约为 20%～40%。）

4）The bottle could/might break when dropped. 这瓶子掉下来时说不定会破碎。

（could/might：可能性很小，概率约为 5%～20%。）

5）The bottle cannot break when dropped. 这瓶子掉下来时不可能破碎。（cannot：几乎没有可能性但仍有极低的概率，概率约为 0%～2%。）

需要注意的是，might 和 could 不是 may 和 can 的过去式，而只是表示可能性更小。can 与 may 和 might 的用法不同，can 用于谈论更为一般性的或“技能上”的可能性，而后者则用于议论具体事例发生的可能性。试比较：

The voltage can be stepped up or stepped down. 电压可以升高或降低。

They may decide to step up the voltage at this substation. 他们也许会将这个变电站的电压升高。

还需注意的是，may not 和 can not 的意思是不一样的。试比较：

This may not be the device you need. = It is possible that this is not the device you need. 这可能不是你需要的仪器。

This can not be the device you need. = It is not possible that this is the device you need. 这不可能是你需要的仪器。

2. 第二组情态动词（will）表示预言或预见

（1）The sea-water corrodes the iron. 这片海水腐蚀这块铁。

（2）The sea-water will corrode the iron. 这片海水将腐蚀这块铁。

其中，（1）是陈述某一地区的海水对某一块铁的作用，而（2）则是预言某一地区的海水对某一块铁的作用。由于做预言就是断言某事将要发生，例如：It will rain tomorrow.（明天将下雨。）因此，（1）和（2）在含义上有差别。再对比：

（3）Sea-water corrodes iron. 海水腐蚀铁。

（4）Sea-water will corrode iron. 海水将腐蚀铁。

其中，（3）是基于对科学定律的了解而做出的普遍陈述，（4）是基于对科学定律的了解而作出的预见。然而，陈述必定发生的某件事与预言这件事实际上是一回事，所以（3）和（4）含义相同。由此得出如下两点结论：

（1）在普遍的、总结式的科学陈述中，will 和一般现在时含义相同。因此，下面两句都成立：

If pure water is heated to 100℃ at sea level，it boils. 如果在海平面将纯水加热到 100℃，它就沸腾。

If pure water is heated to 100℃ at sea level，it will boil. 如果在海平面将纯水加热到 100℃，它就会沸腾。

（2）在有疑问的情况下，陈述（一般现在时）和预言（will）仍然有差别。

Sea-water does not corrode this new alloy. 海水不腐蚀这种新合金。（可能是错误的）

Sea-water will not corrode this new alloy. 海水不会腐蚀这种新合金。（有待证实）

3. 第三组情态动词

should 常用于注意事项和操作说明，must 表示“必须”，have to 表示“不得不”。should 为这三个中最有用的情态动词，例如：

1）Students should be careful when using high voltage electrical equipments. 学生在

使用高压电气设备时要小心。

2) Concrete should contain at least 12% cement. 混凝土至少应含 12%的水泥。

3) The applied voltage should not exceed 300V. 外加电压不可超过 300V。

2.10 含有 must 和 should 的被动语态

含有情态动词的被动结构 "must be+动词-ed" 和 "should be+动词-ed" 通常用来描述应该或不应该做的事。含有这两个情态动词的被动结构在写操作说明、注意事项和通告时用得特别普遍。

All library books should be returned to the library by the end of June. 所有图书馆的书应在 6 月底以前归还。

有许多以-en 结尾的动词常与 must 或 should 组成被动结构。这些动词来源于普通的形容词和名词，见表 2-1 。

表 2-1　　动词、名词、形容词对照表

形　容　词	名　　词	动　　词
tight　紧的		tighten (make tighter) 使紧
loose　松的		loosen (make looser) 使松
weak　弱的		weaken (make weaker) 使弱
deep　深的	depth 深度	deepen (make deeper) 使深
short　短的		shorten (make shorter) 使短
wide　宽阔的	width 宽度	widen (make wider) 使宽
strong　强的	strength 强度	strengthen (make stronger) 使强
high　高的	height 高度	heighten (make higher) 使高
long　长的	length 长度	lengthen (make longer) 使长

例如：The air gap is too narrow; it should be widen. 气隙太窄，应该加宽。(注意两句之间用分号)

另外，说明书必须清楚（如果不清楚，说明书本身就失去意义）。如果用 firstly、secondly、thirdly、then、next、finally 这样的词，说明书会更加清楚。这些词是起承接作用的词。如果由它们连接句子，其后应用逗号。

2.11 定冠词的物主代词用法

在科技英语中，定冠词用在与前面已提到的事物有关系的名词前，表示所有关系，相当于物主代词 its (their)。这种用法的定冠词有如下两种汉译法：

1. 译成物主代词

(1) Ice float on water because the density is less than that of water. 冰能浮于水上，因为它的密度小于水的密度。(句中名词 density 与前面的名词 ice 有所属关系，所以其定

冠词译成“它的”)

(2) Heating a gas causes the molecules to move faster, and to become further apart. 加热气体可以使其分子运动更快，彼此相距更远。(the molecules：其分子)

(3) The solid particles in this liquid are too large for the mixture to be considered a true solution. 由于这种溶液中的固体颗粒太大，它们的混合物很难视为真正的溶液。(the mixture：它们的混合物)

(4) All atoms, except those of hydrogen, contain one or more neutrons in the nucleus. 除氢原子以外，所有原子在其核内都含有一个或一个以上的中子。(the nucleus：原子核)

(5) When a structure is to be built, suitable materials must be chosen for the parts. 要建造一个结构物，就得给它的各个部件选择合适的材料。(the parts：它的各个部件)

(6) When the temperature of a gas is raised, then either the pressure must go up or the volume must go up. 当气体温度升高时，不是其压力增加就是其体积增大。(the pressure：其压力；the volume：其体积)

2. 重复前面有关的名词

(1) The silicon combine with dissolved oxygen in steel, thereby improving the quality. 硅与钢中已溶解的氧化合而改进了钢的性能。(名词 quality 是属于前面的名词 steel 的，所以其定冠词译成“钢的”)

(2) The content of smoke depends on the source. 烟的成分取决于烟源。(the source：烟源)

(3) Every driver ought to know this fact: the distance required to stop a car depends on the square of the velocity. 每个驾驶员应当知道这一点：汽车制动所需要的距离取决于车速的平方。(the velocity：车速)

(4) When a beaker of water is heated from below, the water at the bottom becomes less dense (assuming it is above 4℃) and rises. 当从下面对一烧水杯加热时，烧杯底部的水，由于密度变小（假定在 4℃以上）而上升。(the bottom：烧杯底部)

(5) Fast neutrons from the fission of U 235 escape into the surrounding graphite, where they collide with the atoms. (the atoms：石墨原子) 铀 235 裂变产生的快中子进入周围的石墨中，在那里同石墨原子发生碰撞。

(6) Proper application of these machine elements depends upon a knowledge of the force on the structure and the strength of the materials employed. 正确使用这些机械零件取决于对作用在零件结构上的力的了解以及对所用材料的强度的了解。(the structure：零件结构)

2.12 in that

“in that” 是一种较古老的用法，意思是“由于”“因为”“在……方面”，只用在较正式的文体中。例如：

(1) This insulation material differs from that one in that it has better performance. 这

种绝缘材料之所以异于那种，是因为这种材料的绝缘性能较好。

(2) Vinyl film is similar to cellophane in that it is transparent. 乙烯薄膜在透明性方面同玻璃纸相似。

(3) Electrical engineering is unique, as compared with essentially any other division of engineering or similar enterprise, in that it is dealing with, controlling, and utilizing something invisible and utterly intangible—the electrical phenomena of nature, commonly spoken of as electricity. 电力工程与任何别的工程部门或类似的企业比较起来是独特的，因为它分配、控制和利用的东西是看不见和根本摸不着的，即自然界与电有关的现象。

注意：此例句最后五个字省略不译，从而使整个译文简洁而流畅。

2.13 among others

"among others"可以用来表示两个意思：①其中包括；②……等。例如：

(1) The argument was voiced by Hans J. Morgenthou, among others, outside the engineering itself. 在工程界之外，有人提出这种论据，其中包括汉斯.J. 摩根索。

(2) The kernels allow the calculation of device parameters of great concern for the microwave designer, i. e. in the case of a PA, among others, nonlinear gain, 3rd order harmonics and intermodulation。利用核可以计算微波器件设计者关心的器件参数，即对于功率放大器，有非线性增益、三次谐波和相互调制等。

2.14 不定冠词 a (an) 和数词 one 在意思上的差别

不定冠词a (an) 和数词one有时在意思上稍有差别。不定冠词a (an) 往往偏重于类别上的"一个"或"任何一个"；而数词one则偏重数量上的"一个"，强调的不是"两个"或"三个"。例如：

(1) It took more than a month to design the circuit. 设计这个电路花了一个多月时间。

(2) It took more than one month to design the circuit. 设计这个电路花了不止一个月时间。

(3) A force is a push or a pull. 力是一种推力或拉力。

(4) The desk will collapse at one push or one pull. 这个课桌推一下或拉一下准垮。

第2篇

电力工程科技英语教程

高电压与绝缘技术与专业英语语法

3

高电压与绝缘技术

3.1 What Engineers in Industry Should Know about the Response of Grounding Electrodes Subject to Lightning Currents

3.1.1 Introduction

MOST ground terminations of electrical systems in industrial plants are designed based on concerns related to the flow of slow electric currents, such as short-circuit currents. Nevertheless, these terminations are frequently stressed by lightning-related currents, and, in these cases, their response is quite different from that shown for slow currents.

Malfunctions of electric and electronic systems in industrial plants due to inappropriate response of their grounding electrodes subjected to lightning-related currents are relatively frequent. In order to minimize such malfunctions, it is important to promote adjustments in the design to improve the lightning response of electrodes, through preserving the main features of the designed ground terminations.

In this respect, it is essential to know well the fundaments of the transient behavior of grounding electrodes and to disseminate the corresponding technical culture among electrical engineers working in industrial plants. This paper addresses this issue, discussing the basic knowledge along with recent advances in this field, following the approach indicated next.

Section Ⅱ summarizes the main features of lightning currents. Fundamental concepts of the transient behavior of grounding electrodes are discussed in Section Ⅲ, and the parameters that represent this behavior are introduced. In Section Ⅳ, the simulated response of some typical arrangements of electrodes of industrial plants is analyzed. Finally, based on the considerations of the previous sections, useful practical remarks and recommendations are suggested in Section Ⅴ, for application in industrial environment.

3.1.2 Influent Parameters of Lightning Currents

Lightning-related stresses of industrial electric systems are yielded by direct or nearby strikes. The resulting current $i(t)$ is the source of deleterious effects, and it can exhibit different features according to the specific kind of lightning event. In industrial plants, such effects might cause damages and insulation failures, responsible for economic losses and disoperation of the plant.

Damages result from heating losses associated to the flow of currents through the systems' components. The parameter that quantifies this effect is the so-called specific energy, given by the time integral of the square instantaneous current

$$\int_T i(t)^2 \mathrm{d}t \tag{3.1}$$

Note that multiplying this parameter by the equivalent resistance of the component crossed by the current results in the dissipated energy. Insulation failures result from lightning over voltages established across insulators of electrical systems due to the pulses of current of first and subsequent return strokes, and the major parameters that influence on this occurrence are the peak current and the current front time and waveform.

Direct strikes are responsible for the severest effects since they impress huge impulsive currents (of several tens of kiloamperes) on electrical systems. The electromagnetic field yielded by nearby strikes illuminates electrical systems, inducing voltages of several tens to a few hundreds of kilovolts among their components, depending on the lightning current parameters and the distance between the incident point and the illuminated system. Although much more frequent, the effects of such strikes are rarely able to impress currents as high as 1kA on nearby systems, since their induced voltages are lower than 300kV and typical surge impedances of electrical systems have the order of 400Ω.

In lightning protection studies, only cloud-to-ground (CG) lightning has been considered. Most of such events are negative (about 90%) and consist of multiple-stroke flashes. About 80% of flashes exhibit subsequent return strokes, and their average multiplicity is about 3.

Only part of the subsequent strokes terminate to the same point in the ground. Most events strike to other points, being the average distance to the original termination of around 2 km.

First-return-stroke currents in single- and multiple-stroke flashes exhibit similar patterns. A typical first-stroke current starts with a concave front, followed by an abrupt rise around the half-peak that leads to the first peak. This is typically followed by a second peak, usually higher than the first, and then, the current decays slowly with some subsidiary peaks, ceasing typically after 1 to 3ms. Figure 3.1 illustrates such patterns, referring to the original record of a first-return-stroke current measured recently at Morro do Cachimbo Station (MCS), Brazil.

According to the database by Berger, the median parameters of first-stroke pulses for maximum current, front time (td30), time to half-peak, transferred charge, and specific energy are 31.1kA, 3.83μs, 75μs, 5.2 C, and $5.5 \times 10^4 \mathrm{A^2s}$, respectively.

The currents of subsequent return strokes also show certain homogeneity of parameters, including their waveforms. Most currents exhibit a single peak, shorter front time, time to half-peak, and lower transferred charge compared with first strokes. Figure 3. 2 illustrates a typical waveform of such type of event.

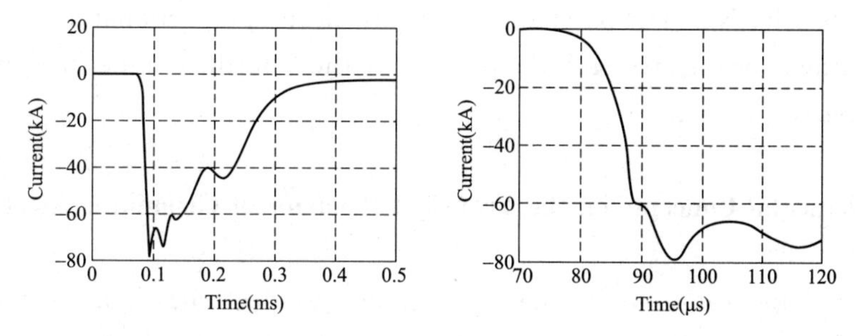

Figure 3. 1 Original unfiltered current of a negative CG first return stroke. Record of a multiple-stroke flash measured at MCS on December 11, 2013.

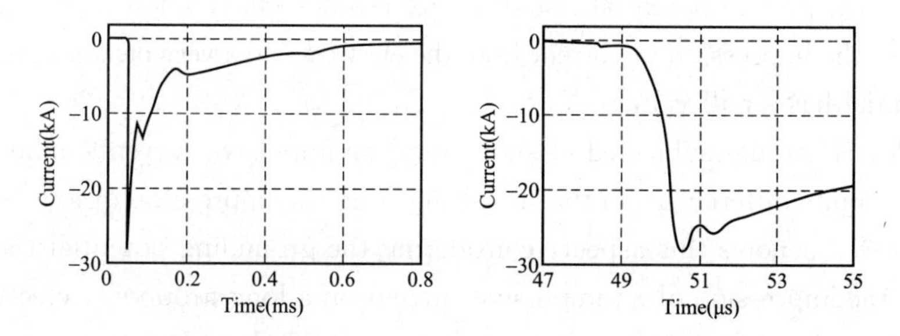

Figure 3. 2 Original unfiltered current of a negative CG subsequent return stroke measured in MCS on December 11, 2013 (second stroke).

According to the database by Berger, the median values of subsequent-stroke pulses show a peak current, front time, time to half-peak, transferred charge, and specific energy of 12kA, 0. 67μs, 32μs, 1. 4 C, and $6.0 \times 10^3 A^2 s$, respectively.

The peaks of the currents of first and subsequent return strokes measured at MCS have median value about 50% higher than those measured by Berger, though the other parameters have similar median values.

Most of the energy of the first- and subsequent-return-stroke currents is associated to components of around 10 - 15kHz. However, the frequency content exceeds 400kHz and 2MHz in the first and subsequent strokes, respectively.

Although currents of positive CG lightning exhibit higher peak value and specific energy in addition to longer time parameters, their occurrence is rare.

Eventually, mainly in temperate regions, upward lightning can be initiated from very high structures in industrial plants, such as chimneys or communication towers. The current of such events is very low (in the range of a few hundred ampere), although their duration is very long (several hundreds of milliseconds). Therefore, although it is not able to produce insulation failures, it is able to transfer large charges and to cause severe damages due to heating. Part of the upward lightning events have their initial stage of low-amplitude current, followed by return strokes, whose currents are still low, comparable with those of subsequent return strokes.

It is worth mentioning that the most important lightning protection parameter is indeed the flash density Ng. The parameter Ng is practically proportional to the expected frequency of lightning events stressing the plant and, thus, to the number of damages and insulation failures in the plant.

3.1.3 Fundamental Concepts on the Transient Behavior of Grounding Electrodes

In industrial applications, the quality of grounding electrodes is measured in terms of their low-frequency grounding resistance R_{LF}, given by the ratio of the instantaneous values of voltage and current in the low-frequency range ($R_{LF} = U_{LF}/I_{LF}$). In these cases, voltage means the ground potential rise of the electrodes (in relation to the remote earth) resulting from the impression of current from the electrodes to a very distant point, where the electric potential has a null value.

However, electrodes subjected to short-duration impulsive currents exhibit a response that is usually quite different from that resulting from the impression of low-frequency currents. Figure 3.3 denotes this aspect, considering the grounding potential rise (GPR) resulting from the impression of an impulsive current on a long grounding electrode.

Frequently, this response is represented in a simplified way by means of the so-called impulse impedance Z_P, given by the ratio of peaks of the GPR and impressed impulsive current ($Z_P = U_P/I_P$). Usually, this ratio has a value different from that of R_{LF}, and the occurrences of the voltage and current peaks are not simultaneous. The attractive aspect of this parameter is that it allows promptly calculating the GPR peak by simply multiplying Z_P by the peak current.

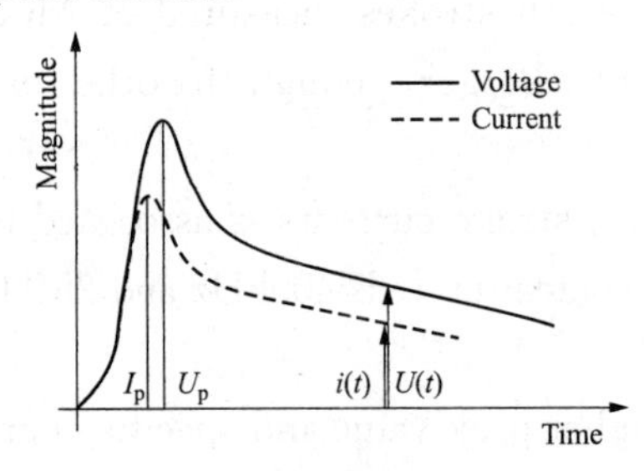

Figure 3.3 Response of a grounding electrode subject to an impulsive current.

Note that, the value of R_{LF} can be also obtained from the impulsive waves in Figure 3.3, as the ratio between the instantaneous values of voltage and current at the tail of the waves, since both voltage and current vary very slowly at this region.

Analyzing the behavior of buried electrodes in frequency domain allows improving the understanding of their impulse response. In this case, it is useful to refer to the harmonic impedance $Z(\omega)$ of the electrodes, given by the ratio of voltage (potential rise developed at the current injection point) and current at same frequency. Figure 3.4 shows the diagram of this impedance for a 10- and a 30-m-long electrode, buried in a low- and a high-resistivity soil. This impedance was calculated in the range of frequency components of lightning currents, using an electromagnetic model (HEM), considered in Section IV. It also shows a simplified representation of grounding electrodes by a distributed-circuit model, to allow exploring interpretations for their frequency response.

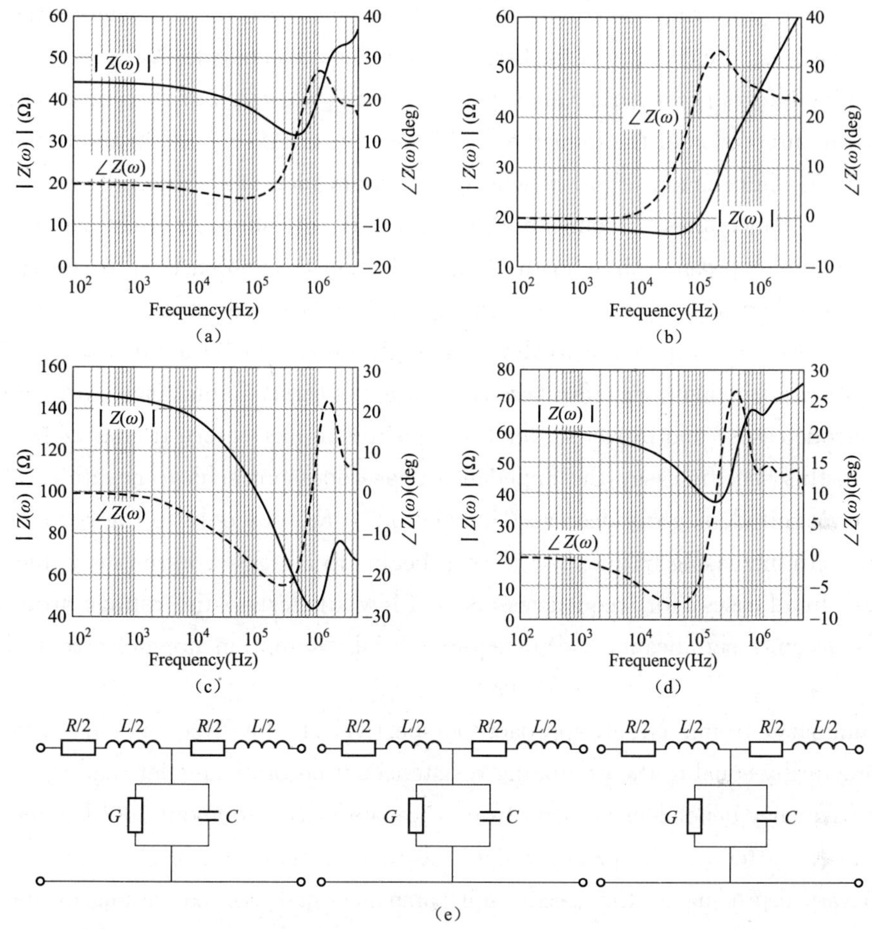

Figure 3. 4 Harmonic grounding impedance of a horizontal electrode of length L, buried 0. 6 m deep in a soil of resitivity ρ_0. $\rho_0 = 300\Omega \cdot m$:
(a) $L = 10$ m; (b) $L = 30$ m, $\rho_0 = 1000\Omega \cdot m$; (c) $L = 10$ m; (d) $L = 30$ m;
(e) Simplified representation of electrodes by a series of connected distributed circuits (R、L、G、C are the per-unit-length longitudinal resistance and inductance and parallel conductance and capacitance of the electrode).

At the low-frequency range, all reactive effects are negligible and thus the internal resistance of the metallic electrodes. Therefore, the harmonic impedance is practically constant and equal to the grounding resistance R_{LF} (see the null angle in all the diagrams of impedance). The impedance (or resistance) of the longer electrodes has lower values (about 2. 5 times lower), and the impedance of the electrodes buried in the soil of higher resistivity is about 3. 3 times larger, the same value of the ratio between the resistivity values. Furthermore, at the low-frequency range, the electric potential is the same all along electrodes.

As frequency rises, above a few kilohertzs, two effects begin to influence the electrode response: the capacitive currents in the soil (see the negative impedance phase) and the decrease of soil resistivity (due to a frequency-dependence effect). Both effects, included respectively in the capacitive and conductive parallel branches of the equivalent circuit in Figure 3.4 (e), cause the decrease of the harmonic-impedance magnitude. As frequency further rises, inductive effects become relevant (see the positive variation of the phase magnitude). Initially, the interaction of the inductive and capacitive effects results in additional decrease of the harmonic impedance. $Z(\omega)$ reaches a minimum magnitude around the frequency in which the impedance phase becomes null. After that, the inductive effect begins to prevail (see the positive impedance phase in the diagrams), and the impedance increases continuously until a few megahertzs. Only the curves in Figure 3.4 (b) consist of an exception to this rule. In this case, corresponding to the long electrode (30m) buried in the low-resistivity soil (100Ω·m), the inductive effect becomes relevant at earlier frequencies (in relation to the short electrode cases) and prevails over the capacitive effect before the latter becomes important. In this specific case, the impedance phase does not achieve negative values: the impedance magnitude slightly decreases in relation to R_{LF} due only to the reduction of soil resistivity with increasing frequency and then begins to increase continuously due to inductive effect. In all cases, for frequencies above a few kilohertzs, the voltage drop along the electrodes becomes significant, and the equipotential assumption along electrodes is no longer valid.

In summary, from a circuit approach perspective, the low frequency harmonic impedance is practically equal to the grounding resistance; it becomes smaller than R_{LF} due capacitive and frequency-dependence effects from a few kilohertzs to about 100kHz and becomes larger than R_{LF} after this frequency range due to inductive effects. The intensity of this effect can vary depending on the specific soil parameters and electrode arrangements. For instance, the capacitive effects are stronger in high-resistivity soils, causing larger reduction of the impedance in the intermediate frequency range (see the diagram of the 1000Ω soil). In addition, the transition of the impedance from a value lower to higher than R_{LF} can vary (it occurs earlier for low-resistivity soils). Nevertheless, this described qualitative behavior of the harmonic impedance holds true for soils and electrodes in general.

Considering the frequency components of impulsive currents, this frequency-domain behavior allows understanding the nonsimultaneous occurrence of the peaks in voltage and current waves and the possibility of Z_P to be either equal, lower, or larger than R_{LF}.

Still, the simplified circuit approach is useful to illustrate that any grounding electrode arrangement can be considered a nonuniform and lossy transmission line. The pulse of current impressed to an extremity of the buried electrode propagates along it. While propagating, this current, along with the associate voltage pulse, is subjected to attenuation and distortion due to the losses in the ground. The most conductive the soil is, the larger the losses yielded by the current dispersed to this medium and, therefore, the larger is the attenuation of the current and voltage waves propagating along the electrode. This attenuation is also stronger at higher frequencies.

This strong attenuation explains a very important parameter in the context of the transient response of grounding electrodes, i. e. , the effective length L_{EF}. This parameter, first introduced by Gupta and Thapar and discussed in an outstanding work by He et al. , corresponds to a threshold length. While the electrode length is increased, the impulse impedance is decreased. However, the longer the electrode is, the larger is the attenuation of the propagating waves and, notably, of their high frequency components. For a limiting length, this attenuation is so strong that the impulse impedance no longer decreases with further increase of the electrode length, although the low frequency grounding resistance continues to decrease, since it is not subject to attenuation effects. The value of L_{EF} depends on several factors. It increases with decreasing soil resistivity (attenuation is stronger at soils of higher conductivity). It also increases with pulses with shorter front time, since they present higher frequency components (attenuation is stronger at higher frequencies).

The ratio of the impulse impedance to the low-frequency resistance, known as impulse coefficient I_C ($I_C = Z_P/R_{LF}$), is another useful parameter, since it allows determining Z_P from a measured resistance R_{LF}.

Finally, it is important to briefly refer to two effects that might influence the transient response of electrodes: soil ionization and frequency dependence of soil resistivity and permittivity.

Soil ionization occurs when very high currents are impressed to short electrodes. Discharges resulting in the soil around the electrodes cause a reduction of the electrode impedance while the process occurs. A large number of studies have addressed the impact of this effect, although all of them are subject to questionability related to the assumption adopted for the critical electric field in the soil in which the ionization process incepts.

Due to this aspect, the so-calledWeck's formula [see (3. 2)] is still used frequently to determine the soil ionization impact, in terms of the reduction it promotes on the grounding resistance R_{LF}. Thus

$$R_I = \frac{R_{LF}}{\sqrt{1 + I/I_{PI}}} \tag{3. 2}$$

In the preceding equations, R_I corresponds to the decreased grounding resistance resulting from the ionization process applied to an electrode arrangement whose original resistance was R_{LF}, assuming a critical inception electric field of E_0 (frequently assumed as 400kV/m). I and I_{PI} are the impressed peak current and the peak current required for the inception of ionization process, respectively, as given by (3. 3).

$$I_{PI} = \frac{E_0 \rho}{2\pi R_{LF}^2} \tag{3. 3}$$

Based on laboratory experiments, the frequency dependence of soil parameters has been known for a long time. Recently, it has been demonstrated based on measurements developed in field conditions that soil resistivity and permittivity are strongly frequency dependent. Both soil resistivity and permittivity ε decrease with increasing frequency. From the results of a large number of measurements, the following conservative expressions were proposed to take this frequency dependence into account for soils in general, in the range of 100 Hz to 4MHz, from the low-frequency resistivity measured by conventional instruments:

$$\rho = \rho_0 \{1 + [1.2 \cdot 10^{-6} \cdot \rho_0^{0.73}] \cdot [(f - 100)^{0.65}]\}^{-1} \tag{3.4}$$

$$\varepsilon_r = 7.6 \cdot 10^3 f^{-0.4} + 1.3 \tag{3.5}$$

In the so-called Visacro-Alipio expressions, ρ and ε_r are the soil resistivity and relative permittivity at frequency f in hertzs, respectively; ρ_0 is the soil resistivity at 100 Hz.

3.1.4 Response of Electrodes Subjected to Lightning Currents

Section Ⅲ addressed the qualitative impulse response of electrodes in general. Here, the quantitative response of electrodes subject to lightning currents is considered, particularly that of two very common arrangements of electrodes in industrial plants, i. e. , the horizontal electrode and the grounding grid. Such response serves as reference for the analyses and recommendations of this work.

The results presented here were obtained by simulation using the hybrid electromagnetic model (HEM). In particular, the application of this model to provide accurate results of the impulse response of electrodes is addressed, as demonstrated by the comparison of measured and simulated GPRs, for instance, those of a grounding grid, illustrated in Figure 3.5.

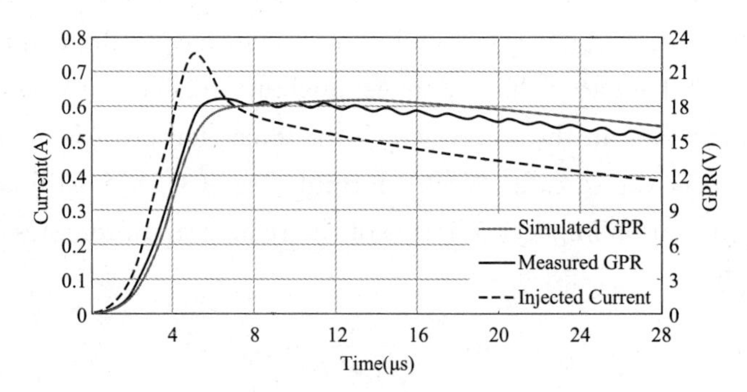

Figure 3.5 Comparison of measured and simulated GPRs of a 20 m×16 m grid composed of 2016m² meshes buried 0.5m deep in a 2000Ω · m soil and subjected to an impulsive 4μs front-time current impressed at the grid center. Steel-galvanized-alloy electrodes of 0.5cm radius. Variation of the relative resistivity and permittivity taken into account from the experimental results provided by the Visacro-Alipio methodology.

As aforementioned, Z_P varies with the current waveform. In lightning protection applications, the interest is focused on the response of electrodes subjected to currents of first and subsequent return strokes. In this respect, it is possible to assess their response and, notably, Z_P, considering the representative lightning current waveforms in Figure 3. 6.

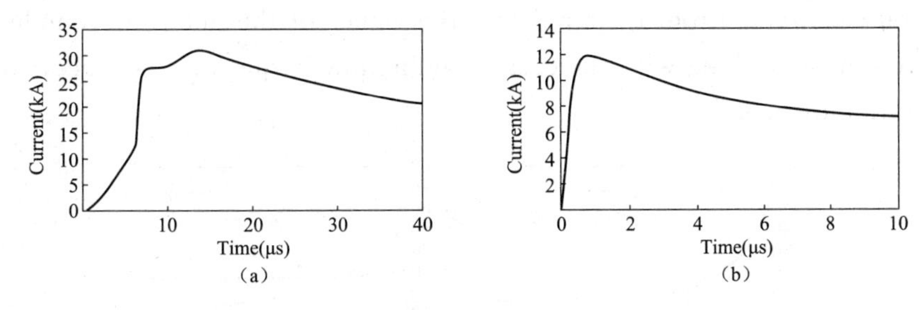

(a) (b)

Figure 3. 6 Representative current waveforms of (a) first and (b) subsequent strokes impressed on the electrodes. Median peak currents and front times: (a) I_p= 31. 1kA, T_d30 = 3. 8μs; T_d10 = 5. 6μs; (b) I_p= 11. 8kA, T_d30 = 0. 67μs, T_d10 = 0. 75μs.

These waveforms reproduce the median parameters of first and subsequent-stroke currents measured by Berger at Mount San Salvatore: peak current, front times, and maximum time derivative and duration (time to half-peak).

Based on the results developed, Figure 3. 7 shows the responses of a 30-m-long horizontal electrode buried in a low- and a high-resistivity soil, subjected to the representative lightning currents in Figure 3. 6 and assuming the frequency dependence of resistivity and permittivity given by the Visacro-Alipio expressions.

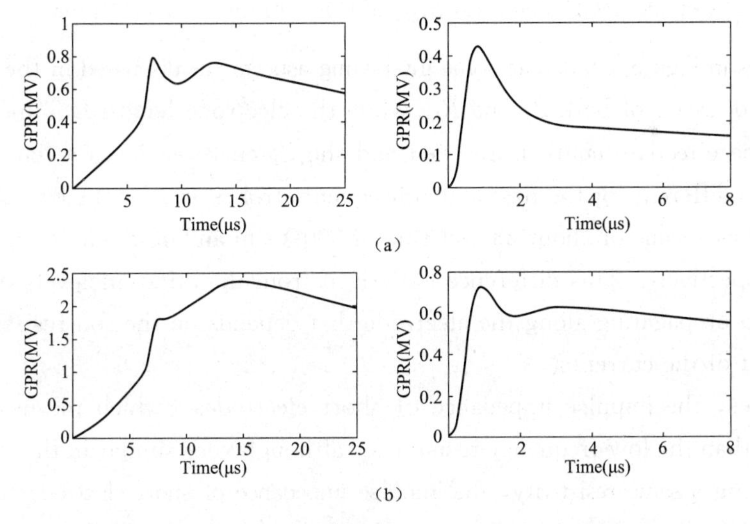

Figure 3. 7 Simulated GPR of a 30m horizontal electrode buried 0. 5 m deep and subjected to representative currents in Figure 3. 6 for (left) first and (right) subsequent return strokes. Soil resistivity ρ_0 : (a) 300Ω • m ; (b) 1000Ω • m.

Note that the responses of the electrode and the values of Z_P are quite different for the first and subsequent return strokes (27 against 36Ω and 80 against 62Ω for the 300 and 1000Ω • m soils, respectively). They are also different from the low-frequency resistance (of about 20 and 60Ω for the 300 and 1000Ω • m soils, respectively). On the other hand, the impulse impedance has almost a proportional increase with soil resistivity.

Simulations of the same type were developed, considering the variation of the length of the horizontal electrode. From their results, the values of the impulse grounding impedance were determined, along with the corresponding low-frequency resistance, to compose the curves in Figure 3.8.

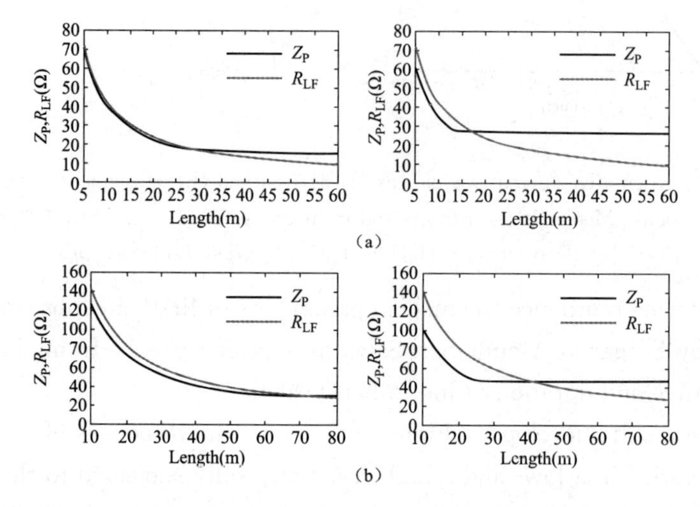

Figure 3.8　Simulated Z_P of horizontal electrodes buried 0.5 m deep (r = 0.7 cm) and subjected to representative currents of (left) first and (right) subsequent strokes as a function of electrode length. Soil resistivity ρ_0 of (a) 300Ω · m; (b) 1000Ω · m.

The curves in Figure 3.8 denote some interesting aspects, as discussed in the following.

Note the decrease of both Z_P and R_{LF} while the electrode length L is increased. This occurs until the effective length is reached and the Z_P curve achieves a flat region. Note that also L_{EF} is different for the first and subsequent strokes and for the low- and high-resistivity soils. It has a value of about 23 and 12m at 300Ω · m and of about 48 and 23m at 1kΩ · m soils, respectively. This difference is derived from the different levels of attenuation of the currents propagating along the electrode that depends on the soil resistivity and frequency content of the currents.

In all cases, the impulse impedance of short electrodes (which means shorter than L_{EF}) is lower than the low-frequency resistance, although very similar in the low-resistivity soil. Considering a same resistivity, the impulse impedance of short electrodes is significantly smaller for subsequent strokes. Inductive effects are not important for short electrodes, and the effect of the capacitive branch in the equivalent circuit in Figure 3.4, responsible for decreasing the impedance, is stronger for subsequent strokes, due to the superior frequency content of their currents.

Z_P becomes larger than R_{LF} for long electrodes (which means longer than L_{EF}), and this trend is enhanced with increasing electrode length, once the resistance decreases continuously, whereas the decrease of Z_P is saturated at a length L_{EF}. In this condition, Z_P is larger for subsequent strokes. For long electrodes, inductive effects prevail, and the effect of the series inductance in the equivalent circuit in Figure 3.4, responsible for increasing the impedance, is stronger for subsequent-stroke currents, due to their superior frequency content.

Curves of Z_P similar to those in Figure 3.8 were developed for horizontal electrodes buried in soils of different resistivity values, and the approximate values of the effective length of electrodes, shown in Table I, were estimated. Indeed, L_{EF} can be determined in different ways. A defined length derivative of the impedance Z_P/d_L is adopted to calculate it. The values of L_{EF} in Table 3.1 were identified as the electrode length corresponding to the crossing of the curves of Z_P and R_{LF} (see Figure 3.8). Although the impulse impedance is the parameter of major interest in lightning protection applications, the measurement of Z_P is rarely feasible in industrial environments, in spite of the availability of specific instruments to perform this kind of measurement, such as the grounding impedance meter described.

Table 3.1 Estimated effective length of horizontal electrode (7mm RADIUS, 0.5m DEEP).

ρ_0 (Ω·m)	100	300	600	1000	2000	4000
First (m)	12.5	22.5	35	48	80	130
Subsequent (m)	7.5	12	17.5	23	36	60

In this respect, a resource to determine this parameter for defined electrode arrangements consists of calculating it from the measured low-frequency resistance. Z_P is given by simply multiplying R_{LF} by the impulse coefficient, which has specific values for first and subsequent return strokes, as illustrated in Figure 3.9 for the 1000Ω·m soil. Note that, in the range of electrode length shorter than or equal to L_{EF}, the impulse coefficient I_C is approximately constant (I_{C0}). The value of I_{C0} depends on the soil resistivity ρ_0 and decreases for increasing soil resistivity ρ_0, as both capacitive and frequency-dependence effects contribute to reducing the impulse impedance. In the range of electrodes longer than L_{EF}, there is an almost linear increase of the impulse coefficient with the electrode length.

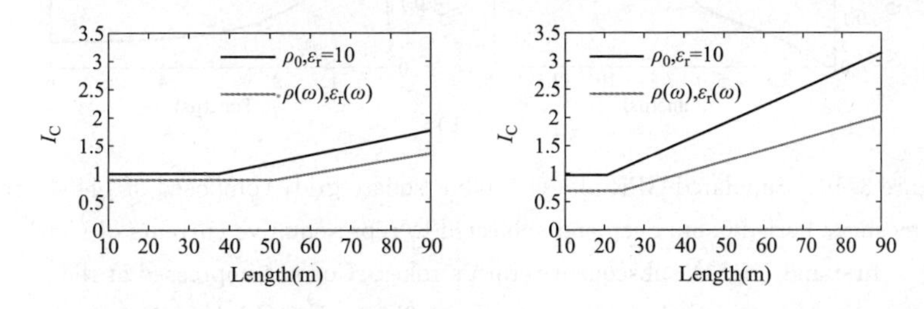

Figure 3.9 Simulated impulse coefficient IC of horizontal electrodes buried 0.5m deep and subjected to representative currents of (left) first and (right) subsequent strokes as a function of electrode length. Soil resistivity ρ_0 of 1000Ω·m.

In terms of the grounding electrode design, the response of electrodes to first-stroke currents is the only focus of interest. Even in those cases in which the impulse impedance of subsequent strokes is larger, the amplitude of the first-stroke currents (median values about three times larger) makes their effects much more severe.

Useful expressions for determining I_C of horizontal electrodes for the first return strokes are presented and are summarized next. The expressions are valid in the range of ρ_0 from 100 to 4000Ω · m. The impulse coefficient is given by (3.6) and (3.7), respectively, for electrodes shorter (I_{C0}) and longer than L_{EF}, being this parameter calculated according to (3.8). Thus

$$I_{C0} = -0.00086 \times (\rho_0)^{0.686} + 0.992 \tag{3.6}$$

$$I_C = \alpha L + \beta \tag{3.7}$$

$$L_{EF} = \frac{I_{C0} - \beta}{\alpha} \tag{3.8}$$

In (3.7) and (3.8), α is equal to $2\times10^{-3} + \exp(-1.55 \cdot \rho_0^{0.162})$, and β is equal to $0.5 + \exp(-0.00046 \cdot \rho_0^{0.83})$.

A similar approach was followed to analyze the lightning response of grounding grids. Figure 3.10 shows this response for a 60×60m² grid, buried 0.8m deep in soils of low and high resistivity, assuming the frequency dependence of soil parameters given by the Visacro-Alipio expressions.

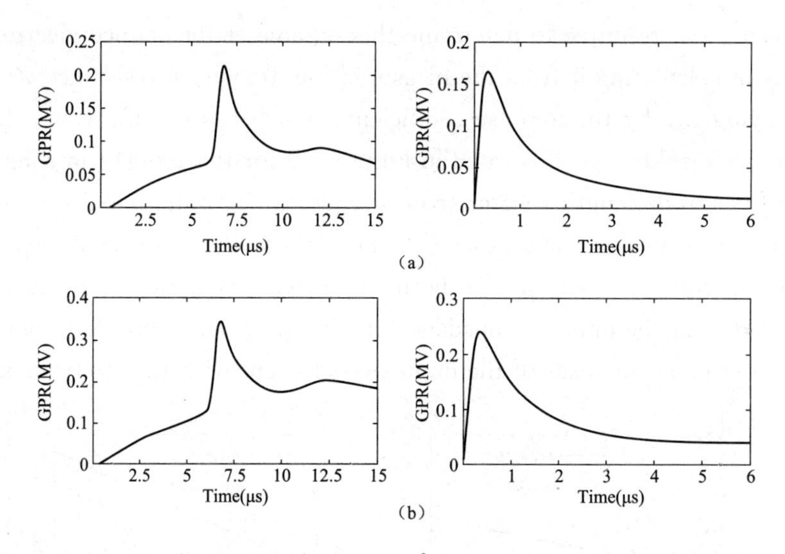

Figure 3.10　Simulated GPR of a 60×60m² square grid, composed of 1445×5m² meshes, buried 0.8m deep and subjected to representative currents of (left) first and (right) subsequent return strokes. Current impressed at the grid corner. Soil resistivity ρ_0 of (a) 300Ω · m; (b) 1000Ω · m.

Naturally, the transient grid response is quite different from that of the 30-m-long horizontal electrode. The lower values of the grid GPR are evident, and the waveforms are very different, mainly in the wave tail, due to the lower values of the grid grounding resistance. However, except for these differences, the qualitative analyses developed for the response of horizontal electrodes still hold true.

In addition, in this case, simulations were developed varying the electrode length to determine corresponding values of Z_P and R_{LF}, as depicted in the curves in Figure 3.11.

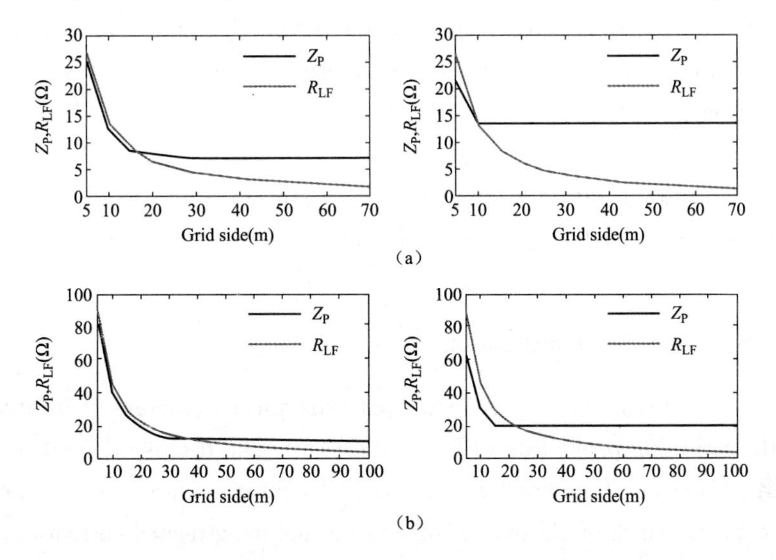

Figure 3.11 Simulated Z_P of square grids with side length L, meshes of $5\times5m^2$, buried 0.8m deep and subjected to representative currents of (left) first and (right) subsequent strokes, for soil resistivity ρ_0. Impression of current at the grid corner: (a) 300Ω • m; (b) 1000Ω • m.

Note that the horizontal scale corresponds to the grid side L_{SIDE} and not to the electrode length, as in the horizontal electrode case. Considering that the increase of the side of a square grid corresponds approximately to a double increase of the electrode length, the curves show that the behavior of the impulse impedance is very similar to that of the horizontal electrode case. Using the grid side as variable compresses the relative length and contributes to the impression that the difference between the Z_P and R_{LF} curves are not so significant as in the horizontal electrode case. If this compression is taken into account, one can see that the relative behavior is indeed quite similar. This includes the value of the effective length of the electrodes, which is practically the same in both cases.

Useful expressions were also developed for determining the impulse coefficient I_C of grounding grids subjected to first-stroke currents. Such expressions are valid in the range of ρ_0 from 100 to 4000Ω • m and are summarized next. They are similar to those developed for horizontal electrodes, although L_{EF} is replaced by L_{SAT}. The latter corresponds to the threshold of the grid-side length above which the decrease of the grid impedance resulting from increasing the side length is saturated.

For a grid side shorter than L_{SAT}, given by

$$L_{SAT}=\exp(0.9381\times\rho_0^{0.1865}) \tag{3.9}$$

the same expression (3.6) allows calculating the impulse coefficient.

For a grid side longer than L_{SAT}, the impulse coefficient is given by

$$I_C = \alpha L_{SIDE} \tag{3.10}$$

In (3.10), α is equal to $2.9 \cdot 10^{-3} + \exp(-0.8935 \cdot \rho_0^{0.2129})$.

Thus, from a known low-frequency grounding resistance R_{LF}, the impulse impedance Z_P of grids or horizontal electrodes can be promptly determined from (3.6) - (3.11). Equation (3.11) is as follows

$$Z_P = I_C \cdot R_{LF} \tag{3.11}$$

Expressions are also provided for the impulse impedance of some other arrangements of electrodes of interest in industrial plants, including those for grids with currents impressed at their center.

3.1.5 Conclusion and Recommendations

This paper has summarized the relevant patterns and parameters of lightning currents required for the lightning protection of industrial plants and discussed fundamental aspects of the lightning response of grounding electrodes. By means of a conceptual approach, the differences on the response of electrodes subjected to lightning-related currents and that resulting from the impression of low-frequency currents were addressed, focusing on the requirements for applications in industrial environments.

Based on the discussions of previous sections, major concluding remarks are highlighted next to support some practical recommendations.

A. Concluding Remarks

While electrodes subject to low-frequency currents, such as those of short circuits, exhibit an equipotential condition along them, this condition vanishes when electrodes are subjected to lightning-related currents, except for very short electrodes. The electric potential yielded by such impulsive currents might vary a lot along the electrodes, and this has important implications.

Reactive and propagation effects are responsible for such behavior of the electrodes subjected to lightning currents, and representing the electrodes simply by a grounding resistance R_{LF} is not recommended. Even in a simplified way, representing the lightning response of electrodes requires more elaborate parameters, such as the impulse grounding impedance Z_P.

The concise representation of the electrodes by Z_P is very attractive in the lightning protection perspective, once it allows promptly determining the maximum GPR yielded by impulsive currents from the product of Z_P by the peak current.

The impulse impedances Z_P of the first and subsequent return strokes are different due to the different frequency contents of corresponding currents. Thus, determining the maximum GPR yielded by the first- and subsequent-return-stroke currents requires multiplying the specific value of Z_P of each type of stroke by the corresponding peak current.

In most applications, low values of grounding impedance are pursued, and the most effective practice for achieving such low values consists of increasing the area covered by the electrodes, although this practice is limited by constraints related to propagation effects. Specifically, when there sistivity of deeper layers of soil is much lower than that of the superficial layer, using long vertical electrodes can be very effective to achieve low impedance values.

To take the attenuation of the current surge propagating along the electrodes into account, a specific parameter is defined, consisting of the effective electrode length L_{EF}. The knowledge of the value of this parameter under different conditions of soil resistivity, as indicated in Table 3. 1, is very important in applications involving long electrodes. The value of L_{EF} decreases with decreasing soil resistivity and increasing frequency. Thus, L_{EF} is shorter for the subsequent-return-stroke currents compared with those of first ones.

The impulse impedance of electrode arrangements can be either smaller or larger than the low-frequency grounding resistance. Electrodes shorter than L_{EF} have a value of Z_P typically smaller than R_{LF}, and this trend becomes more pronounced with increasing soil resistivity. In electrodes longer than L_{EF}, the effect is the opposite. Z_P is typically higher than R_{LF}, and this trend becomes stronger with increasing electrode length.

Short electrodes ($L < L_{EF}$) have their subsequent-stroke impulse impedance smaller than that of first ones, whereas electrodes longer than L_{EF} have their subsequent-stroke impulse impedance larger than that of first ones.

Different from the low-frequency response of electrodes, due to attenuation effects, their lightning response depends on the position of the connection of the downward conductor to the electrode.

The impulse impedance of an electrode arrangement can be determined from the product of its low-frequency resistance by the impulse coefficient I_C, using expressions available in the literature, such as those mentioned in Section Ⅳ for horizontal electrodes and grids, i. e. , (3. 6) - (3. 10).

Ionization effects might improve the performance of concentrate electrodes subjected to large lightning currents, reducing their impedance. However, this effect is not important in arrangements involving long electrodes, such as substation grids and counterpoise wires of transmission lines.

Differently, the frequency dependence of soil resistivity and permittivity improves the lightning response of both concentrate electrodes and arrangements involving long electrodes buried in soil of moderate and high resistivity, independently of the current amplitude. The effect is not important for electrodes buried in soils of resistivity below 300Ω • m. The literature provides expressions for taking this effect into account in engineering applications, such as (3. 4) and (3. 5).

B. Practical Recommendations

Lightning protection applications always require a concise representation of grounding electrodes in industrial environments.

- This work advocates the use of the impulse impedance in such applications. Indeed, the first-stroke impulse impedance is the recommended parameter, due to the much more pronounced effects of the first-return-stroke currents.

Most standards concerned with lightning protection applications recommend a low grounding resistance value to ensure an improved lightning response of electrodes.

- With only a few exceptions, qualifying the grounding performance by means of the grounding resistance should be avoided. A low grounding resistance value does not imply a low impulse impedance value, unless the electrodes are very short. Very long electrodes ($L \gg L_{EF}$) always have a value of R_{LF} much lower than that of Z_P. In this case, the low value of R_{LF} obtained by means of long length of electrodes yields a false expectation of an improved lightning response of the grounding electrode.

The effects of both soil ionization and frequency dependence of soil parameters (ρ, ε) are beneficial, in terms of improving the lightning response of electrodes, although the first is effective only in the case of concentrate electrodes.

- Considering the typical large arrangements of electrodes of industrial plants, adopting a conservative approach in the calculation of the lightning response of electrodes is recommended. According to this approach, only the frequency dependence of soil resistivity and permittivity would be taken into account, using, for instance, expressions (3.4) and (3.5). The effect of soil ionization would be ignored, except for short electrodes. In this latter case, expressions (3.2) and (3.3) could be used to estimate the impact of the effect.

In most cases, the measurement of the impulse grounding impedance is very difficult in industrial environments.

- It seems prudent to determine this parameter indirectly from the low-frequency resistance R_{LF} and the impulse coefficient I_C. In this case, the recommendation for estimating the first-stroke impulse impedance of the electrode arrangement consists of measuring the grounding resistance using conventional low-frequency techniques and using expressions for I_C provided in the literature, such as those in Section Ⅳ.

Due to propagation effects, the position of the down conductor connection to the electrodes becomes relevant.

- Considering electrical components with a single termination to the ground, connecting the down conductor to central positions of long electrodes is preferable, instead of connecting to their extremities. The impression of a lightning current at the center of a large grid (sides much longer than L_{EF}) results in an impulse impedance, whose value is less than half of that resulting from the impression of current at the grid corner.
- In addition, in case of large electrode arrangements (very common in industrial

plants), it is recommended to use multiple downconductors, whose connections are distributed at distant positions along the electrodes. This ensures the decrease of the impulse impedance in relation to that of a single connection, when electrodes are longer than L_{EF}. Section Ⅳ provides expressions for estimating the effective length of horizontal electrodes and grounding grids.

The electric potential of electrodes subjected to lightning currents varies a lot along them, due to propagation effects.

- It is not recommended to ground the terminals of linked equipment at distant positions along electrodes. For instance, connecting the two extremities of shielding cables at different points may allow destructive currents to circulate through such aerial loop.
- In particular, it is not prudent installing very high structures, which consists of preferential points for lightning incidence, close to areas from which control cables are derived to command the operation of equipment installed at distant positions, even if a large grounding grid covers the whole area containing the structure and these positions. In case of lightning strike to the structure, these cables can transfer huge voltages, leading to damages or equipment malfunctions. For instance, damages of components or unexpected trips of switchgears have been reported in substation yards of industrial areas, due to lightning strikes to communication towers installed beside the substation's control houses. The transference of lightning current surges by control cables has been responsible by such effects.
- If preventing the condition described above is not possible, at least adopting mitigation practices is recommended. In cases of cables connecting two different positions at the grounding grid (or connected to a position and coupled with another distant one), the transference of lightning surges can be minimized by installing a metallic aerial or insulated bar connected to the grid at both positions (for instance, at the tower base and at equipment grounding position). This practice is able to reduce enormously the voltage between such positions of the grid, practically eliminating the source of damages and malfunctions.

In industrial plants, it is relatively common to connect different grounding grids, working toward a low value for the resulting grounding resistance. This practice is not effective for improving the lightning performance of the set of grids, if they are distant (distance longer than L_{EF}) and the connection is done by means of buried electrodes.

- In this case, connecting the grids using aerial conductors is very effective to improve the lightning performance of the set of grids.

Several standards, for instance, those addressing lightning protection systems, recommend a limiting value of the grounding resistance of electrodes.

- This recommendation is considered questionable. First, as stated before, a low value of R_{LF} does not imply necessarily a low value of the first-lightning-stroke impulse impedance. Furthermore, the source of damages and malfunction is the flow of current. This current only flows between two points, if there is a difference of potential (or voltage) between them.

- Thus, practices aimed to ensure the same potential between such points under lightning current stress are very effective to reduce damages and malfunctions.
- In addition, the use of optical links to connect equipment with components distributed at distant positions at the industrial plants, such as programmable controllers, is very effective. Each component is electrically isolated from the others. It has a local energy supply and a single local connection to the ground.

A last comment concerns the need of measurement of the grounding impedance at the industrial areas consisting of large interconnected grids and buried metallic components. Sometimes, it is not feasible to measure even the low-frequency resistance.

- In most cases, it is not so important to know the value of this impedance. Due to the huge area covered by the interconnected electrodes, R_{LF} has certainly a low value. Using multiple downconductors or aerial cables to connect such distributed electrodes ensures a minimized value of their impulse impedance, in addition to decreasing the difference of electric potentials along them.

3.2 Insulation Contamination of Overhead Transmission Lines by Extreme Service Conditions

3.2.1 Introduction

THE problem of contamination of overhead transmission line insulators emerged almost immediately after these lines were first built. For instance, M. R. Shariati et al give information on the problem of pollution on the external high-voltage insulation back in 1912. This is quite logical as, from the very beginning, power engineers have chosen to build lines near or within the territories of industrial enterprises, which did not only manufacture their goods but also generated and still generate environmentally unfriendly waste. Despite considerable pressure on the industry coming from environmentalists and the continuous improvement of dust and gas trapping technology, this problem has still no solution. Moreover, there is a problem of natural pollution, when transmission lines run across steppes and deserts (especially with saline soils), near seas and salty lakes. A combination of natural and industrial pollution poses the greatest hazard to insulation, especially when the contaminants are prone to solidification. In this case, the harm exceeds the sum of separate factors. Moreover, the pollution combination for each specific factor is unique and affects the insulator efficiency in its own way.

Unfortunately, all the attempts to create a mathematical model of insulator pollution and its subsequent behavior have failed: the process depends on too many factors. Models that do appear are limited to the approximation of experimental data and do not provide much of a predictive force.

In view of the above, the problem of experimental research into the behavior of linear insulation in the polluted environment remains topical. At the same time, the experiments need to be repeated for each specific usage environment, since, for example, the research findings for the pollution of the insulators used in forest regions are ineffective for the seaside or industrial area conditions.

3.2.2 Statement of Work

The authors have examined a power system of a coastal industrial area with the prevailing chemical enterprises on an alkaline soil.

The specific number of annual outages on the transmission lines with the voltage of 110kV exceeded 25 per 100km by the fifth year of service. The outages were caused by the polluted insulator flashover at the operating voltage. The lines included porcelain PF-70A or glass PS-70E (U70BS) normaltype (Figures 3.12) or two-winged PFD-70V, PSD-70E (U70BSP) (Figures 3.12) insulators, which formed tension strings, as well as normal-type or dust-proof bellshaped porcelain insulators with massive upper surface PFG-5A (it is no longer in production but it was widely used in the power system) in suspension strings (Figure 3.12). Figure 3.12 also shows sketches of insulators that were never used in the power system but did participate in the experiments due to their good aerodynamics-PFK-70A and PSK-70A (U70AD) (Figures 3.12). The identifications of insulators in Figure 3.12 are in accordance with GOST 27661-88 and, where it is possible, with IEC 305-78.

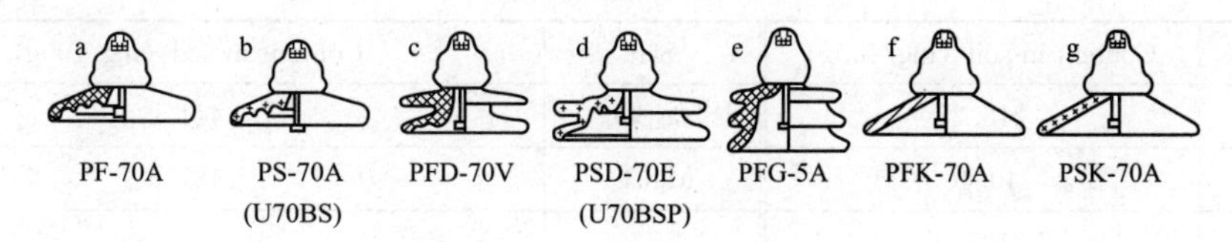

Figure 3.12 Insulator sketches.

See Table 3.2 for the geometrical parameters of insulators. Here:

H is the insulator construction height;

L is the leakage distance;

D is the diameter of insulating element ("disc");

S is the surface area of insulating element;

F is the insulator form factor.

$$F = \int \frac{l}{\pi D(l)} \mathrm{d}l \tag{3.12}$$

Table 3.2 **Geometrical parameters of insulators.**

Insulator (Figure 3.12)	H (cm)	L (cm)	L/H	D (cm)	S (cm^2)	F
a	14.0	32.4	2.31	27.0	1745	0.78
b	13.0	29.5	2.27	25.5	1435	0.76
c	19.8	40.0	2.02	27.0	2420	1.04
d	13.0	40.0	3.08	27.0	2265	1.00
e	19.8	45.0	2.27	25.0	2525	1.00
f	12.0	31.2	2.60	33.0	1840	0.74
g	12.0	30.5	2.54	32.0	1729	0.72

During the first years of exploitation, there was largely alkaline dust pollution on the linear insulation. With the launch of fertilizer and sulphuric acid manufacturing enterprises, almost all the transmission lines were fully or partially in the area of entrainments of industrial plants.

The authors of this research received a task of formulating recommendations on replacing the insulation of the transmission lines in the area of the energy provider's responsibility with a view to reduce the frequency of unscheduled outages. At the same time, an increase in the construction height of strings was extremely undesirable, since it would lead to a change in the pylon dimensions, which would involve non-allowable capital expenses.

3.2.2.1 Insulation Operating Conditions

Soils along the overhead transmission line routes are severely cured and gypsified; vast areas are occupied by saline and alkaline soils as well as sands.

The content of soluble salts in the soil ranges from 1.0%- 1.2% to 5%- 5.4%. The predominating saltsin the soils are sulphates and chlorides (Table 3.3).

Table 3.3 **Content of salts in soil.**

Salt	Content in soil (mg/100g)	Salt	Content in soil (mg/100g)
$Ca(HCO_3)_2$	216.27	Na_2SO_4	441.9
$NaCl$	40.40	$MgCl_2$	161.9
$CaSO_4$	2135.6		

The district under study has an arid desert climate with very hot summers and short but cold winters. The average temperature in January is −1.4℃, in June +29℃. In summer, the temperature may reach +45℃ (absolute maximum is 48.6℃), and in winter, plummet down to −15℃ (absolute minimum is −30℃); the ground heats up to +70℃ in summer. The daily temperature gradients are on average 6 - 8℃, with the maximum of 19℃. The average air humidity varies from 56% in summer to 75% in winter. However, the average daily humidity reaches 80%- 99% on up to 140 days and 100% on up to 20 days annually. The average wind speed is 7.2m/s in January and 5.6m/s in July. The maximum wind speed reaches 30m/s. When the wind speed exceeds10 - 12m/s, it causes dust storms (on average, 40 - 50 dust storms a year). The dominant wind direction is south-easterly in winter and northerly in summer (see Figure 3. 13).

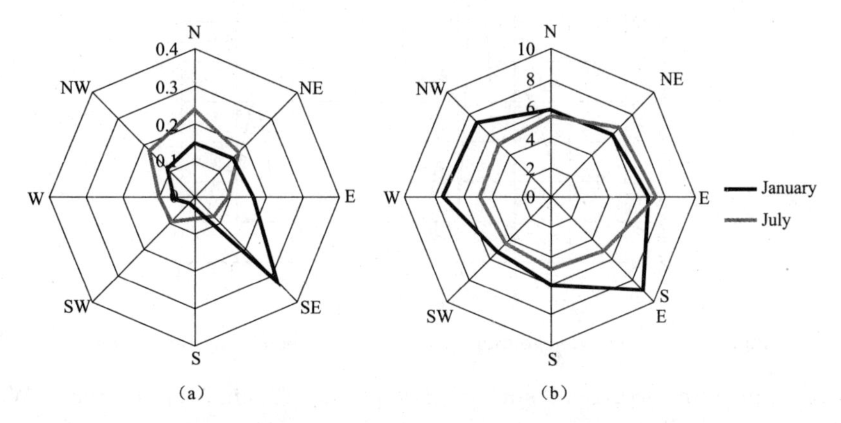

Figure 3. 13 Wind rose on the territory under study:
(a) wind frequency by direction; (b) wind speed by direction.

Precipitation in the area under study is scarce; the average amount is 100 - 110 mm per year. At the same time, drastic temperature fluctuations within 24 hours, an extreme continental climate, and the proximity of the sea cause frequent fogs and dews.

The duration of fogs ranges from 1. 5 to 24 hours and more, and the average duration of dews is more than twice as high as that of fogs. The insulation flashovers mostly happen at night and in the morning (Figure 3. 14 and Figure 3. 15), in other words, at the time of fogs or dews or both.

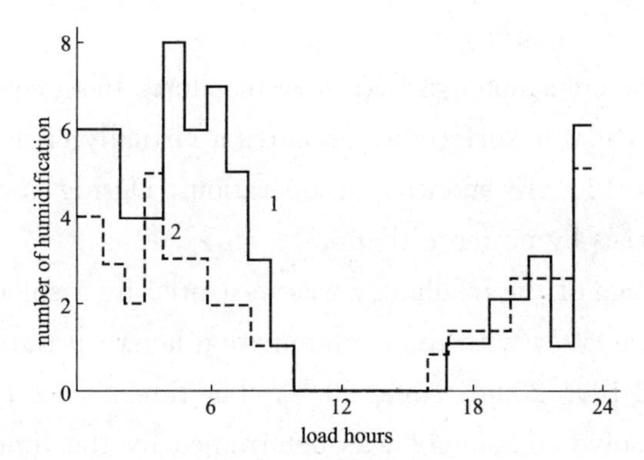

Figure 3. 14 The average annual distribution of the number of insulator wetting cycles by time of day: dew (1), fog (2).

3. 2. 2. 2 Test Procedure

The dynamic pattern of insulator contamination was studied on a test bench without energizing. The test set-up was located in the area of simultaneous contamination by entrainments of a fertilizer plant and alkaline dust. There was also a production facility of complex fertilizers at the distance of about 500 m. The insulator strings were suspended at

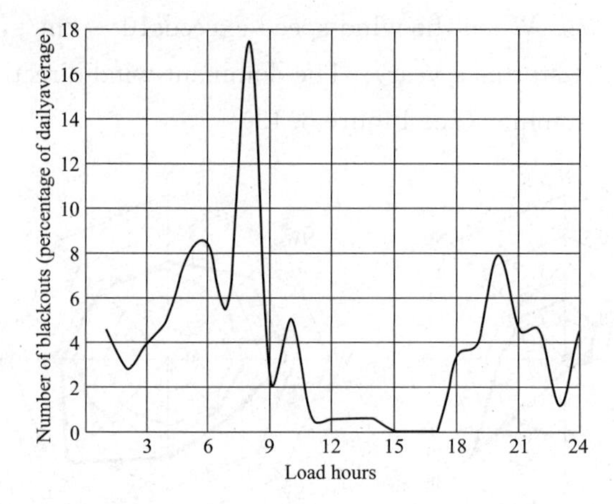

Figure 3. 15　The distribution of insulation flashovers by time of day.

the height of 12 m from the ground surface, comprising 10 similar units each. We tested at least six insulator strings of each type.

During the first six months of exposure, we removed the insulators (one string of each type) from the test bench every month and then twice a year, at the end of drought seasons.

Two insulators of each string were designated for the study of pollution layer parameters. The remaining ones served for the laboratory determination of discharge characteristics.

The pollution layer parameters were determined separately for the top and bottom surfaces of insulators. We used a scraper to remove the contamination mechanically from each of the given surfaces and then weighted and analyzed its chemical composition.

The total duration of the study was 32 months.

The contamination accumulation dynamics were monitored this frequently during the initial exposure because the insulator surface accumulates a virtually critical amount of contaminants after as little as 12 - 18 months of operation. During further tests (up to 32 months), it only increases by no more than 20%- 40%.

The flashover voltage of the insulators was measured by a smooth increase of voltage in accordance with IEC 60507, with the contamination layer saturated by water until reaching the conductivity of 180 - 200μS/cm at 20 ℃. The time needed for the salts of the contamination layer to dissolve completely was determined by the time of surface resistance settlement and varied from 35 to 55 minutes. The experiment was held on a high-voltage test bench with the output voltage of 100 kV and the rated power of 100 kVA (high-voltage transformer IOM 100/100 powered by a voltage regulator in this case Norris transformer AOMKT 100 / 0. 5). The bench power was increased by a series capacitor compensation, which provided the solid fault current of more than 3 A at a voltage of 20 - 40 kV. To protect the high-voltage transformer from overloads, the capacitor was shunted by a discharge gap, which activates after the output current of the setup reaches 1. 15 A. The flashover voltage was measured using an electrostatic kilovoltmeter that was connected directly in parallel with the insulators being tested. See Figure 3.16 for the flow sheet and Figure 3.17 for the general view.

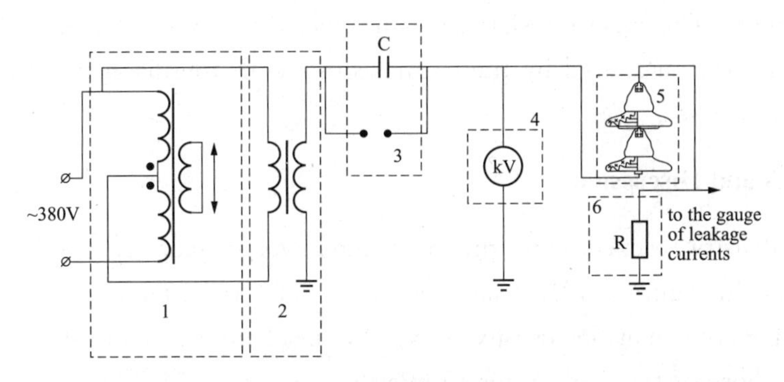

Figure 3.16 Test set-up flow sheet: Norris transformer (1), high-voltage transformer (2), series capacitance with an arrester (3), kilovoltmeter (4), test object (5); leakage current shunt (6).

Figure 3.17 Test set-up general view. Designations are the same as in Figure 3.16 and Figure 3.15 are off screen.

For the insulators of each removal period, the average flashover voltage was determined as the average of the minimum values obtained in each of the three measurements during the test of four strings (two elements in each). The number of insulators in a string was based on the capabilities of the test bench, contaminations on the insulators being spread almost uniformly.

The voltage distribution along the creepage path during polluted and wet insulators loading with high voltage is completely determined by the leakage current density, resulting voltage drop has maximal value near the insulator cap and pestle, while on the plate edge (where is the maximum diameter of the insulating parts) it is minimal. Accordingly partial discharges that precede a full overlap insulator occur. Since all the tested insulators have a variable diameter along the leakage path, there is much the same distribution pattern of voltage, and a special study of this distribution was not carried out.

The surface resistance of insulators was measured using a megohmmeter with the voltage of 2,500V and monitored by oscillographic testing of high voltage (80% of the average flashover voltage) and leakage current. The control oscillographic test was carried out under impact loading. The surface resistance values measured by the megohmmeter agree satisfactorily with those measured by the oscilloscope, with the discrepancy reaching 10%-20% in just a few cases.

3.2.3 Results and Discussion

Pollution density reached its maximum on the lower surfaces of type '*a*' and '*c*' insulators (identified in Figure 3.18). It amounted to 10.59 and 10.3 mg/cm^2 respectively.

The maximum contamination density on type '*a*' insulator was recorded after 27 months of testing and on those of type '*c*', after 15 months.

The minimum contamination density by the end of the testing was recorded on the upper surfaces of '*g*' and '*f*' type insulators: 0.83 and 0.99 mg/cm^2 respectively and on the lower surface, on *g*, *f*, and *c*: 1.94; 2.04 and 2.06 mg/cm^2. The research has shown an insignificant difference between the contamination density on the upper and lower surfaces of smooth bell-shaped insulators (*f* and *g*). This difference has been the smallest throughout the research period and never exceeded 3-4-fold. The insulators with massive lower surface (*a*, *b*) had a considerably higher contamination density on the lower surfaces than on the upper ones throughout the course of the research. This proportion, for instance, reached 1 : 13 for type '*a*' insulators (10.59 and 0.77 mg/cm^2 respectively). The content of soluble salts was recorded to be 30% higher in the deposits on the upper and middle surfaces of some insulators (*c*, *d*, *f*, *g*) than on their lower surfaces (15%-20%).

The peak amount of soluble salts was registered after 20 months of tests on insulator *f* : 60.7% (upper surface, equivalent NaCl salinity 43%). By the end of the exposition period, the content of soluble salts reduced by 40%-50% on all the insulators due to meteorological factors.

It takes 12-18 months of exposure for insulators to accumulate the limit amount of pollution. Subsequently, the amount of pollution collected fluctuates around this level under climatic factors combined with industrial entrainments. The contaminations are prone to solidification.

Substances contained in fertilizer plant entrainments dominate the pollution composition. The pH level of the aqueous solution of contaminations from the insulators lies within 6 ± 0.5 (acidic medium), which also suggests the domination of the substances from entrainments of fertilizer plants in the contamination composition. The specific bulk conductivity of the insulator contamination solution reached 1.5 - 1.6μS/cm.

During the starting period of exposure, the strings were removed from the test bench every month, which allowed us to detect a pronounced seasonality in the contamination of the upper surface on all the seven insulator types (Figure 3. 18). After as little as 40 days since the installation on the test bench (first removal, July-August), pollution accumulated on the upper surface of the insulators in significant amounts-0. 564 mg/cm^2 on insulator b to 1. 185 mg/cm^2 on insulator *d*. Later on, until October-November, the upper surface contamination density was decreasing down to the minimal values-0. 483 mg/cm^2 on insulator *b* and 0. 238 mg/cm^2 on insulator *g*. The upper part contamination density went up again in November to December.

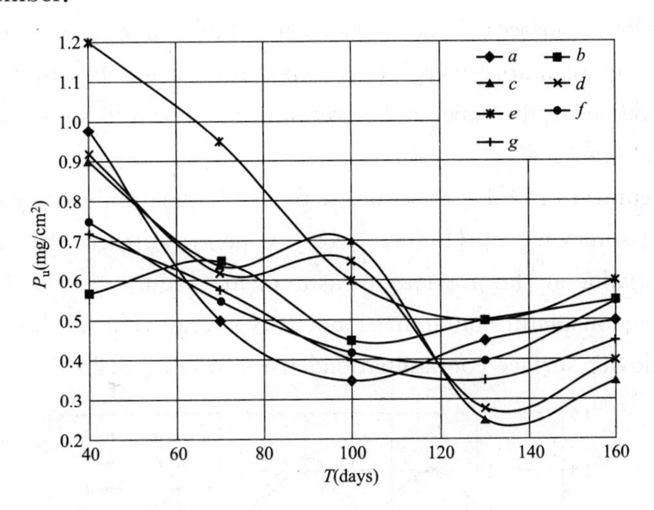

Figure 3. 18 Upper insulator part contamination density.

The inhomogeneity of insulator pollution on the test bench was assessed by the value of the ratio

$$\alpha = P_u / P_l \tag{3.13}$$

where P_u and P_l are the contamination density of the upper and lower part of the insulator.

In the first lot, all the insulators have $\alpha > 1$, i. e. the contamination density on the upper disc surfaces is higher than on lower ones. On the contrary, in the second and all the other lots, the lower insulator part appeared to be more polluted (Figure 3. 19, Y-axis is logarithmic).

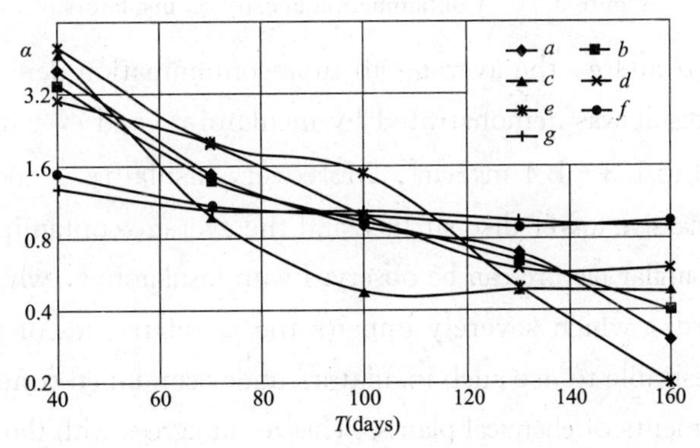

Figure 3. 19 Inhomogeneity of insulator contamination chart.

The upper surface of the insulators is more exposed to environment, including pollution. Therefore, depending on the ambient conditions and wind direction, it "gains" contamination as easily as it "loses" it. The lower surface, on the contrary, is more sluggish in this respect.

It accumulates pollution at a slower rate but the self-purification is severely hindered; the contamination is virtually not removed from the lower surface at all.

During a short period between the installation and the first removal of the insulators (40 days of exposure), their lower surfaces did not accumulate pollution in considerable amounts.

Progressively, as insulators were being exposed on a test bench, the difference in the self-purification conditions of the upper and lower surfaces began to show in all the lots starting with the second one, $\alpha < 1$.

At the fifth removal (160-day exposure), the contamination density of the upper insulator surfaces went somewhat up (Figure 3. 18), which resulted in the increase of factor α as compared to its value at the previous measurement (Figure 3. 19). The pattern of the change in the average insulator contaminationdensity overall reflects the dynamic pattern of the change in the lower surface contaminationdensity (Figure 3. 20).

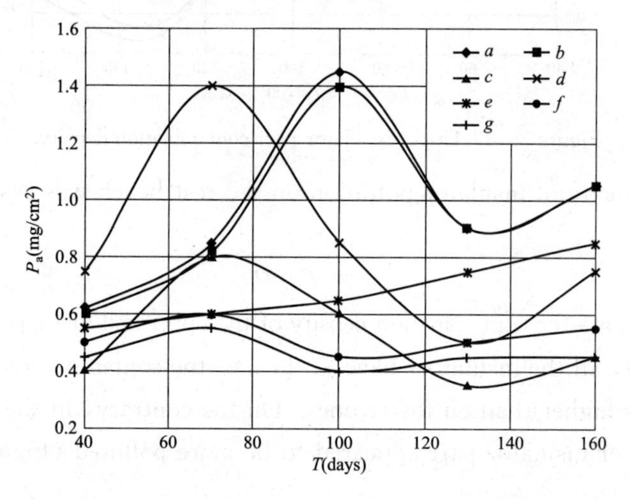

Figure 3. 20　Contamination density on insulators.

With insulators *d* and *g*, the average all-time contamination density did not exceed 0. 9 mg/cm^2. The worst result was demonstrated by insulators *c* and *e*, whose peak contamination density amounted to 1. 3 - 1. 4 mg/cm^2. Moreover, insulators *a* and *b* largely accumulate contaminants on the massive lower disc surface and the average contamination density is constantly growing. A similar picture can be observed with insulators e, whose insulating element is extended downwards, which severely impairs the aerodynamics of the insulator string. Therefore, it is undesirable to use such insulators under combined contamination with alkaline dust and entrainments of chemical plants. This result agrees with the data obtained by other authors in similar conditions.

The flashover voltage of most contaminated insulators goes down by 73%- 77% after as little as 40 - 70 days of exposure vs. the flashover voltage of clean wet insulators. High average pollution density (0. 5 - 0. 8 mg/cm^2) and high conductivity of the layer (7 - 14 μS) associated with it reduce the flashover voltage down to such values (12 - 16 kV) that make it impossible to provide the fail-safe operation of insulators without an urgent insulation enhancement or purification (washing).

The measurements did not allow detecting a clear connection between the discharge voltage and surface conductivity. The leakage currents measured during the last half-cycle of the applied voltage before the flashover also did not correlate with the discharge voltage (Table 3. 4). The authors have found this correlation but in their case, the inhomogeneity of the contamination was considerably lower, similar to that of artificial contamination.

At the same time, we can see the connection between the discharge voltages and the contamination surface density, which follows from the simultaneous change in these parameters in time. We can assume that the surface conductivity is no universal characteristics for the insulator contamination, at least when the contamination is not uniform.

Table 3. 4 shows the lowest flashover voltages of contaminated insulators for all exposure. In the table:

U_d is the insulator flashover voltage;

I_l is the leakage current of the insulator in the last half cycle of the applied voltage before the flashover;

E_H is the insulator flashover voltage relative to its construction height;

E_L is the insulator flashover voltage relative to its leakage current path length;

$И = F/R$ is the conductivity of the insulator pollution layer saturated by water, whereas R is resistance measured by the megohmmeter.

The increase of the leakage current path length$L$ reduces its utilization efficiency, which is supported by the lower specific flashover voltage along the leakage path length E_L. Thus, making the insulator shape more sophisticated does not increase its operation reliability under the conditions considered.

However, the ratio of the insulator stray path length$L$ to its construction height H directly affects the specific flashover voltage of insulators along the construction height E_H. According to Tables 3. 2 and 3. 4, the increase of the parameter L/H leads to the increase of E_H. This, in turn, makes it possible to improve the insulation reliability without increasing the construction height of the strings.

The specific flashover voltage of insulators along the leakage current length E_L does not depend much on L/H ratio. It is, however, strongly dependent on the pollution layer conductivity: increasing the conductivity up to 8 - 12 μS/cm for all the seven insulator types leads to a sharp decrease of the flashover voltage; a consequent increase of the conductivity leads to a slight alteration of the voltage (Figure 3. 21).

Table 3.4 **Minimum flashover voltage of tested insulators.**

Insulator	U_d (kV)	I_b (mA)	E_H (kV/cm)	E_L (kV/cm)	и (μs)
a	9.5	580	0.68	0.29	61.6
b	7.2	270	0.55	0.24	37.0
c	14.4	500	0.73	0.36	34.9
d	13.8	350	1.06	0.34	25.5
e	14.5	320	0.73	0.32	22.3
f	10.7	270	0.89	0.34	25.2
g	11.8	280	0.98	0.39	23.7

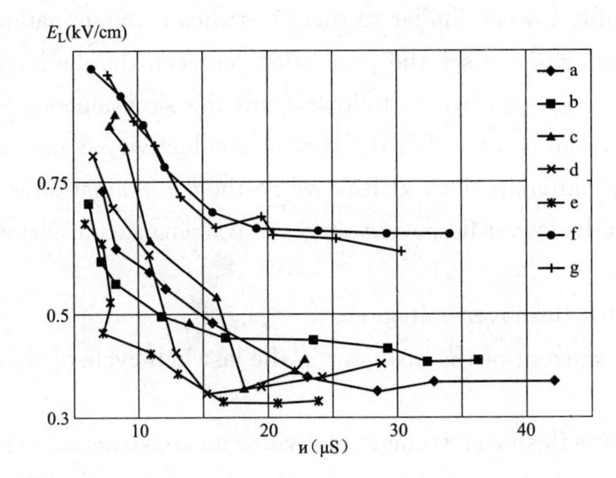

Figure 3.21 Dependence of specific flashover voltage of insulators along the leakage current length E_L on the conductivity of the insulator pollution layer и.

3.2.4 Conclusion

Combined natural and industrial contamination of high intensity leads to a nearly critical decrease of insulator flashover voltage during the first 1.5 – 2.5 months of operation.

Thus, during the bench tests, the recommended period of the insulator removal at initial stages of the study is 1.5 – 2.5 months. The total duration of the tests can be reduced down to 24 months.

The insulators with a massive lower surface, which are normally the first choice in the contamination-intensive regions, demonstrate an alarming trend to accumulate pollution and decrease flashover voltage even after short exposure periods. At the same time, insulators with a smooth bell-shaped insulating element with fine aerodynamic properties demonstrate excellent self-cleaning capabilities both individually and as part of a string. They also possess the best specific discharge characteristics. This enables us to recommend these insulators for usage in the regions that are similar to the one under study.

4

专业英语语法(2)

4.1 否定结构

在科技英语中，除了使用一般否定句外，较为常用的否定结构还有全部否定、部分否定、双重否定、意义否定以及否定的转移。下面分别予以讨论。

4.1.1 全部否定

英语句子中用全否定词 not，no，none，no one (nobody)，nothing，neither，nor，neither …nor …，nowhere，nohow，never 等词来表达全部或完全否定。这些否定词不论作什么成分，其所在的句子都要译成全部否定。所谓全部否定是指否定整个句子的全部意思。例如：

(1) We usually regard energy, including electromagnetic energy, as not occupying space. 通常认为能量，包括电磁能不占有空间。

(2) No place is absolutely safe from the lightning threat; however, some places are safer than others. 没有一个地方是绝对不会遭到雷击威胁的，但是有些地方比另一些地方更安全些。

(3) There are no moving currents of air. 空气不流通。(否定主语转移否定谓语)

(4) None of the textbooks available have mentioned this special phenomenon. 现有的教科书均没有提到这一特殊现象。

(5) Some people believe that there may be aliens, but no one is sure. 有些人认为可能存在着外星人，但谁也没有把握。

(6) Nothing in this world moves faster than light. 世界上没有什么能比光跑得更快。

(7) So far, nobody has ever carried out such experiments. 迄今，还没有任何人做过这类实验。

(8) Neither of the two transformers is satisfactory. 这两台变压器都不令人满意。

(9) It is well-known that glass can not conduct electricity , nor can rubber. 众所周知，玻璃不能导电，橡胶也不能导电。(nor 连接两个并列分句，第二分句为倒装语序省略句)

(10) There exist neither perfect insulators nor perfect conductors. 既没有完全的绝缘体，也没有完全的导体。

(11) Never is aluminium found free in nature. 铝在自然界从不以游离态存在。(句子

以 Never 开头，采用倒装语序）

4.1.2　部分否定

英语中有些否定句所表示的意义是部分否定，即非全部否定，指句子的意思部分否定、部分肯定，相当于汉语“不全是”“并非全是”“不都是”“不是都”“不是所有都”“不是每个都”“未必都”“不都”“不是两者都”“不总（全）是”“不全是”“不常”等之意。其构成方式是由：①某些含全体（整体）或多数意义的不定代词，如 all、everything、everybody、both、each、many、much 等；②某些含全体或多数意义的形容词，如 every、whole，total、complete、entire 等；③某些程度、方式、频率副词以及表示全体的副词等，例如 always、very、often、quite、usually、wholly、entirely、wholly、completely、totally 等与表示否定词 not 连用而成，常见的有以下三种句式：

（1）all … not …，everything … not …，both … not …，each … not …，many … not…，much … not …等。

这种结构由表示全体或多数意义的不定代词与否定词连用，由于否定词 not 与谓语在一起构成谓语否定，所以形式上很像全部否定，但实际上却是部分否定。因此，汉译中应当特别注意，译为“并非所有”“不是每个都……”“并非每一个”“并非全是……”“未必都是……”“不是两者都”“不是许多”等含义。例如：

1）All that can conduct electricity is not metal. 能导电的未必都是金属。

2）Everything is not possible. 并非一切均有可能。

（2）all … not…，every … not …，both … not …等。

这种结构由表示全体或多数意义的形容词与否定词连用，也是部分否定，译为“不是每个（所有）都……”“并非全是……”“未必都是……”“不是两者都”等含义。例如：

1）It should be noted，however，that all electrical systems are not conservative.

然而，应当注意，并非所有的电（气）系统都是守恒系统。［不能译为：所有的电（气）系统都不是守恒系统。这一意义应是：None of the electrical systems is conservative.］

2）All the chemical energy of fuel is not converted into heat. 并非所有燃料的化学能都能转变成热量。

不过，这种“all … not …”结构有时并非意为“并非都……”，而是指“即使全部……也不……”。例如：

3）All the achievements in automation cannot make man free from taking part in management of modern industries. 即使全部实现自动化，也不能把人从现代企业的管理中完全解放出来。

4）Every subject is not treated in the same way. 不是每个题目都同样处理。

5）Every color is not reflected back.（=Not every color is reflected back.）并非每种色光都会反射回来。

6）Both of the substances do not dissolve in water. 不是两种物质都溶于水。

7）Both of the substances are not involved in the chemical reaction. 这两种物质并非

都参与了化学反应。

8）Both transistors are not of the N-P-N type. 并非两个晶体管都是 N-P-N 型的。

9）Each transistor in the box is not out of order. 盒里的晶体管并非每一个都是坏的。

10）In a thermal power plant，all the chemical energy of the fuel is not converted into heat. 在火力发电厂里，燃料的化学能并未全部转化为热量。

英语中这种表达部分否定的形式有两种：

①否定词 not 置于谓语处，形式为“all … not …”；

②否定词 not 放在 all、every、both 这些词之前，形如 not all、not every、not both 等，这时其部分否定的意义就很明显，而不会译错。前者是传统说法，虽然不合逻辑，但习惯上这样用。现代英语的流行趋向是采用后者，即将否定词直接置于被否定的那部分主语之前，这样从逻辑和语法上看都比较合理，不至于造成误解，所以正在为越来越多的人采用，尤其在美国书刊中更为常见。例如：

11）Not all substances are conductors. 并非所有的物质都是导体。

12）Not all matter is visible. Air is not visible，but it is matter. 并非一切物质都是看得见的。空气是看不见的，但是它是物质。

13）Not every minute difference is noted. 并非每一个细微的差别都被人注意到了。

14）Not many of the machine tools are self-regulating and self-controlling. 许多机床不都是自动调节和自动控制的。

15）Not much of waste water is utilized by them. 他们利用的废水不多。

（3）not always、not often、not altogether、not entirely 等。

这种结构由否定词 not 和某些表示程度、方式、频度以及全体的副词连用，可以翻译为“并不一定……”“并不总是……”“并不完全是……”。这种用法也属于否定转移。例如：

1）An engine may not always do work at its rated horse power. 发动机并非始终以其额定马力做功。

2）Force do not always produce movement. 力并不一定能引起运动。

3）A cubic foot of air does not always weigh the same amount. 1 立方英尺空气的重量并不总是一样的。

4）Materials are not always useful in the form in which they are found. 处于原始形态的物质并非总是有用的。

5）The electrons within a conductor are not entirely free to move but are restrained by the attraction of the atoms among which they must move. 导体中的电子运动并不是完全自由的。电子必须在原子之间运动，从而要受到这些原子的引力的束缚。

6）Sounds are produced by vibrating bodies. The vibrations are not often visible but they occur none the less. 声音是由震动产生的。震动并不是经常（不常）能看得见的，但仍然发生了震动。(none the less 仍然、还是)

7）Yet the views which have supplanted this theory are not altogether suitable for pictorial representation. 可是，已经取代这种学说的观点并不完全适于用图表示。

“主语＋not＋all（both，every，…）”这种结构汉译为“并不是都……”“不都……”“并非总是……”等时，all、both、every 等词在句中的位置是作宾语，包括作介词宾语或修饰宾语，而（1）、（2）中 all、both、every 等词在句中的位置是作主语或修饰主语，但它们均为部分否定。例如：

1）I don't know everything about the course. 我并不了解这门课程的全部内容。

2）He doesn't know both of the experts. 这两位专家他并不都认识。

3）The NC lathe is not the best lathe for all jobs，it is more economical for certain kinds of work and quantities. 数控车床并非适用于所有加工，它只对某些加工种类和数量而言是比较经济的。

对于全部否定和部分否定可作如下比较：

Every result is not correct. 并非每个结果都对。（有对有错）

No result is correct. 所有结果都是错的。（都错）

I don't remember all the formulas. 这些公式我并不都记得。（有的不记得）

I don't remember any of these formulas. 这些公式我都记不得。（全不记得）

All these transformers are not made in China. 这些变压器不都是中国制造的。（有的是）

None of these transformers are made in China. 这些变压器都不是中国制造的。（都不是）

Both of the installations went wrong. 两台设备没有都坏。（有一台坏了）

Neither of the installations went wrong. 两台设备都没坏 .

every … not … = some（其全部否定应当用 nobody、no、none、no one）

all … not … = some（其全部否定应当用 none）

both … not … = one（其全部否定应当用 neither）

not often … = sometimes

not always = sometimes

4.1.3 双重否定

英语的双重否定结构通常是由两个否定词（no、not、never、nothing 等）出现在同一个句子中或一个否定词与某些表示否定意义的词语或短语结合连用而构成的，即否定之否定，意思是为了委婉地表达肯定意思，或者是表示强意肯定，其目的在于加强语气。英汉两种语言都有这种否定结构。前者称为弱化的双重否定，后者则称为强化的双重否定。

1. 弱化双重否定

这种双重否定大多包含两个否定词，其中一个常由否定前、后缀构成。因为两个否定词中有一个为另一个所否定，这就使否定的语气抵消了一部分，因而产生弱化的效果，实际上等于一个语气委婉温和的肯定词。类似的情况在汉语中也有，例如，“难免错误”“无可厚非”“不无理由”“未免过分”等。弱化双重否定的特征是其句中两个否定词一般仅紧紧连接在一起，在汉译时多保持原文双重否定形式而直接译成双重否定，也可转移为肯定句。例如：

（1）Variations in excess of the ranges given are not infrequent. 变化超出规定范围者并非罕见。（译为双重否定）

（2）It is not impossible that a breakthrough in condition assessment of high voltage insulation in power system equipment will be attained in the foreseeable future. 电力系统设备高压绝缘的状态评估技术在可预见的将来会取得突破，这不是不可能的。（译为双重否定或译为肯定：电力系统设备高压绝缘的状态评估技术很可能在可预见的将来会取得突破。）

（3）The great advances in high voltage engineering was not unexpected by some engineers. 对于高电压工程方面的巨大进展，某些工程师并非未曾料到。（译为双重否定）

（4）We do not deny that there is much room for the perfection of our research work. 我们不否认我们的研究工作大有改进的余地。（译为双重否定）

（5）A radar screen is not unlike a television screen. 雷达荧光屏跟电视荧光屏一样。（译为肯定）

（6）The flowing of electricity through a wire is not unlike that of water through a pipe. 电流过导线就像水流过管子一样。（译为肯定）

（7）There is nothing unexpected about it. 一切都在意料之中。（译为肯定）

（8）We have to lose no time to install the steam turbine. 我们必须抓紧时间安装汽轮机。（译为肯定）

（9）But unlike common radio waves，nuclear radiation is not harmless to human beings and other living things. 但核辐射不同于普通的无线电波，它对人类以及其他生物是有害的。（译为肯定或译为双重否定：……对人类以及其他生物不是无害的。）

（10）The factory boss has not been unmindful of these requirement. 厂长并不是不关心这些要求。（译为双重否定）

以上例句中的弱化双重否定，大多数是由否定词 not 与一个含有否定意义的形容词（多由词缀法构成）或动词组成。事实上，间或也还有其他形式，如 lack of disadvantage（并无不利）、free from instability（没有不稳定性）。总之，弱化的双重否定产生的效果是否定语气或程度的削弱。

2. 强化双重否定

两个否定词并用能产生加强整个句子语气的作用。但应注意，这里所指的语气是肯定的、正面的语气，而不是否定的、反面的语气。从结构形式上看，这种双重否定的两个否定词一般并不紧紧连接在一起。因此，不存在弱化双重否定中一个否定词否定另一个否定词的情况，故而多按原文结构直接译为双重否定，强化双重否定中两个否定词所否定的分别是句子的不同部分。这种否定句也有两种译法：

（1）根据上下文的需要保持并译成双重否定：“不会不”“无不”“未尝不”“并非不”；

（2）译成肯定句，译词用“没有……不（没）……”等。这种否定形式最常见的有：no（not）…no（not）…没有……，没有……不……；no（not）… but … 没有……不……；no（not）…without … 没有……不……，除……不……；not（none）… the less … 不因……就不……；no（not）… unless … 没有……就不……，除非……才……；

not … until (till) … 不到……不，直到……才……；no (none) … other than … 不是别的，正是……；not … but that … 并非……不……，虽然……；not … a little … 不少、很多、大大。例如：

(1) No machine ever runs without some friction. 从来没有一台机器运转时是没有摩擦的。(译为双重否定)

(2) No machine can be made completely frictionless. 没有摩擦作用的机器是造不出来的。(译为双重否定)

(3) There is no steel not containing carbon. 没有不含碳的钢。(译为双重否定)

(4) Sodium is never found uncombined in nature. 自然界中从未发现不处于化合状态的钠。(译为双重否定)

(5) No flow of water occurs unless there is a difference in pressure. 没有压差，水就不会流动。(译为双重否定)

(6) It is impossible for heat to be converted into a certain energy without something lost. 热转换成某种能而没有什么损耗是不可能的。(译为双重否定)

(7) There can never be a force acting in nature unless two bodies are involved. 在自然界中不牵涉到两个物体，就不能有作用力。(译为双重否定)

(8) Bearings should not be unboxed or unwrapped until the moment for fitting has arrived. 不到安装的时候，不应开启包装箱或包装纸。(译为双重否定)

(9) There is no law that has not exceptions. 凡是规律都有例外。(译为肯定)

(10) One body never exerts a force upon another without the second reacting against the first. 一个物体对另一物体施作用力必然会受到另一物体的反作用力。(译为肯定)

(11) In making experiments you can not be too careful. 做实验时，越小心越好。(谨慎总不嫌过度。)(译为肯定)

(12) It is not until meteors strike the earth's atmosphere that they can be seen. 直到流星进入地球的大气层时，才能看得到它。(译为肯定)

(13) Not until the invention of the high altitude rocket did the direct study of upper atmosphere become possible. 直到高空火箭发明之后，直接研究上层大气才成为可能。(译为肯定)

(14) You can do nothing without energy. 没有能量，你就什么也做不成。(译为双重否定)

(15) Without substance, there could be no world. 没有物质就没有世界。(译为双重否定)

(16) You can not keep a gas in one place at all unless you imprison it in a bottle, or a ballon, or something like that. 除非你把气体封闭在瓶子里、气球里或类似的容器里，否则你根本不可能把气体保持在一个地方。(译为双重否定)

(17) In our factory not a generator but what was designed and made in China. 在我们厂没有一台发电机不是中国设计和制造的。(译为双重否定)

(18) The atomic furnace will *not* work *unless* it has enough fuel. 原子反应堆若没有足

够的燃料是不能运转的。（译为双重否定）

（19）No flow of water occurs unless there is a difference in pressure. 没有压差，水就不会流动。（译为双重否定）

（20）In the absence of electricity，large scale production is impossible. 没有电，大规模生产就不可能。（译为双重否定）

（21）All these electric protective devices should not be unboxed or unwrapped until the moment for fitting has arrived. 不到安装这些电气保护装置时不应打开包装箱或包装纸。（译为双重否定）

（22）With the introduction of the electronic computer，there is no complicated problem but can be solved in a few hours. 有了电子计算机就没有几小时算不完的复杂算题。

（23）There is no material but will deform more or less under the action of forces. 在力的作用下，没有一点也不变形的材料。（或译成肯定：一切材料在力的作用下多少会有些变形。）

注意："There is no … but"是一种双重否定结构，其中 but 是关系代词，引出定语从句，相当于"that … not"，从句与主句都具有否定意义。汉译时，可译成双重否定"没有……不"，也可以译成肯定。

4.1.4 意义否定

所谓意义否定是指句子在形式上是肯定句，但由于句子中用了含有否定意义的词或短语，所以实际为句意否定的否定句。构成意义否定的词有动词、名词、副词、介词、连词、形容词及词组。例如：

too…to … 太……不……；太……没有……

in place of… 代替……

too…for … 太……不……

out of… 缺乏……；脱离……；在……外

far from … 一点也不……，远非……；没有……反而

in lieu of… 代替……

free from … 没有……；免于……

in vain… 无效；徒然……

safe from … 免于……；没有……

short of… 缺少；没……

keep from …使……不受；避开；阻止；免于

rather than… 而不是；与其……不如……

save from…使……不受；避开；阻止

miss 失败；没有

protect from… 使……不受；避开；阻止

refuse 拒绝

no more… 也不；一样不……

ignore　忽略；不顾

but for…　若非；如果没有……

overlook　忽略；忽视

anything but…　绝不是……

exclude　排除；除去

instead of　而不（是）……；代替……

fail　失败；缺乏

prevent from…　不……；阻止……

few　几乎没有；很少

little　几乎没有；很少

neglect　忽略；忽视

absence　缺乏；缺少

exclusion　排除；除去

but that　若非；要不是

without　没有；不必

except　除……之外

下面按照词义否定程度的不同将意义否定分为两类，即分别由半否定词和准否定词构成的意义否定。

1. 由半否定词构成的意义否定

英语中根据否定程度以及否定语气的强弱，可以将否定词分为全否定词、半否定词和准否定词。顾名思义，全否定词（组）表示的是否定语气，例如 no，not，never，none，nobody，nothing，nowhere，neither，nor，neither … nor … 等。所谓半否定词是与全否定词相对照、语气较委婉、程度较温和的一组词，它们从字面上由于没有 no、not，但具有否定语气，因而称为半否定词，例如 little、few、hardly、scarcely、barely、seldom、only a little（few）或 but a little（few）= little（few）等。由这些半否定词可以构成意义否定，因而要译成否定语气。例如：

（1）An insulator is a material that has few free electrons and that resists the flow of electron. 绝缘体是一种几乎不含自由电子的物质，因而它阻止电子的流动。

（2）Metals, generally, offer little resistance and are good conductors. 通常，金属几乎没有（或译：没有什么）电阻，因而是良导体。

（3）Pure iron has few useful properties. 纯铁没有什么有用的特性。

但是，not a little、no little 的意思是“许多”和“很（颇）”，同样，not a few 的意思是“许多……”“相当多……”，都是肯定语气；而 not a bit、not in the least 的意思是“一点也不”“一点也没有”，则是否定语气，汉译时应加以区别。此外，little 和 few 都可以用作名词，但 little 还可以用作副词。

hardly、scarcely 和 barely 这三个词均含有否定的意味，前两者一般可以互换，但严格地说，在用法上是有区别的。hardly 着眼于“困难”，常用于修饰表示能力的词来对能力加以否定，而 scarcely 往往强调不足，常与 enough、sufficient、any 等表示程度的词连用；barely 一般与上述两词意思相近，往往表示“仅仅”“勉强”“刚刚”等意思。简言之，hardly 往往强调困难，scarcely 和 barely 则着眼于多少，是对数量的否定。但在一般书面语中，特别是指“就这么一些，一点也不多”时，这三个词可以说完全同义，可互换使用。例如：

（4）A separate development in hydraulic controls, though hardly a trend, is an outgrowth of the hydrofluidic amplifier. 液压控制系统的另一项发展（虽然说不上是一种趋势），就是液体射流放大器的崛起。

（5）There are scarcely any power plant which can convert more than 55% of the fuel energy into electric energy. 几乎没有（或译：简直没有）什么发电厂能将燃料中 55%以上的能量转换为电能。

（6）Sometimes a very simple electronic device can be availed to enable a human being to be relieved from working in an environment which is so dirty as to be barely tolerable. 有时采用非常简单的电子装置，就能使人从一种脏得几乎不能忍受的工作环境中解脱出来。

seldom 这个词的否定语气远不及 never，但与 often 和 always 一对照，否定含义却十分明显。例如：

（7）Though the design engineer seldom has the ultimate responsibility for quantity production of the parts or machines he designs, he must be familiar with different manufacturing process. 虽然工程设计人员对于自己设计的零件或机器的成批生产很少负有主要责任，但是关于各种不同的生产工艺却必须熟悉。

seldom 的否定语气也可通过与其他副词相结合的方式而加强，如 seldom if ever 与 seldom or never 就相当于汉语的“绝无仅有”“极其难得”“简直不……”等。

（8）The robot seldom if ever tires or losses interest in its task. 机器人对于所承担的工作简直不知疲劳或失去兴趣。

表 4-1　　半否定词构成的意义否定的含义

few 几乎没有 little 几乎没有	a few 有几个（可数） a little 有一点（不可数）
hardly, hardly any, hardly possible scarcely, scarcely possible barely	表示几乎不、几乎没有、几乎不可能 等于 almost nothing, almost no, almost not at all, nearly no
rarely, rarely any, seldom, hardly ever	等于 almost never
too … to …　太……以致不能 enough … to　足够……以致可以，正好 not … enough … to　不……不能……	too=excessively　表示否定 enough=sufficiently　表示肯定

（9）Certainmaterials whose resistivity is not high enough to classify them as good insulators, but is still high compared with the resistivity of common metals, are known as semi-conductors. 某些材料的电阻率不够高，不足以划为良绝缘体；但是与普通金属相比，

其电阻率还是高的，故称为半导体。

2. 由准否定词（组）构成的意义否定

所谓准否定词（组）是指其含义否定，但形式是肯定的词（组），有动词、名词、副词、介词、连词、形容词等，其中多数是单独的词，有些则是词组。

（1）动词与动词词组。

1）The motor refused to start for lack of maintenance. 由于没有保养，马达发动不了。（动词 refuse 含有否定意义）

2）The first law of thermodynamics denies the possibility of creating or destroying energy. 热力学第一定律否认有创造或消灭能量的可能性。（动词 deny 含有否定意义）

3）The machine is provided with a punched-tape reader which is mounted in such a conspicuous position that the operator cannot miss it . 这种机器装有穿孔读出器，其位置非常显眼，操作者不会看不见。（动词 miss 含有否定意义）

4）The value of loss in this equation is so small that we can overlook it. 这方程中的损耗值小得可以忽略不计。（动词 overlook 含有否定意义）

5）Neglecting friction，the output is equal to the input. 如果不计摩擦，输出就等于输入。（动词 neglect 含有否定意义）

此外，还有很多能表达否定意义的动词，例如 avoid、decline、defy、ignore 等。

6）The meeting was put off because we failed to obtain the necessary information. 因为我们没有获得必要的资料，致使会议延期举行。

7）If the follower loses contact with the cam，it will fail to work. 随动件如果与凸轮脱开，就不能工作。

8）Space lacks for an overall illustration of it. 篇幅有限，不能对之加以全面说明。

9）As rubber prevents electricity from passing through it，it is used as a insulation material. 由于橡皮不导电，所以用作绝缘材料。

10）It is gravity that keeps us from falling off. 正是重力使我们不会从地球上掉出去。

11）The system of power supply in the city leaves much to be desired. 该市的供电系统极不完善。

（2）名词。

1）Failure to satisfy these requirements could be catastrophic for users. 不能满足这些要求，会使客户感到分外困难。

2）In computers the tubes do not have to control all the shades of brightness in an image or be carefully tuned to receive one frequency to the complete exclusion of all others. 在计算机中，电子管无须控制图像的亮度或仔细地调得只接受某一个频率，而将其他频率全部排除在外。

3）The absence of a corridor could lead to lower cost housing being built under OHLs. 没有回廊可以使得建在架空线下的房屋成本降低。

4）There is an excess or deficiency of neutrons in any combination，the isotope will be unstable. 中子过多或不足，形成的同位素就不会稳定。

5）Lack of markets makes a product useless. 没有销路（市场）就使产品归于无用。

（3）形容词。

1）The ripple or harmonics in the direct voltage are objectionable in many cases. 在许多情况下，直流电压中的纹波或谐波是不利的。（形容词 objectionable 含有否定意义）

2）The whole system is kept unchanged. 整个系统保持不变。（过去分词 unchanged 含有否定意义）

形容词而具有否定含义的，往往还借助于与一定的介词搭配，才能将否定意思表达出来。

3）The moving body is kept free from the action of forces. 该运动的物体未受外力的作用。（形容词短语 free from 含有否定意义）

4）If short of hydraulic system pulsations associated with on-off operation，proportional control is capable of reducing wear and tear on hydraulic hoses and gearboxes. 如果没有开关操作所连带的液压系统脉动，比例控制就可降低液压软管和齿轮箱的磨损。

5）Hydraulic systems devoid of noise is feasible，the issue is only the cost and time. 没有噪声的液压传动系统是可能实现的，问题仅仅在于资金有无和时间迟早而已。

6）It is the last type of machine for such a job. 这是最不适合于此种作业的机器。

7）High set-up cost should be the last to blame for the fiasco. 决不应将这次惨重失败归咎于创建费用过高。

“the last＋名词＋动词不定式（或定语从句）”中的 the last 意思往往是“最不值得”“最不合适”“最不可能”“决不会”等，这是一种修辞手段，与该词的具体意义不同，再如：

That's the last thing you should do. 你万万不该干那种事。

试比较：The last thing he did that night was to talk about the accident. 他那晚做的最后一件事就是讨论事故。

（4）副词。

1）Far from increasing the reaction rate，high temperature decrease it. 高温不但没有增加这个反应速率，反而减低了反应速率。

2）The temperature is too high for the boiler to withstand. 温度太高，这锅炉经受不住。（too＝excessively，其否定含义是从“太”“过”这一本义引申出来的）

3）Silver is too costly to be used for general contacts. 银太贵重，不常用来做一般的电器触头。

从上面的例句可以看出，too 后面的不定式或 for 的介词短语都是用来修饰 too 的，因而 too 具有否定含义。倘若 too 后面的不定式不修饰 too 而修饰紧接 too 后的词语，那么 too 就无否定含义；此外，若这个不定式修饰谓语，表示目的或结果，自然也不会使 too 产生否定含义。也就是说，句型“too＋apt、ready、eager、easy、inclined、willing、happy 等形容词＋ to ＋不定式”表示“很”“十分”“太……”等意，而不是“太……不”之意，应该严格区别开来。例如：

4）Men are too apt to forget，but computers do not. 人们往往（或译：太容易）健

忘，而计算机绝非如此。

5）He was too eager to meet his friend to sit out the meeting. 他渴望会见友人，等不到散会就离去了。（第一个不定式短语修饰 eager，无否定含义；第二个不定式短语修饰 too，具有否定含义）

6）Which do you think is the better alternative：to aim too high to fall of most of the things aimed at，or to aim too low to win a success which is but commonplace? 下面两者之间你以为何者为佳：把目标定得过高，以致大部分不能完成，还是把目标定得太低，结果只赢得平凡的成就?（两个 too 后面的不定式都是修饰前面不定式的，都是结果状语，故 too 无否定含义）

（5）介词与介词短语。

1）Atomic electric batteries can operate without being recharged for decades. 原子电池组可以连续工作几十年，而不必再充电。

2）So instead of using weight as a standard we use mass，because that doesn't change. 因此我们用质量作为标准而不用重量，因为质量不会变化。

3）But for air and water，nothing could live. 没有空气和水，什么也活不了。

4）Fluids flow under stress instead of being deformed elastically as solids do. 流体在应力作用下流动，不像固体那样会发生弹性变形。

5）When a metal is exposed to a corrosive medium and is subjected to a load，fracture may occur even though the stress is less than that required to cause fracture in the absence of corrosion. 金属零件接触到腐蚀介质并承受负荷时，即使应力比在无腐蚀情况下造成断裂所需要的来得小，断裂也仍然可能产生。

6）This method is beyond doubt superior to that one. 这一方法无疑优于那种。

介词 beyond 含有否定意义，这个介词常与许多名词结合成为表示某一否定含义的介词短语。例如：beyond question（不成问题），beyond compute（＝immeasurable）（不可胜数），beyond conception（＝inconceivable）（不可设想），beyond controversy（无可争议），beyond dispute（＝indisputedly）（毋庸争辩），beyond measure（＝excessively）（非常，格外），beyond doubt（毫无疑问），beyond expectation（意料不到），beyond reach（力所不及），beyond [one's] capacity（power）（为……力所不及），beyond recognition（无可辨认），beyond expression（无法形容）等。

由 out of 构成的表示否定的短语也很多，如 out of order（坏了），out of alignment（不对准的），out of contact with（不与……接触），out of control（不加控制的，失去控制），out of doubt（无疑），out of focus（不聚焦，散焦），out of operation（不运转），out of step（不同步，不一致），out of round（不圆），out of touch（不接触），out of true [不精（正）确]，out of use [不（能）使用的] 等。

（6）连词。在很多情况下 rather than、more than 或 more … than 等形式表达了否定的意思。

1）rather than。连词 than 与副词 rather 连用时，具有“而不”之意，着重说明客观上的差别。这一结构起连词作用，连接两个并列成分，用以肯定前者而否定后者，因而一般译为“而不是”。

Note that we talk about mass，rather than weight. 注意，我们讲到的是质量而不是重量。

2）but。but 或作并列连词，或作从属连词，有否定含义；但另一分句（并列成分）或主句往往也是否定的，汉译为“而不”“若不”“除非”。例如：

a. No machine tool here but it is automatized. 这里没有一台机床不是自动化的。（but 在此句中为并列连词）

b. The temperature would have continued to go up but that feedback took place. 若不是发生反馈，温度还会继续升高。（but 常与 that 连用，引出条件状语从句，相当于 unless 或 if … not，主句常为否定句，其谓语往往采用虚拟语气）

c. No question is so difficult but it can be settled. 没有什么问题如此困难，以致不能解决。（but 引出结果状语从句，主句常为否定语气）

d. Who knows but (that/what) there may be other factors that brought about the abrupt transformation? 谁能说不会有其他因素导致这一突变呢？（but 有时与 that 或 what 连用，表示否定语气，相当于 that …not，引出宾语从句）

e. It is anything but a high precision grinder. 这绝不是一台高精密磨床。（以 anything 置于 but 前面，加强了 but 的否定语气，anything but 形成了一个固定词组，此外还有 everything、anybody、anyone、everybody、everyone、anywhere、everywhere 等＋but，它们意为“除……以外的任何事情（东西、人、地方），根本不……，决不……”）。

3）more … than …。more … than …结构常用于对不同人或物的同一性质或方面进行比较，有三种译法：

a. “比……更”或“A 多于 B”：表示对同一性质来比较其不同的程度。

b. “与其说……不如说……”：对于两种不同性质加以比较而有所取舍。它连接的是两个并列的表语或宾语，但在意义上偏重前者，而压低后者。汉译时要注意两个表语或宾语的顺序，例如，可把“more A than B”顺译为“是 A 而不是 B”，或逆译为“与其说是 B，不如说是 A”。

c. “more A than B”：仍是对于两种不同性质加以比较而有所取舍。它连接的是两个并列的表语或其他成分，译为：“不是 B 而是 A”“没有 B 而是 A”。后两个用法即是这里的意义否定。例如：

a）The transformer is more efficient than that one. 这台变压器比那台效率更高。

b）The division between the former and the latter is more apparent than real. 前者和后者的区别是表面的而不是实际的。

c）A river that carries so much silt per cubic meter (590 kilograms - over half a ton!) is more like liquid land than water. 一条河每立方公尺夹带了那么多泥沙（590kg——0.5t 多!），与其说是河水，倒不如说是很像液状陆地。

d）The stages of development of the ship from the first tentative sea-going examples to the various types in use today relate more to development in the means of propulsion than to particular changes in the shipwright's art，the shape and purpose of a ship being limited more by its means of propulsion than by any other single consideration. 船舶从最初试验性航海样品船到今天各种类型的船只，其发展进程与其说同造船人特有的技术改革有关，还不如说同动力设备的发展有关，因为一艘船的动力设备对其外形和用途的影响比其他任何因素都大。

e) A model mounted between centers is more often used than a flat template and this is positioned on a beam at the rear of the machine. 常常使用的是装在两个顶尖之间的模型，而不是扁平样板，这个模型安置在机床背部的横梁上面。

f) Technically, it is a better safety device than regulator. 从技术上讲，这不是一种调节器，而是一种安全装置。

4) before。连词 before 也隐含有否定意义。例如：

Please stay at the substation before I come back. 我没回来之前请你呆在变电站。

5) unless。在不少场合也可译成“不”。例如：

a. Don't remove the cover unless steam escape from the valve. 蒸汽不从阀门逸出，就不要移开盖。

b. A body will remain at rest unless something is done to change that state. 物体静止时不施加外力，那种静止状态就将保持不变。

(7) 插入语。一些插入语也可以表示意义否定，例如 let alone、much less [=not to mention (say)、not to speak of、to say nothing of] 都是作为插入语使用的。其含义相当于汉语的“别说”“更毋论”“更不用说”……。但是它们用法上有区别，let alone 后面可接谓语动词、名词、动名词、介词短语或过去分词；much less 后面可接谓语动词、名词、动名词；而 to say nothing of、not to speak of 和 not to say (not to mention) 后面只能接名词或动名词。它们都是一种准否定语气。例如：

1) They had never seen such a massive transformer, let alone made it. 这样大的变压器他们连看都没有看到过，更不用说制造过了。

2) I can hardly understand his books about power systems, much less his lectures. 他关于电力系统的书我简直看不懂，更不用说他的讲课了。

3) There was hardly any machine-building industry then, to say nothing of a motor industry. 那时连机器制造业都没有，更不用说电机制造业了。

4.1.5 两种否定转移

英语和汉语在表达否定概念时所使用的语法规则和词汇手段都存在着较大的差别，因而在翻译否定句时经常会遇到两种否定转移现象，但是这种转移只是语言表达上更加流畅汉化，避免了洋化汉语表述，而语义并没有改变。否定转移主要表现在两个方面：

(1) 语法结构上的否定转移，即对于非主从句而言，英语原文句中否定某个语法成分在汉译时却转移到否定另一个语法成分；或者对于主从句而言，主句中的否定词形式上是否定主句中某个成分，但在汉译时却转移到了否定从句中的另一个成分。

(2) 句子内容的否定转移，即英语中的否定形式译成汉语时可用肯定形式，英语中的肯定形式译成汉语时可用否定形式。

1. 语法结构上的否定转移

（1）非主从句语法结构上的否定转移。这种否定转移主要包括三种情况，即由否定主语或宾语可转移至否定谓语；由否定谓语转移至否定状语；由否定系词可转移至否定表语等。

1）由否定主语或宾语（或介词宾语）转移至否定谓语。用否定词 no 否定某个作主语或作宾语的名词时，在汉译时通常可以转移至否定谓语，即英原文否定主语，汉译文否定谓语，即由否定主语转移至否定谓语。例如：

a. Matter must move，or no work is done. 物质必须运动，否则就没有做功。（由英原文否定主语转移至否定谓语）

b. There is no perfect conductor and no perfect insulator. 既没有理想的导体，也没有理想的绝缘体。（由英原文否定主语转移至否定谓语）

c. No smaller quantity of electricity than the electron has ever been discovered. 从来没有发现过比电子电荷更小的电量。（由英原文否定主语转移至否定谓语）

d. No body can be set in motion without having a force act upon it. 如果不让力作用于物体上，就不能使物体运动。（由英原文否定主语转移至否定谓语）

e. It is guaranteed that no penetrating coolant will destroy the spindle bearings. 保证渗透性冷冻剂不会损毁主轴承。（原文否定主语，译文否定谓语）

f. But for the heat of the sun，nothing could live. 要是没有太阳的热，什么也不能生存。（由英原文否定主语转移至否定谓语）

g. In general，no new substance forms in a physical change. 一般说来，物理变化不形成新的物质。（由英原文否定主语转移至否定谓语）

h. Neutrons carry no charge. 中子不带电荷。（由英原文否定宾语 charge 转移至否定谓语）

i. Hydroelectric power generation uses no fuel. 水力发电不用燃料。（由英原文否定宾语 fuel 转移至否定谓语）

j. Liquids are different from solids in that liquids have no definite shape. 液体和固体的区别在于，液体没有一定的形状。（由英原文否定宾语转移至否定谓语）

k. A transformer provides no power of its own. 变压器本身不产生电力。（由英原文否定宾语转移至否定谓语）

l. The previous chapter tells us nothing about what electricity is. 上章没有告诉我们，电究竟是什么。（由英原文否定宾语转移至否定谓语）

m. The moon has no atmosphere and certainly carries no life. 月球上没有大气，因而一定不会有生命。（由英原文否定宾语转移至否定谓语）

n. There are very few applications of metals in which it is necessary for part to deform appreciably in service. 金属零件的应用极少要求其在工作中作出明显变形。（由英原文否定主语 applications 转移至否定从句谓语）

o. The straight line passes through none of the points. 这条直线不通过任何一点。（由英原文否定介词宾语转移至否定谓语）

2）由否定谓语转移至否定状语。英语句子取正常语序时，表示程度、方式和频率的状语往往是否定的重点。因此，在否定谓语的否定句中含有程度或方式状语（这个状语可能是副词或介词短语或不定式短语），汉译时可由否定谓语转移至否定状语。例如：

a. Electric current cannot flow easily in some substances. 电流不能顺利地在某些物质中流动。（cannot flow 从语法上来说是否定谓语，但从逻辑上看讲不通。不是“电流不能流动”，而是“不能顺利地流动”。从逻辑上看是否定状语 easily。）（由英原文否定谓语转移至否定状语）

b. The decision makers do not agree frequently. 决策人员的意见不是经常一致的。（由英原文否定谓语转移至否定状语即频度副词 frequently）

c. He does not distinguish clearly between red and purple. 他不能清楚地把红色和紫色区分开。（由英原文否定谓语转移至否定状语）

d. Insulators prevent heat from escaping, because their molecules do not conduct heat easily and quickly from one to the next. 绝缘体材料能防止散热，因为这种材料的分子不能轻易而迅速地把热从一个分子传递给相邻的分子。（由英原文否定谓语转移至否定状语）

e. The planets do not go around the sun at a uniform speed. 行星并不是匀速地绕着太阳运行。（由英原文否定谓语转移至否定状语）

f. These electrons do not revolve around the nucleus in a disorderly fashion. 这些电子并不是无规则地绕着原子核旋转。（由英原文否定谓语转移至否定状语）

g. Annealing is often specified for casting that did not cool uniformly. 退火常常规定用于那些冷却的不均匀的铸件。（原文否定谓语动词 cool，译文否定状语 uniformly）

h. They did not adopt the method for that reason. 他们不是为了那个原因才采纳这个方法的。（由英原文否定谓语转移至否定原因状语 for that reason）

i. As we know, electricity cannot be conducted by means of insulators. 我们知道，电不能靠绝缘体来传导。（原文否定谓语，译文否定状语 by means of insulators）

j. They have not sponsored the symposium to compare notes on the controversy. 他们主办的这次专题讨论会，不是为了对那项争议交换意见。（由英原文否定谓语转移至动词不定式 to compare 引起的目的状语）

3）由否定系动词转移至否定表语。

a. Kirchhoff's laws do not seem obvious except to those who are familiar with them. 除了那些熟悉基尔霍夫定律的人以外，基尔霍夫定律看起来似乎并不一目了然。（do not seem obvious … 语法上是否定系动词 seem，若译成“不似乎清楚”就不符合汉语习惯，因此汉译时应由否定谓语转移至否定表语）

b. Green plants cannot grow strong and healthy without sunlight. 没有阳光，绿色植物就长不结实、长不好。（由英原文否定系动词转移至否定表语）

c. If we compare the weight of air with the weight of stone or metal, air does not seem to be very heavy. 如果我们比较一下空气与石头或金属的重量，空气似乎不是很重。（由英原文否定系动词转移至否定表语）

（2）主从句语法结构上的否定转移。这种否定转移主要包括两种情况，即由否定主句中的谓语转移至否定状语从句；由否定主句谓语或主语转移至否定宾语从句中的谓语等。

1）由否定主句中的谓语转移至否定（整个）状语从句。

a. not … so … as …，not … as … as …，not … as …。

这种否定主句中谓语的句子中，带有 so…as 连接的比较状语从句或由 as 连接的方式状语从句，应译成“不像……那样”，而不应直译成“像……那样不”。例如：

a）Sound does not travel so fast as light. 声音不像光那样传播快。（does not travel 从语法上看来是否定主句的谓语，但逻辑上讲不通，不是“不传播”，而是“不像光那样快”，即否定由 as 引出的状语从句。）

b）The sun's rays do not warm the water so much as they do the land. 太阳光线使水增温，不如它使陆地增温那样高。（由否定英原文主句中的谓语转移至由 as 引出的否定状语从句。）

c）In a gas the molecules cannot pass along their kinetic energy or heat from one to another as easily as molecules of a liquid or a solid can. 气体中的分子不能像液体或固体分子那样很轻易地把动能即热能从一个分子传递给另一分子。（cannot pass along 从语法上看来是否定主句的谓语，但这样讲逻辑不通，即不是“不能传递动能”，而是“不能像液体……那样容易……”，否定由 as 引出的状语从句。）

d）Now oxygen does not boil away from liquid air so easily as nitrogen. 现在氧不像氮那样易于从液态空气中汽化出来。（由否定英原文主句中的谓语转移至否定 as 引导的省略状语从句）

e）Although computer is one of the wonders of our time，its operation is not as difficult as one might think. 尽管计算机是当代奇迹之一，但是它的操作并不像人们想象的那样困难。（由否定英原文主句中的谓语转移至否定由 as 引出的状语从句）

f）Friction is not always a bad thing as you might think. 摩擦并不像你所想的那样总是件坏事。（不可直译为：摩擦像你所想的那样并不总是件坏事。）

b. not …，because … 和 not … because …。这是一个用法特殊的否定结构。连词 because 与否定词 not 同用时，可能有两种不同的含义，因为 not 有时否定主语的谓语动词，有时却否定 because 从句。因此，汉译时必须从上下文的逻辑意义上来判断。

a）not …，because，译为：“因为……所以没有（不）……”。例如：

Pure iron is not used in industry because it is too soft. 纯铁因为太软而不用在工业上。（否定主句谓语动词）

I did not go to the lecture，because I was not interested. 因为我不感兴趣，所以没去听讲座。

b）not … because，译为：“不是因为……才（而）……”。例如：

I did not go to the lecture because I was interested. 我并不是因为感兴趣才去听讲座的。（由否定原文主句中的谓语转移至否定 because 引导的状语从句）

The mountain is not valued because it is high. 山的价值并不在于它高。（由否定原文主句中的谓语转移至否定 because 引导的原因状语从句）

The engine did not stop because the fuel was finished. 发动机不是因为燃料用完而停

止的。（不能译成：发动机因为燃料用完而不停止。由否定原文主句中的谓语转移至否定 because 引导的状语从句。）

The version is not placed first because it is simple. 这个方案并不因为简单而放在首位。（由否定原文主句中的谓语转移至否定 because 引导的状语从句）

We do not resort to robots for productive labor because we indulge a taste for exotic. 我们采用机器人从事生产劳动，并不是由于我们喜欢标新立异的缘故。（由否定原文主句中的谓语转移至否定 because 引导的状语从句）

Aeronautic industry does not use aluminum because it finds the metal a material with rich deposits. 航空（飞机）工业之所以用铝作原料，并不是由于该金属蕴藏丰富。（由否定原文主句中的谓语转移至否定 because 引导的状语从句）

2）由否定主句谓语或主语转移至否定宾语从句中的谓语英语中表示判断、信念或推测等意义的动词（如 believe、think、suppose、expect、imagine 等）为否定式并带有由 that 引出的宾语从句（that 可省略）或动词不定式表示的宾语补语等，这时，翻译时可由否定主句中的谓语或主语转移至否定宾语从句中的谓语，即应把这类动词的否定式译成肯定，而把其后从句中的谓语或宾语补语译成否定。例如：

a. At first people would not believe that what he had said was true. 开始，人们认为他说的并不是事实。

b. We do not consider melting or boiling to be chemical changes. 我们认为熔化或沸腾不是化学反应。

c. Nobody has believed for a long time that the moon is a comfortable place to live in. 长期以来，人们认为月球并不是一个能够舒适生活的地方。

d. Electrical and electronics engineers did not believe this was possible but they agreed they would study the matter. 电气和电子工程师都认为这是不可能的，但他们同意研究这种现象。

e. Physicists did not think that a death-ray could be produced. 物理学家都认为不可能产生死光。

f. Seeing a ball flying, we don't expect the ball to fly forever. 我们看到球飞滚时，认为它不会永远飞下去。

g. Ordinarily one does not believe air to have weight. 人们通常认为空气没有重量。

h. Ordinarily we don't consider air as having weight. 我们通常认为空气没有重量。

i. I don't suppose he will come to the lab tomorrow. 我认为他明天不会来实验室。

j. They do not think it necessary to recharge these cells. 他们认为不必将这些电池重新充电。（not 的否定重点不是 think，而是作为宾语补足语的 necessary）

k. I did not quite remember the client complaining about the newly delivered machine's failure to comply with the specifications. 我的确记得买方对新交货的机器没有完全符合规格表示不满。（not 的否定重点不是 remember，而是作为宾语补足语的 complaining。不要译为：我不大记得买方对新交货的机器没有完全符合规格表示过不满。意思与正确的译文就大有出入。）

l. Some operators believe that the claimed increase in reliability with more complex bus arrangement is not all realized. 有些运行人员认为，以复杂的母线排列来取得高可靠性并非都能实现。

可见，上面这一类表示对某一问题持否定见解的句子，在英语习惯中，否定词往往放在主句谓语动词之前，如直译就为“不认为……”“不觉得……”，而汉语表达习惯常常与此不同，因此应译为“认为不……”“觉得……不是……”。

2. 句子内容的否定转移

所谓句子内容的否定转移，即是将英原文句子中看似否定的句式（其意实为肯定）转移至肯定内容或反之。

（1）否定句式转移至肯定意义。有些英语句子，从所用词汇和语法结构上看是否定句的形式，但实际为肯定意义，因此汉译时要特别注意。比较常见的有以下几种结构：

1）can not …＋too …（无论怎样……也不过分）

a. can not …＋too＋形容词。从英文字面上，这一结构的意思是“不能太”。但其意思却是“it is impossible（difficult）to＋be＋over 形容词/副词……”（不可能是……过分的）。其中的 can 表示可能性，加 not 就是“不可能”的意思，too 有“over-”（过分）之意。汉译时切勿译成否定句，而要译成肯定句，通常译成“无论怎样……也不过分”“应尽量”“越……越好”等。例如：

a）You cannot be too careful in doing high voltage engineering experiments. [＝It is impossible（difficult）to be over careful in doing high voltage engineering experiments.] 你在做高电压工程实验时越小心（仔细）越好。

b）You can not pay too much attention to your experimental skills. 越注意你的实验技能越好。这个结构中的 not 可以改用 scarcely 或 hardly，too 也可以改用 over、enough、sufficient 等词，意思不变。例如：

You cannot be too careful. ＝You can not be overcareful.
＝You can not be careful enough.
＝You can not be sufficiently careful.
＝You can not take enough care.
＝You can not take sufficient care.
你越小心（仔细）越好。

b. cannot＋（realize、emphasize，exaggerate，etc.）＋too＋副词。从英文字面上，这一结构的意思也是“不能太”。但其意思却是“it is impossible to overdo …”（不可能是……过分的）。其中的 can 仍表示可能性，加 not 就是“不可能”的意思，too 仍为“over-”（过分）之意。汉译时也切勿译成否定句，而要译成肯定句，通常译成“无论怎样……也不过分”“应尽量……”“越……越好”“非常”等。例如：

a）We can not estimate the value of modern science too much.（＝It is impossible to overestimate the value of modern science.）现代科学的价值，无论如何重视也不算过分。（不能译成：我们不能过高地估计现代科学的价值。）

b）Young scientists can not realize too soon that existing scientific knowledge is not nearly so complete，certain，and unalterable as many textbooks seem to imply. 青年科学家们应尽快地认识到，现有的科学知识并不像很多教科书所写的那样完整确切而不可改变。

c）We cannot emphasize too strongly that the principles of chemistry derive from experiments；chemistry is an experimental science. 化学的原理都是由实验得出来的这一点怎样强调也不过分，所以化学是一门实验科学。

d）It cannot be too strongly emphasized that electrical circuit theory is the basis of electrical engineering. 电路理论是电气工程的基础，这一点无论怎样强调都不算过分。

e）It cannot be too much emphasized that mathematics is the basis of all other sciences. 应当尽量强调数学是其他一切科学的基础。

c. cannot＋overemphasize，overrate，overestimate 等词。由于“cannot …＋too＋形容词”“cannot＋（realize，emphasize，exaggerate，etc.）＋too＋副词”两结构中 too 均有“over-”（过分）之意，所以“cannot＋over-＋动词原形”也属于这一结构，其意不变。例如：

a）The importance of scientific researches can not be overvalued. 科研的重要性，无论怎样估计也不会过高。

b）So the importance of circuit breaker can not be overemphasized. 因此，断路器的重要性再强调也不过分（或译为：因此，断路器的重要性无论怎样强调也不过分）。

c）Newton's contribution to modern science can hardly be overrated. 牛顿对现代科学的贡献非常大。(或译为：牛顿对现代科学的贡献是无论怎样评价也不会过高的。)

d）The importance of designing techniques to conserve materials as well as that to raise productivity can hardly be emphasized too much. 不但设计一些提高生产率的技术，而且设计一些节约材料的方法，对于这一点，无论怎样强调也不会过分。

e）I can not read enough of his books on power system stability. 我非常爱看他写的关于电力系统稳定方面的书。(总也看不够)

2）not …＋too …＋to＋不定式这是 too …＋to＋不定式句式的否定形式，意为“完全可以”“不致太……而不”。例如：

This device is not too complicated for us to assembly. 这台仪器并不复杂，我们完全可以装配。

3）nothing（nobody，no one，none，nowhere，etc.）＋but …。这个结构中，but 为介词，意为“除了……之外（except）”，它与上述否定词连用表示“除了……之外，什么（谁，一个人，什么地方）也不是（也没有）……”。nothing ＋ but …、no one、(none) ＋ but …根据句子具体内容，等于 only、the only 或 alone 的意思，常译为“仅仅是，只不过是”。例如：

a. Heat energy is nothing but the energy of motion of molecules. 热能只不过是分子运动的能量。(nothing but ＝only)

b. I saw nothing but a transformer. 除了一台变压器外，我什么也没看见。(nothing but＝only)

c. None but he knew what happened at the substation. 除了他，谁也不知道变电站发

生了什么。(或译为：只有他才知道变电站发生了什么)。(none but =alone)

d. Write nothing but the requirements on this side. 在这边只写要求。(nothing but = only)

e. No transformer is truly useful here but the one which has no defects. 只有没有任何缺陷的变压器才真正可以用在这里。(No transformer … but = the only)

f. You can find that sort of sophisticated testing instrumentation nowhere but (=only) in America. 你只有在美国才能找到那种先进的检测仪器。

nothing but 之后不限于接名词或代词，还可以跟动词不定式。例如：

g. He does nothing but carry out experiments all day. 他整天只是做实验。

h. There is nothing for it but to wait for the experimental results. 对于这个问题只能等待实验结果。

(2) nothing else but。此结构中 nothing else 意为"没有别的事""不是别的"，but 为介词，意为"除了……之外"，因此整个结构意为"不外乎是，正好是"。nothing else but 在句中起副词作用，相当于副词 only。例如：

They are nothing else but the elementary particles which constitute matter. 它们不外乎是构成物质的基本粒子。(或译为：它们只是构成物质的基本粒子。)

(3) nothing other than。此结构中的 other than 意为"除了……之外"，故 nothing other than 意为"除了……之外，什么也不……""只不过是""无非是""仅仅是""只有"。例如：

a. It's nothing other than the usual resonance. 这只不过是寻常的谐振。

与 nothing other than 同义的词组还有 nothing more than、nothing else than。例如：

b. She wanted nothing more than study. 她只想学习。

c. The electrical "fluid" that early scientists talked about is nothing more than electrons flowing along a wire. 早期科学家谈论的电"流体"，只不过是沿着导线流动的电子。

d. The contribution of the Greek was nothing less than the creation of the very idea of science as we know it now. 希腊人的贡献只不过是建立了正如我们今天所了解的"科学"这个概念。

(4) no more than。这一结构与上一结构中的 nothing more than 等结构相似，不过它起副词作用，相当于副词 only，一般译为"只有""仅仅"，主观上认为"少了"，仍有否定的含义。例如：

The bridge can bear no more than 6 tons of weight. 这座桥只能承受 6t 重量（重量不算大）

(5) not less+形容词+than。这一结构表示"不亚于""不比……差"。例如：

This course is not less necessary than math to students. 这门课程对于学生的必要性并不亚于数学。

(6) not+终止性动词+until 或 not until …。这一结构译为"直到……才"。例如：

a. This fundamental equation in electrical engineering did not receive an adequate answer until little more than 30 years ago. 这个电气工程上的基本方程式直到 30 年以前才得

出合适的答案。

b. Not until the early part of the nineteenth century was it realized that all the o-called natural organic compounds contain carbon as a constituent element. 直到 19 世纪初人们才认识到所有这些所谓的天然有机化合物都含有碳这一组成元素。

c. It was not long until transistors were being manufactured in great quantities and in as many different configurations as were vacuum tubes. 不久，人们就生产出大批晶体管，而且型号同晶体管一样多。

(7) hardly (scarcely) … when (before) … 或 no sooner … than …这一结构译为"刚一……就""一……就"。例如：

Hardly was Resistor-transistor Logic well launched when Diode-transistor Logic appeared. 电阻—晶体管逻辑电路刚发展成熟，就出现了二极管—晶体管逻辑电路。以上这种句子结构也是形式否定、意义肯定。

4.2 带有情态动词的否定句

情态动词后面用否定词时，情态动词的含义一般都在否定范围以内。例如：

(1) Being radioactive, the used fuel from a nuclear power station cannot be left lying around. 核电站用过的燃料有放射性，不可到处乱放。

(2) These materials need not be water-tight. 这些材料无需具有防水性能。

(3) The temperature must not exceed 1000℃. 温度不得超过 1000℃。

倘若用 may 来表示可能性时，一般都不在否定范围之内，应汉译为"也许不会"或"可能……"，而绝对不要译为"不可能……"。如果在汉语中要表示"不可能……"，则对应的（用否定词来表示的）英语应当用 cannot。例如：

(1) Frictional forces may not accompanied by motion. 产生摩擦力的同时，可能并无运动。

(2) These materials may not require further treatment. 这些材料可能不需要进一步处理。

(3) The materials that cannot be made into magnets are called non-magnetic materials. 不可能制成磁铁的物质称为非磁性物质。

(4) Water and alcohol cannot be separated by distillation. 蒸馏法不可能分离水和酒精。

4.3 否定句中应注意几个问题

(1) 对于不定代词和某些词汇的否定，应当采用相对应的否定不定代词或者英语习惯用法。如 and、all、both、every 等，不采用对它们直接否定的方法，而可以采用 or、any、neither、none 等否定不定代词和 neither … nor、not either … or 等。例如：

1) 肯定句：I know all these transformers.

对应否定句的错误形式：I don't know all these transformers.（形成部分否定形式而非全部否定）

对应否定句的正确形式：I don't know any of these transformers.

对应否定句的正确形式：I know none of these transformers.

2）肯定句：Both controller and I/O interface are correct.

对应否定句的错误形式：Both controller and I/O interface are not correct.

对应否定句的正确形式：Neither controller nor I/O interface are correct.

（2）对否定视情况可以加上辅助词。例如：

肯定句：I understand now. 对应的否定句：I don't understand yet.

肯定句：It's still raining. 对应的否定句：It's not raining any more.

加上辅助词 at all 的否定句：The COMS contents couldn't remain at all without battery.

（3）注意否定词含义持续的时间，下面两例含义不同。

1）He got asleep until two o'clock.

错译：他一直睡到 2 点。

正确：他直到 2 点还在睡。（行动 A 一直持续到行动 B 开始，B 一开始，A 即停止）

2）He didn't get asleep until two o' clock.

错译：他直到 2 点还未睡着。

正确：他直到 2 点才睡着。（行动 A 一直持续到行动 B 开始，B 一结束，A 才开始）

（4）注意否定句转换时相应的副词变化。例如，always、often 可对应于使用 never、nearly、seldom、hardly 等。例如：

1）He always passwords his computer. 他总是对其用的计算机加上口令。

错译：He always doesn't passwords his computer.

错译：He not always passwords his computer.

正确：He passwords his computer not always.

正确：He never passwords his computer.

2）I nearly saw everything now. 我几乎看到了所有情况。

3）I hardly saw anything yet. 我几乎什么也没看到。

4.4 在表示“用……”时 with，by 及 by means of 的区别

按汉译英的一般习惯，表示“用”工具时，用 with；表示“用”什么方法或手段时，多用 by 或 by means of。但英译汉时，似乎看不出什么差别。例如：

（1）The road was cleared by（or by means of）a bulldozer.

The road was cleared with a bulldozer.

那条路曾是用一台推土机清出来的。

（2）Greater speeds can now be attained by modern aircraft with（or by）the new metals which are now being developed.

用正在研制的新材料制成的现代化飞机可以获得更大的速度。

但是，当名词表示的是交通工具或动名词时，一般只能用 by，而不用 with。例如：

(1) I usually go to work by bus/train/car/boat.

我通常是乘公共汽车（火车、小汽车、船）上班。

(2) Production will be greatly increased by introducing the new machinery.

引进新机器一定会大大地增加生产。

当名词所表示的工具意义较强，且暗含“使用”的意思时，通常只用 with，而不用 by。如：

(1) Frequent measurements of the bar were made with a micrometer. 人们过去常用千分尺测量棒料。

(2) Someone had broken the window with a stone. 有人用石头打破了窗子。

此外，在表示工具意义时，虽可以用 by，但一般仅限于被动句，往往相当于动因，主动句则只用 with。例如：

(1) The work is firmly held in the lathe by the centres. 用顶尖可以将工件牢牢地夹在车床上。

(2) The lathe firmly holds the work with the centres. 车床用顶尖将工件牢牢地夹住。

(3) The ball was caught with/by his left hand. (He caught the ball with his left hand.) 他用左手接球。

第3篇

电力电子与专业英语语法

5

电 力 电 子

5.1 Failure Identification in Smart Grids Based on Petri Net Modeling

5.1.1 Introduction

THE rapid developments of information and communication technology (ICT), and power monitoring and metering technologies are having a significant impact on power systems. In fact, the integration of new technologies allows utilities and suppliers to rethink network design and operation in order to improve efficiency, quality, and reliability. These emerging capabilities open the doors to a new concept for power networks: the smart grid. Smart grids are based on the installation of infrastructures and the implementation of data integration with automated analysis of the event data provided at various substations. The infrastructure includes communication and monitoring equipment with most of the monitoring functions provided at the substation via intelligent electronic devices.

This vision of power systems allows the implementation of new network management strategies in order to reduce costs by applying demand side management, enabling grid connection of distributed generation (DG), incorporating grid energy storage for DG load balancing, and identifying, eliminating, or containing failures, such as widespread power grid cascading failures. The latter problem is particularly acute in emerging power systems and makes quick fault diagnosis and identification essential for pinpointing the faulted elements and their position in the power grid. Furthermore, the fault diagnosis and identification problem under the "smart grid philosophy" must address both the possibility of communication failures and the correctness and consistency of the information conveyed to control centers by ICT systems, whose development is strongly encouraged by policy and regulatory initiatives. For example, in 2009, the U. S. government announced the largest single energy grid modernization investment in history, funding a broad range of technologies that will spur the nation's transition to a smarter, stronger, more efficient, and reliable electric system.

In recent years, much research work has been devoted to fault diagnosis and identification, and a variety of models, procedures, and algorithms have been proposed. Expert system techniques have been considered; in particular, presents a decision support system that automatically creates rules for knowledge representation and develops an efficient fault diagnosis procedure. A Bayesian network for fault diagnosis on the distribution feeder was built on the basis of expert knowledge and historical data, expertise was represented by logical implications and converted into a Boolean function. Different methods for more general systems have been pointed out: An intelligent supervisory coordinator for process supervision and fault diagnosis in dynamic physical systems was presented, a mode identification method for hybrid system diagnosis was introduced, the authors proposed a scheme of applying wireless sensor networks for online and remote energy monitoring, and for fault diagnostics for industrial motor systems.

Fuzzy logic approaches have been applied to power system fault diagnosis. These techniques offer the possibility to model inexactness and uncertainties created by protection device operations and incorrect data. Min propose a method for fault section estimation that considers the network topology under the influence of a circuit breaker tripped by a preceding fault. To deal with the uncertainties due to the protection systems, fuzzy set theory was applied to the network matrix in order to examine the relationship between the operated protective devices and the fault section candidates. A fast diagnosis system for estimating the faulty section of a power system by using a hybrid cause——effect network and fuzzy logic was presented. Applications for fault detection and diagnosis in motor drives were also presented.

In recent years, Petri net (PN) and fuzzy PN (FPN) techniques have received significant attention by researchers. The fault clearance process was modeled by a PN, and a reverse PN was introduced to estimate the fault section. An FPN technique was used to deal with incomplete and uncertain alarms generated by protective relays and circuit breakers. PNs were applied to a supervisory control scheme for human - machine interface systems.

Up to now, the fault diagnosis methods have not considered the significant influence of ICT in power systems, and few approaches have been developed to integrate methods to detect both faults in distribution networks and failures in communication systems. Furthermore, the presence of DG is typically not taken into account when anomalous situations arise.

In this paper, we develop a method based on PN theory for performing failure diagnosis in smart grids, where the term failure is reserved for both power and communication system failures. The proposed method exploits the underlying network topology in order to carefully design a PN, both to detect failures in data transmission and to identify faults in the distribution network. The use of a PN model allows the description of events that occur simultaneously. Since, in power networks, a fault can activate simultaneous events (trip of the protection, data transmissions, etc.), a PN proves to be a very suitable modeling tool for this purpose. The design of the PN model is carried out by systematically composing multiple PN models for single protection systems and current detectors: Such an approach allows also the identification of the faults in the distribution system despite possible strong penetration of DG. This makes the proposed approach substantially different from those adopted, where only the failure detection algorithm has been considered without giving the opportunity to identify the fault location. In order to validate the proposed method, simulations on typical distribution systems are carried out.

5.1.2 PN Background

In order to describe a chain of events (e.g., the tripping of a protection system), it is often necessary to represent activities that occur in parallel with each other but not independently: For example, a given event may not be allowed to occur or activate unless other activities conclude or certain conditions hold. In power systems, PNs have already been used to evaluate reliability and security, and to simulate the behavior of the protection systems.

A PN is a mathematical model that allows not only the representation and description of an overall process but also the modeling of the process evolution in terms of its new state after the occurrence of each event. PNs can be represented by a graph that can be used as a visual communication aid with the ability to simulate the dynamic and concurrent activities of the underlying system. As a mathematical tool, a PN allows the specification of state evolution equations, together with algebraic relations governing the behavior of the system.

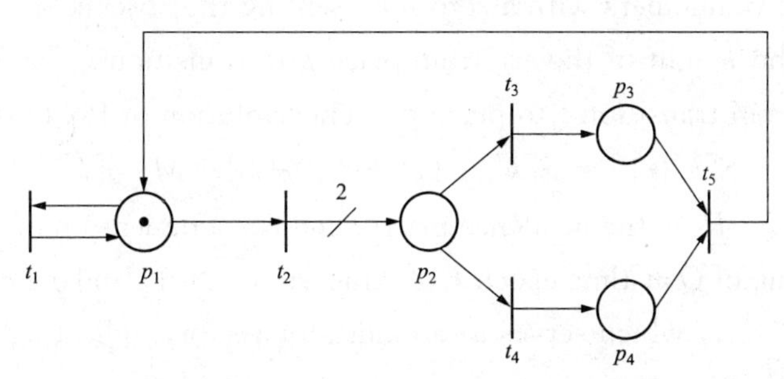

Figure 5.1 PN example.

Graphically, PNs can be identified as a directed bipartite graph with two sets of nodes, (circles) places and (bars) transitions, and (arrows) directed arcs connecting places to transitions or transitions to places. An example of a PN is shown in Figure 5.1. It consists of four places, represented by circles; five transitions, depicted by bars; and directed arcs connecting places to transitions and transitions to places. Places, transitions, and arcs model the process state, the events that occur, and the evolution laws of the process, respectively. The actual process state is identified by the position of tokens in the places. The number of tokens at places changes when some event happens (transition firing). Typically, input places represent the availability of resources, transitions represent their utilization, and output places represent the release of resources.

In the net of Figure 5.1, for instance, place p_1 is both an input and an output place of transition t_1 and an input place of transition t_2. Place p_2 is the output place of transition t_2, the input place of transitions t_3 and t_4, and so forth. Note that, from transition t_2 to place p_2, there are two arcs, which is sometimes denoted by a single arc of weight 2. When transition t_2 fires, it removes one token from place p_1 and deposits two tokens to place p_2; the tokens at place p_2 then become available to transitions t_3 and t_4.

In order to analyze the dynamic behavior of the modeled system by considering its state evolution (token changes at places), we need to keep in mind that each place may hold a nonnegative number of tokens, pictured by small solid dots, as shown in Fig. 1. The presence or absence of a token in a place indicates whether a condition associated with this process state is true or false or, for a place representing the availability of resources, the number of tokens in this place indicates the number of available resources. At any given time instance, the distribution of tokens in places is called the PN marking and defines the current state of the system.

Formally, we let $Q=(P, T, P^-, P^+)$ be a PN with a place set P of n places $P=\{p_1, p_2, \cdots, p_n\}$, a transition set T of m transitions $T=\{t_1, t_2, \cdots, t_m\}$, and $P^-: P\times T\rightarrow N$ and $P^+: P\times T\rightarrow N$ representing the matrices that specify the arc weights from places to transitions and the arc weights from transitions to places, respectively (N is the set of nonnegative numbers with a zero representing the absence of an arc). Specifically, $P^-(i, j)$ is the weight of the arc from place p_i to transition t_j, and $P^+(i, j)$ is the weight of the arc from transition t_j to place p_i. The evolution of PN Q is given by

$$S[k+1]=S[k]+(P^+-P^-)\sigma[k]+M\cdot\sigma[k] \tag{5.1}$$

where $\underline{M}\equiv\underline{P}^+-\underline{P}^-$ is the incidence matrix and has a dimension of $n\times m$, the vector $S[k]$ is the marking of Q at time epoch k of dimension $n\times 1$, and $\sigma[k]$ is the firing vector of dimension $m\times 1$, which serves as an indicator vector, indicating the transition that fires at time epoch k.

According to PN theory, transition $t_j\in T$ is enabled at time epoch k if and only if each input place p_i to transition t_j has a number of tokens greater than or equal to the weight of the arc between p_i and t_j; this statement translates to $\underline{S}[k]\geqslant\underline{P}^-(:, t_j)$, where $\underline{P}^-(:, t_j)$ indicates the t_j column of $\underline{P}^-$. The $\underline{P}^+$ and $\underline{P}^-$ matrices for the example in Fig. 1 are the following:

$$P^-=\begin{bmatrix}1&1&0&0&0\\0&0&1&1&0\\0&0&0&0&1\\0&0&0&0&1\end{bmatrix}\qquad P^+=\begin{bmatrix}1&0&0&0&1\\0&2&0&0&0\\0&0&1&0&0\\0&0&0&1&0\end{bmatrix}$$

where the columns correspond to transitions and the rows to places. The incidence matrix is

$$M = P^{+} - P^{-} = \begin{pmatrix} 0 & -1 & 0 & 0 & 1 \\ 0 & 2 & -1 & -1 & 0 \\ 0 & 0 & 1 & 0 & -1 \\ 0 & 0 & 0 & 1 & -1 \end{pmatrix}$$

At the time epoch $k = 0$

$$S[0] = [1 \quad 0 \quad 0 \quad 0]^{T}$$

and the firing of transition t_2 causes

$$S[1] = S[0] + M - [0 \quad 1 \quad 0 \quad 0]^{T} = [0 \quad 2 \quad 0 \quad 0]^{T}$$

Depending on the underlying modeling assumptions, the firing of transition t_2 may indicate the occurrence of an event (e. g. , trip of a circuit breaker) that causes the system to transition to another state (e. g. , opened circuit breaker). In our examples later on, we will use the PN to capture and interpret activity that occurs in the system. Due to faults that may occur in the system, the number of tokens in places can become negative. This is not a violation of any physical constraints but rather a result of the interpretive nature of our PN models.

5.1.3 PN Model of Distribution Network for Fault Diagnosis

In this paper, we do not explicitly take into account timing information (because we employ untimed PNs), but the order of activation of different relays (main, primary, secondary, etc.) is captured by the PN model. One can potentially extend these ideas in a context that includes timing information by employing timed PNs.

The behavior of protection systems (relays with circuit breakers) can be modeled by PNs in which relays and circuit breakers are modeled by places of the PN. The number of places of the proposed PN model depends on the number of trip thresholds that characterize a given protection system. Figure 5.2 shows the protection PN (PPN) model for a three-threshold protection system, which can be generalized to a different number of trip thresholds in a straightforward manner. In the remainder of this section, we describe the PN model of a protection system in detail.

The PN model is built assuming the following operational hypotheses for each protection system:

1) Each protection system is capable of detecting the current direction;

2) it is possible to obtain from relays information regarding the status of "ready to send the main, primary, or secondary trip signal to the circuit breaker";

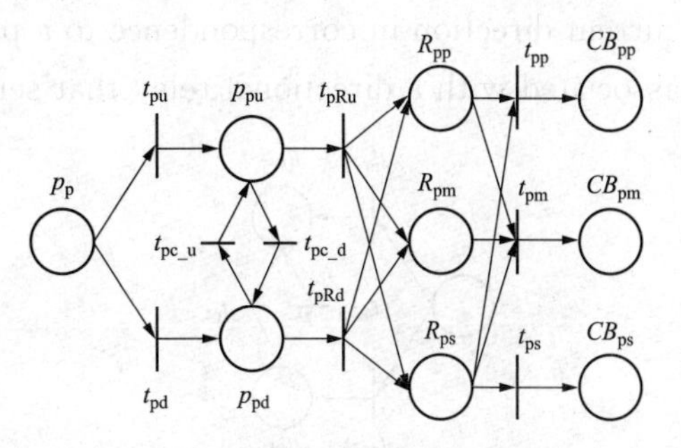

Figure 5.2 PN model of protective relay.

3) it is possible to obtain information regarding the tripped threshold at which the cir-

cuit breaker opened;

4) all events described before are observable, can be sensed, and can be conveyed by a communication system.

Given the advent of ICT, the aforementioned hypotheses become increasingly realistic. For example, sensors and communication devices that are available at relays and circuit breakers may be tasked with providing the information required in items 2) and 3).

In terms of the PN model, the available information from the sensors (as current directions, trips, etc.) is associated to events described by transitions, and the status of the protection system is described by places. In particular, place p_p represents the initial state of the protection, when its state is that of monitoring the current direction. Places p_{pd} and p_{pu} correspond to the status of "evaluation of current direction": In particular, if a token is in p_{pu}, the current through the protection is upstream, whereas if the token is in p_{pd}, the current is downstream. Transitions t_{pu} and t_{pd} indicate that the protection system has begun to detect the corresponding current direction. While in state p_{pu}, a change of current direction causes transition t_{pc_d} to fire; similarly, while in state p_{pd}, a change of current direction causes transition t_{pc_u} to fire. When a token is in p_{pu} or p_{pd}, the system has detected current direction, and if the values exceed certain anomalous set points, transition t_{pRd} or t_{pRu} may fire. A token in R_{pm}, R_{pp}, or R_{ps} means that the main, primary, or secondary threshold of the relay R_p is now sensing the fault current and the circuit breaker can be activated. Such an action is represented by the firing of transitions t_{pm}, t_{pp}, and t_{pr}, respectively. Thus, the final marking with a token in place CB_{pm}, CB_{pp}, or CB_{ps} means that the main, primary, or secondary CB has operated. The PPN is designed assuming the following temporal order of CB trips: CB_m, CB_p, and CB_s. This is reflected in the PN model by ensuring that if CB_{pm} trips, no tokens remain in R_{pp}, R_{pm}, and R_{ps}, whereas if CB_{pp} trips, no token remains in R_{pp} and R_{ps}. As mentioned earlier, more general PPN models for protection systems with a different number of threshold levels can be devised in a similar manner.

An important part of the proposed PPN model is the *current detector* which senses the current direction in correspondence to a particular bus (e.g., DG or load bus). It can be associated with a directional relay that sends the information of the current direction by using a communication system. The *detector PN* (DPN) model is shown in Figure 5.3. The places and transitions have the same meaning as described earlier when commenting on Figure 5.2. In the following, the current detector and its PN model will also be used without associating it to a circuit breaker in order to detect only the current direction in the grid.

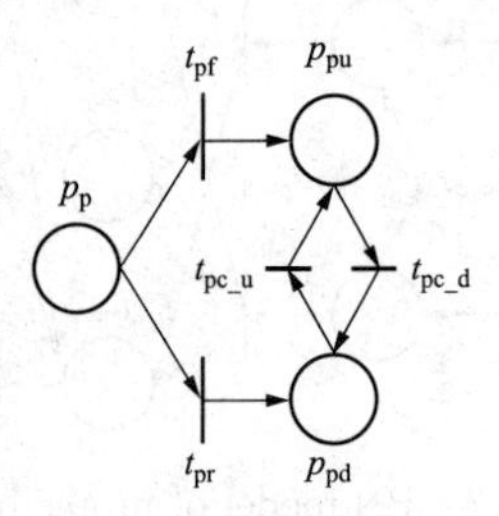

Figure 5.3　General DPN model of current detector.

5.1.4 Methodology for Failure Detection

When one or more failures occur, protection systems and detector currents reach certain values and/or indicate certain activity. We assume that such information is available to a *reasoner*, who needs to assess its coherence. The reasoner can be a SCADA system that, by means of a remote terminal unit (RTU) and communication channels, receives and assesses data. Thc proposed method for failure detection and identification is based on a new strategy that consists of carefully designing a PN model that captures all protection systems and current detectors, as well as their activity. Employing the proposed PN model, the reasoner is able to diagnose the fault position, also determining in the process whether incorrect data were provided to the reasoner.

More specifically, the proposed fault diagnosis method is based on the following three steps:

1) development of a PN model appropriate for failure detection;

2) detection of data transmission failures;

3) identification of power system faults.

The first step allows the design of a PN suited for the analysis of the given power system, the second step identifies failures in data transmission and/or false trips, and the third step, which occurs only if the second one is successful, identifies the actual fault position.

A. Design of PN for Failure Detection

The design of the PN depends on the characteristics of the underlying power network in terms of both connected DG and protection systems. In particular, the expected degree of accuracy depends on the sensing information available at the corresponding DG and protection systems. The number of generators affects the possibilities for fault current direction, and the protection systems determine the complexity of the PN in terms of places and transitions.

The design procedure is now illustrated via an example. Specifically, we consider the section of the distribution network shown in Figure 5.4, assuming a smart grid with adequate sensor availability (as described in Section 5.1.3).

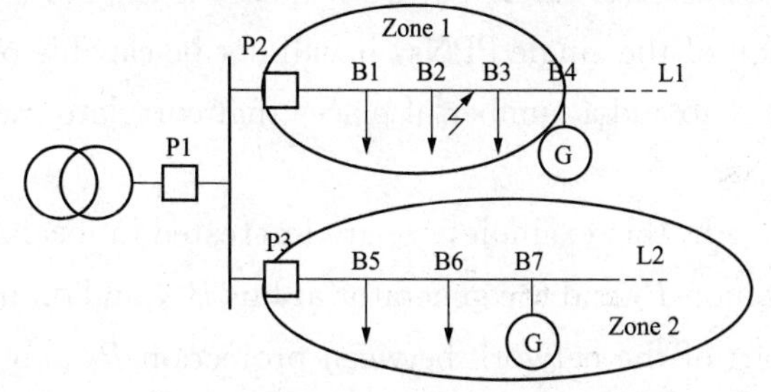

Figure 5.4 Section of power network.

In the first step, each protection system is considered, and its associated PPN model is constructed (refer to Figure 5.2). In this example, the network consists of three protection systems (P_1, P_2, and P_3), so we construct three PPN models. The structure of each PN depends on the number of threshold levels of each single protection system. Assuming that P_1 consists of two threshold levels, and both P_2 and P_3 consist of three threshold levels, we construct the three PPNs according to the rules presented in Section 5.1.3; the resulting scheme is shown in Figure 5.5. The second step inserts the DPN models corresponding to the buses that have sensors for recognizing the current direction. Assuming the existence of a sensor at bus B_4, to account for the presence of the generator, one DPN is added to the PN, as shown in the circled part of Figure 5.5.

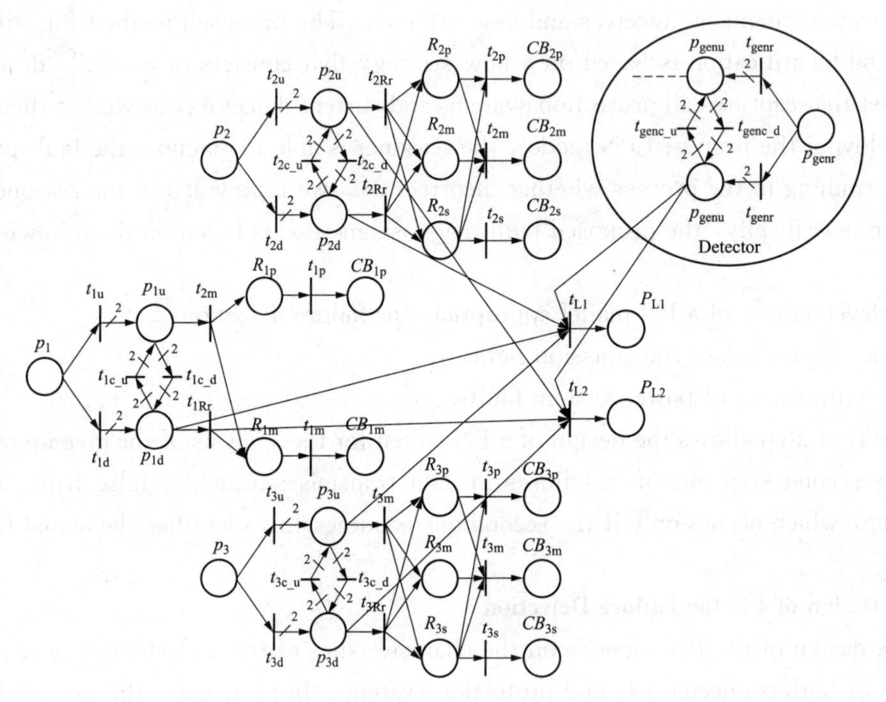

Figure 5.5　PN model for the power system section in Figure. 5.4.

The PN shown in Figure 5.5 has unequal (not unity) weights, unlike the PPNs introduced in Section 5.1.3. The reason is the need to be able to localize faults and it will become clearer later. As we will see, if the PN is constructed according to a simple composition of the single PPNs, it will not be capable of localizing the faults. It will also be necessary to add a number of places that correlate the information obtained from the constructed PNs.

In this example, we are interested in localizing the faults on line 1 (L_1), between protection P_2 and the generator at bus B_4, and on line 2 (L_2), downstream protection P_3. The part of the network between protection P_2 and bus B_4 will be named *zone* 1, and the part between P_3 and the end of the line L_2 will be named *zone* 2. Thus, we need to add places that we call place location faults (PLFs). The first PLF is p_{L1} and accounts for zone 1, and the second is p_{L2} and accounts for zone 2. In order to correlate the information, the connections between the constructed PNs and the two PLFs must be taken into account. Note that a fault between bus B_4 and protection P_2 is supplied from both the generator and the substation, resulting in a downstream current in protections P_1 and P_2, and an upstream current in B_4 and P_3. This observation suggests that we should connect places p_{1d}, p_{2d}, p_{3u}, and p_{gen} to place p_{L1}. The same reasoning can be applied to place p_{L2} (fault in zone 2), keeping in mind that the only PN on line 2 is associated to protection P_3. Obviously, the connections are carried out by means of the two transitions t_{L1} and t_{L2}, which model the sensors at a control center that is able to gather and indicate the fault location. Clearly, the PN can evolve correctly only if we consider two tokens in correspondence of the input and output arcs of the places p_{pu} and p_{pd}, as shown in Figure 5. 5. In the example, with two tokens in p_{2d} (the protection system is detecting the current direction), we can have both the following: 1) Relay 2 senses an anomalous value of the current and transition t_{2Rr} fires (so that the first token is placed in R_{2p}, R_{2m}, and R_{2s}), and 2) (when all sensors detect the current direction with tokens in p_{2d}, p_{1d}, p_{3u}, and p_{genl}, and the control center receives such information) transition t_{L2} fires and the second token placed in p_{L1}. The marking of p_{L1} indicates that the fault is on feeder L_1, in zone 1.

Summarizing, the steps that are followed to systematically design the PN model for each protection system are the following: 1) Insert a PPN model for any protection system with a number of relays and *CB* places equal to the number of threshold levels of the corresponding protection systems. 2) Insert a DPN model in correspondence of a load or generator connection points. A higher number of detectors allow for greater accuracy in the identification of faults. 3) Insert a PLF for any power branch between two protection systems and/or current detectors. Their marking will indicate the position of the fault. 4) Connect the PPN and DPN to the inserted PLF: Connect place p_{pd} (p_{pu}) to *i*-PLF if, in correspondence to a fault in a section modeled by *i*-PLF, the current in the protection system flows downstream (upstream).

It is noted that if the downstream protection system does not posses any generator, the connection specified at step 4) is not carried out.

B. Procedure for Data Transmission Failure

This procedure uses the PN model and the data, to reconstruct its evolved state and identify the failures via parity checking operations, usually based on integer arithmetic or taken modulo p (according to the finite Galois field GF (p), where p is a prime number). The method has already been used and is detailed. Here, only the features required for reproducing the results obtained in subsequent sections are reported.

The existing approach encodes the state of the original PN by embedding it into a *redundant* one in order to enable the diagnosis of failures in PN transitions and/or places. For this purpose, it assumes that the PN activity (transition firing) is observable and the PN state is periodically observable. A related work on fault-tolerant controller implementation using error-correcting codes over Galois fields has been proposed.

In order to detect data transmission failures or/and false trips in the protection system, the concepts of transition and place faults must be recalled. A transition fault occurs at transition t_j if either of the following happens: 1) Transition t_j fires and no tokens are deposited to its output places, even though all tokens from the input places are consumed (postcondition failure), or 2) no tokens are removed from the input places of transition t_j, but all tokens are correctly deposited to its output places (precondition failure). A place fault occurs at place p_i if the number of tokens in the place is corrupted.

In order to identify faults in a PN, a redundant PN Q_H is constructed. We design the redundant PN Q_H via modulo p operations according to a linear error-correcting code over the finite field $GF(p)$. In particular, assuming the PN introduced in (5.1), d places are added to the original PN Q such that

$$S_H[k] = \left[\frac{I_n}{C}\right] S[k] \tag{5.2}$$

for all epochs k. In (5.2), I_n is an $n \times n$ identity matrix and C^* is a $d \times n$ nonnegative integer (modulo p) matrix to be designed.

The evolution (5.1) of the redundant PN can be parameterized as

$$S_H[k+1] = S_H[k] + \left[C^+ \frac{P^+}{P^-} - D\right]\sigma[k] - \left[C^+ \frac{P^+}{P^-} - D\right]\sigma[k] \tag{5.3}$$

where D is a $d \times n$ nonnegative integer (modulo p) matrix, also to be designed. The d additional places, together with the n original places, form the places of the *redundant PN embedding* Q_H. Note that, due to the interpretive nature of the redundant PN (and due to the presence of failures), the number of tokens in its places can become negative.

In order to identify transition and/or place faults, let $\underline{e}_T^+$、$\underline{e}_T^-$, and $\underline{e}_p$ denote the indicator vectors of postcondition, precondition, and place faults, respectively, within the epoch interval [1, 2, …, K]. Incorporating precondition or postcondition faults and place faults, the erroneous PN state at epoch K is given by

$$S_f[K] = S[K] - \Gamma^+ - e_T^+ - e_T^- + e_P \tag{5.4}$$

In order to detect failures, a syndrome vector is introduced and assessed. At epoch K, the syndrome vector is defined as $\underline{S}[K] = F\underline{S}_f[K]$ for $F \equiv [\underline{C}^* - \underline{I}_d]$ and can be shown to satisfy

$$S[K] = D(e_T^+ - e_T^-) + \{-C^+ I_d\} e_P \tag{5.5}$$

and be identically zero if no faults have taken place ($e_T^+ = 0$, $e_T^- = 0$, $e_P = 0$).

Place faults at epoch$K$ are detected first by evaluating the place failure syndrome modulo p, in which case (due to the particular construction) the failure syndrome is verified to be

$$S_p[k] = [-C \cdot I_d] e_p (\mathrm{mod}\, p) \tag{5.6}$$

Supposing no place faults ($\underline{e}_P = 0$) or assuming that all place faults are correctly identified, the transition failure syndrome at epoch K is verified to be

$$S_T[K] = D(e_T^+ - e_T^-) \tag{5.7}$$

Clearly, the syndrome S_T $[K]$ cannot be used to identify multiple faults in which a precondition and a postcondition failure affect the same transition.

Thus, the structure of matrices D and C (which are design parameters) determines the number of faults that can be detected/identified.

Such a procedure can be applied to any power system model previously introduced in order to monitor the PN evolution (as long as the status—number of tokens at each place—becomes periodically available). As mentioned earlier, the number of (place and transition) faults that can be detected/identified depends on the rank properties of matrix $[\underline{D} - \underline{C} \times \underline{I}_d]$.

C. Identification of Faults

The identification of faults is based exclusively on the evolution of the built PN model according to the design rules, as indicated in Section 5.1.4-A. The natural evolution of the PN provides information on the position of the fault when transition t_i, corresponding to i-PLF, becomes enabled.

We illustrate the fault identification procedure by using the network in Figure 5.4. In this case, omitting the PN of the detector at the DG bus (PN in the circle), the protection system is characterized by three circuit breakers with their relays. As mentioned earlier, protection P_1 is assumed to have two trip thresholds, and protections P_2 and P_3 are assumed to have three trip thresholds. The connection between p_{pd} (p_{pu}) and PLF places p_{L1} and p_{L2} is carried out according to the design procedure. If, at the end of the PN evolution, transition t_{L1} can fire (so that we end up with a token in p_{L1}), the fault is on line 1; otherwise, if t_{L2} can fire (so that we end up with a token in p_{L2}), the fault is on line 2.

If a more accurate identification is required, detectors have to be integrated in the scheme. In particular, detectors can be connected in correspondence to buses and can be modeled according to the scheme in Figure 5.3. In the previous example, if a detector is in correspondence to the generator in B_4, the DPN model is inserted into the PN model, as shown in the circle in Figure 5.5: When the final token is in p_{L1}, the reasoner can deduce that the fault originated before bus B_4. It is noted that, without the detector, the system was able to localize the fault, specifying only the faulted line (L_1); with the detector, the system is able to localize the fault on line 2, between protection P_2 and bus B_4.

5.1.5 Case Studies

In order to evaluate the proposed methodology, a typical radial distribution network is used, as shown in Figure 5.6. Even though the method is independent from the type of fault in the network, all short-circuit currents in branches of this example have been calculated by means of the PowerWorld 9.0 software (in order to know the power flow directions and the current values in correspondence of the protection systems, thus presenting a more realistic case study). The considered radial network is characterized by two satellite centers (two substations characterized by one input, an MV feeder, and several protected MV output lines) and six feeders starting from two HV/MV substations. The position of the relay/circuit breaker, which is typical in many countries (e. g. , Italy), requires significant coordination for the protection system to be able to handle two-way power flow in a distribution system with DG. It is assumed that there is a control center (reasoner) that monitors the behavior of the protection system.

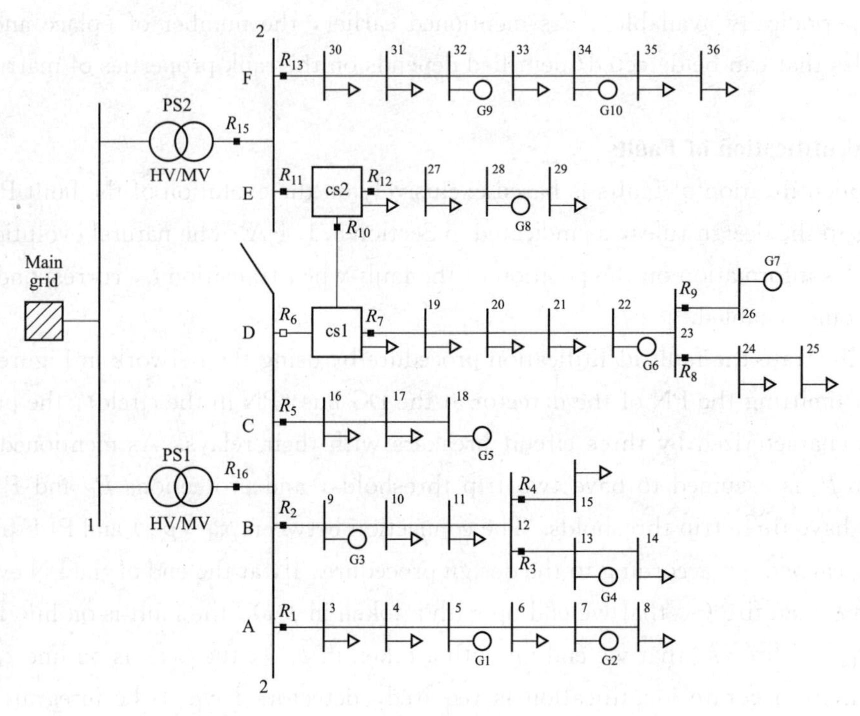

Figure 5. 6　Typical radial distribution system.

Two scenarios are considered: 1) Scenario 1: Incorrect data are transmitted to the control center (SCADA—reasoner) from protection devices (RTUs) and 2) scenario 2: correct data are transmitted to the control center (SCADA—reasoner) but incorrect relay trips are carried out. The two case studies concern a failure in the communication system (reasoner receives partially incorrect information from the protection system—scenario 1) and a failure in the protection trip mechanism (scenario 2). These problems are solved separately only to clarify the discussion. However, they may occur simultaneously and still the scheme would adequately resolve them (as long as enough sensing capability is available). It is noted that, even though the same network for both case studies has been used, the method is quite general and can be applied to any other test case. The considered network is complex in terms of protection system, so that both scenarios can be presented.

A. Scenario 1

A three-phase fault occurs at section bus2—cs2 of feeder E, and protective relays R_{12},

R_{11}, and R_{10}, with corresponding circuit breaker numbers CB12, CB11, and CB10, respond to the fault. These operations are correct because they do not allow the fault to be supplied by both PS2 and G6, and G7 and G8. However, suppose that the following information arrives at the control center: Relays R_{11m} and R_{11p} have sent the trip signal to CB_{11}; CB_{11m} is reported open and CB_{11p} is unknown; in regard to protection 10, current downstream is detected, relay R_{10m} has sent the trip signal to CB_{10m}, and the status of CB_{10m} is reported open; and, regarding protection 12, current upstream is detected, relay R_{12m} has sent the trip signal to CB_{12m}, and the status of CB_{12m} is reported open.

In order to analyze the validity of the aforementioned information, the control center can run the proposed procedure, presented in Section 5. 1. 4-B, having designed the PN according to Section 5. 1. 4-A. In particular, in this case, three protection systems send data to the control center while protection devices 13 and 15, installed in PS2 and at the beginning of feeder F, are crossed by the fault current but do not send any signal.

As far as scenario 1 is concerned, the evaluation of the correctness of the received messages (Section 5. 1. 4-B) is assessed for each protection system that sends a message to the control center. In this case study, the procedure for identifying incorrectly transmitted data is applied only to PPN model 11, which is considered suspicious because it sends two signals to the control center (trip signals from R_{11m} and R_{11p}). The corresponding PPN model, consisting of three threshold levels, is the one already shown in Figure 5. 6 under the condition that the protection subscript number is 11. Here, the PN part related to location fault localization does not come into play because scenario 1 faces only an incorrect data detection problem (this part will be important for scenario 2). For this reason and without loss of generality, all weights are set to unity for simplicity. Furthermore, it is noted that p_{pr11} and p_{pl11} imply a current toward the right side and the left side of Figure 5. 6, respectively.

We now briefly illustrate the procedure of designing a redundant PNQ_H for the PPN in Figure 5. 6, with $n = 9$ places and $m = 9$ transitions. The initial marking of the PPN is

$$S[0] = [1 \quad 0 \quad 0 \quad 0 \quad 0 \quad 0 \quad 00 \quad 0]^T$$

The calculations are carried out modulo p, with $p = 13$. As mentioned earlier, in order to use modulo p operations, it is necessary to assume that all transitions are physically observable, e. g. , there exist sensors that indicate whether transitions t_{11l}, t_{11r}, t_{11c}_l, t_{11c}_r, t_{11Rl}, t_{11Rr}, t_{11p}, t_{11m}, and t_{11s} have fired. Now, two places are added to the original PPN Q, and matrices C^* and D are chosen to be.

The $\underline{D}$ and $\underline{C}^*$ matrices are designed so that the syndrome vectors in (1. 6) and (1. 7) are unique for any single place fault or single transition fault. With these choices, matrices Γ^- and Γ^+ in (3) become

$$
\Gamma^{-}=\begin{bmatrix}
1 & 1 & 0 & 0 & 0 & 0 & 0 & 0 & 0 \\
0 & 0 & 0 & 1 & 1 & 0 & 0 & 0 & 0 \\
0 & 0 & 1 & 0 & 0 & 1 & 0 & 0 & 0 \\
0 & 0 & 0 & 0 & 0 & 0 & 1 & 1 & 0 \\
0 & 0 & 0 & 0 & 0 & 0 & 0 & 1 & 0 \\
0 & 0 & 0 & 0 & 0 & 0 & 0 & 1 & 1 \\
0 & 0 & 0 & 0 & 0 & 0 & 0 & 0 & 0 \\
0 & 0 & 0 & 0 & 0 & 0 & 0 & 0 & 0 \\
0 & 0 & 0 & 0 & 0 & 0 & 0 & 0 & 0 \\
22 & 35 & 50 & 55 & 68 & 89 & 105 & 126 & 128 \\
24 & 63 & 128 & 49 & 166 & 141 & 145 & 181 & 42
\end{bmatrix}
\begin{matrix} p_{11} \\ p_{11d} \\ p_{11f} \\ R_{11p} \\ R_{11m} \\ R_{11s} \\ CB_{11p} \\ CB_{11m} \\ CB_{11s} \\ \\ \end{matrix}
$$

$$
\begin{matrix} t_{11r} & t_{11l} & t_{11cr} & t_{11d} & t_{11Rt} & t_{11Rl} & t_{11p} & t_{11m} & t_{11s} \end{matrix}
$$

$$
\Gamma^{-}=\begin{bmatrix}
0 & 0 & 0 & 0 & 0 & 0 & 0 & 0 & 0 \\
1 & 0 & 1 & 0 & 0 & 0 & 0 & 0 & 0 \\
0 & 1 & 0 & 1 & 0 & 0 & 0 & 0 & 0 \\
0 & 0 & 0 & 0 & 1 & 1 & 0 & 0 & 0 \\
0 & 0 & 0 & 0 & 1 & 1 & 0 & 0 & 0 \\
0 & 0 & 0 & 0 & 1 & 1 & 0 & 0 & 0 \\
0 & 0 & 0 & 0 & 0 & 0 & 1 & 0 & 0 \\
0 & 0 & 0 & 0 & 0 & 0 & 0 & 1 & 0 \\
0 & 0 & 0 & 0 & 0 & 0 & 0 & 0 & 0 \\
16 & 37 & 42 & 63 & 87 & 100 & 93 & 106 & 126 \\
23 & 63 & 127 & 50 & 181 & 155 & 133 & 163 & 44
\end{bmatrix}
\begin{matrix} p_{11} \\ p_{11d} \\ p_{11l} \\ R_{11p} \\ R_{11m} \\ R_{11s} \\ CB_{11p} \\ CB_{11m} \\ CB_{11s} \\ \\ \end{matrix}
$$

$$
\begin{matrix} t_{11r} & t_{11l} & t_{11cr} & t_{11d} & t_{11Rr} & t_{11Rl} & t_{11p} & t_{11m} & t_{11s} \end{matrix}
$$

and the initial marking of the new redundant PPN system is

$$S_H[1]=[1\ \ 0\ \ 0\ \ 0\ \ 0\ \ 0\ \ 0\ \ 0\ \ 0\ \ 0\ \ 9\ \ 11]^T$$

We now follow the evolution of the PPN model associated to the protection system 11. Assume that downstream current direction is detected, so that transition t_{11r} fires. The signal of the current direction is detected by the control center that calculates the state of the system to be

$$S_f[1]=[0\ \ 1\ \ 0\ \ 0\ \ 0\ \ 0\ \ 0\ \ 0\ \ 0\ \ 3\ \ 1\ \ 0]^T$$

Furthermore, assume that the current is recognized as overcurrent, and transition t_{11Rr} fires, making the relay ready to trip. The control center assesses the following state of the PPN

$$S_f[2]=[0\ \ 0\ \ 0\ \ 1\ \ 1\ \ 1\ \ 0\ \ 0\ \ 0\ \ 0\ \ 22\ \ 25]^T$$

In this state, the three unitary values in correspondence of the places R_{11m}, R_{11p}, and R_{11m} indicate that the relay is sensing the current fault and a trip command must be sent to the circuit breaker. Now, when main relay 11 sends the trip signal to the corresponding circuit breaker (fires transition t_{11m}), the switch fails to operate. The control center receives the trip signal but is not informed about the main circuit breaker status and no token is deposed in the place CB_{11m}, so it can deduce that a postcondition fault occurs. The perceived state of the system becomes

$$S_f[3]=[0\ \ 0\ \ 0\ \ 0\ \ 0\ \ 0\ \ 0\ \ 0\ \ 0\ \ 0\ \ -104\ \ -156]^T$$

At this point, the primary relay can operate, transition t_{11p} fires, and the final state of PPN 11 becomes

$$S_f[4]=[0\ \ 0\ \ 0\ \ -1\ \ 0\ \ -1\ \ 1\ \ 0\ \ 0\ \ 0\ \ -116\ \ -168]^T$$

In order to detect failures, the control center evaluates the syndrome vector via $[-\underline{C}^* \ \underline{I}_2] \cdot \underline{S}_f$

$$S_p=[-C*I_2]S_f[4](\bmod 13)=\begin{pmatrix}0\\0\end{pmatrix}$$

$$S_T=[-C*I_2]S_f[4]=13\times\begin{pmatrix}8\\12\end{pmatrix}\neq\begin{pmatrix}0\\0\end{pmatrix}$$

By inspecting matrix D, the eighth column corresponds to the syndrome vector S_T. Therefore, the reasoner can deduce that a postcondition (transition) fault has occurred and a data transmission to the control center was erroneous (because the eighth column of D corresponds to the main relay of protection system 11). The same procedure is repeated for protection systems 10 and 12, but no anomalies are found. When this phase ends, the position of the fault can be determined by using the third procedure. Note that the scheme described in this section requires that t_{11r}, t_{11Rr}, and t_{11m} are observable transitions (so that they can drive the token changes in the redundant places) and that the state of p_{11}, p_{11r}, …, CB_{11p}, CB_{11m}, CB_{11s} eventually becomes available. Both of these requirements are satisfied in this example.

In the aforementioned discussion, we did not take into account two typical phenomena in electrical systems: uncertainty about the measurement data and the noise that can be added to any signal sent or received via the communication system. Our approach has assumed that action against such uncertainty (measurement/noise) is taken individually (by each sensor or receiver) using some type of threshold rule, so that eventually, discrete information is obtained at the control center (e.g., a transition has fired or a circuit breaker has opened). Once this thresholding is accomplished at the sensor/receiver level, any errors due to the uncertainty and the noise are processed automatically by the proposed diagnosis algorithm, as long as the failure detection strategy is designed to be robust to the resulting error. The presented case study and the next one highlight the insensitivities to noisy measurement or uncertain data.

Furthermore, it is noted that the introduced delay to provide information to the reasoner is not critical. In fact, the operations of fault detection and localization require execution times that are perfectly compatible with the classical delays introduced by communication systems. However, an excessive delay is implicitly processed automatically by the system, as it is recognized as "information not received" from the control center or as misoperation of protection systems.

B. Scenario 2

In this case, a three-phase fault is supposed to occur at section cs2—cs1, on the link between the two cs's (between feeders E and D). The protective relays R_{12}, R_{11}, and R_7, with their corresponding circuit breakers CB_{12}, CB_{11}, and CB_7, have responded to the fault, and the information is communicated correctly to the control center. The fault position can be detected by analyzing the evolution of the PN model shown in Figure 5.7, where only the connections to the fault transitions are considered (refer to Section 5.1.4-C). The final state of the PN contains a token in place $p_{R10}\_cs_1$, and we conclude that, even though relays and circuit breakers have operated correctly, there were incorrect trips: Not all loads connected to feeder E have been unsupplied. A correct coordination among relays could have minimized the disservice because relays R_{10} and R_7, and their CBs would have tripped.

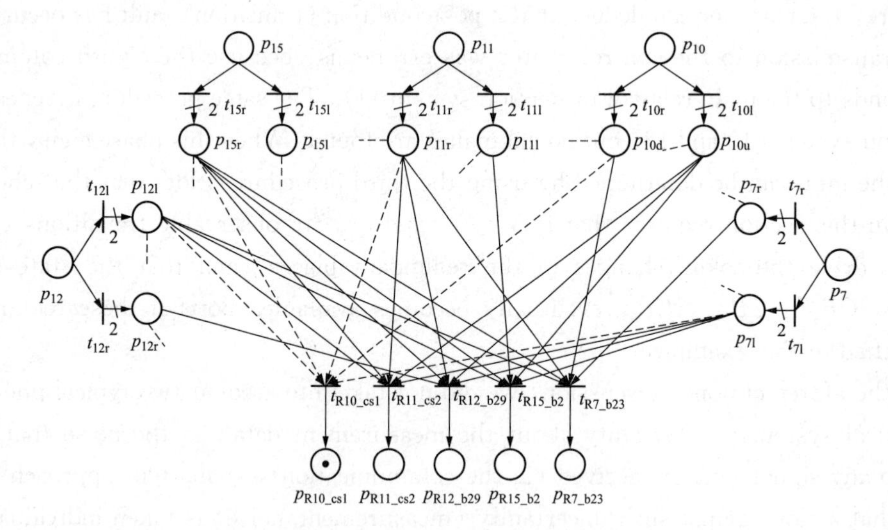

Figure 5.7 PN model for radial distribution system of case 2.

Any network topology changes, due to planning or operational needs (e.g., reconfiguration), are adequately addressed by the introduced method that allows a high degree of modularity in the composition of the PNs. The possible situations can be classified in three cases. 1) In the new network topology, the placement of the protection systems and the direction of the short-circuit currents do not change. 2) In the new network topology, the placement of the protection systems does not change, but the direction of the short-circuit currents changes. 3) The new network topology changes the placement of the protection systems. In the first case, the PN designed before the topology changes occur does not change, and in the second case, only the connections between places p_{pd} and p_{pu} (status of "evaluation of current direction") and PLFs must be changed, but without altering the number of places and transitions; in the latter case, we must take into account the introduced (removed) protection systems by adding (removing) new PPNs to connect (disconnect) to PLFs.

It is noted that, as regards the two presented case studies, generally, it is not possible to associate a data transmission error to a transition fault and a false trip error to a place fault; this has been confirmed by further simulations with different systems, not presented in this paper.

5.1.6 Conclusion

This paper has developed a new model for smart grids in order to be able to rapidly perform fault diagnosis despite erroneous data transmissions. The approach is based on PN theory applied to a carefully developed PN model of the underlying network topology under test. The design of the PN model has been carried out in a modular fashion by composing PN models for individual protection systems along with current detectors. The presented method is capable of dealing with distribution systems with DG, which typically present challenges including false tripping due to coordination loss of protection devices. The method has been validated by means of a case study on a typical radial distribution network, composed of different protection devices that require complex coordination. In particular, the effectiveness of the method has been demonstrated by means of two case studies related to protection systems and DG. From our study, we conclude that the proposed method can remove a lot of complexity in data analysis and allows quick assessment of information while avoiding cascading failures in power system protection.

Clearly, the application of this method favors the creation of a smart grid but requires new investment in ICT and protection systems. From an industrial point of view, the technical feasibility of the proposed solution requires the development of appropriate software and adjustments of protective systems. Currently, incremental investments appear distorted as a result of insufficient decoupling of the industry, and network operators have few incentives to develop the grid in the overall market interest. Nevertheless, considering that both theU. S. and the European Union have planned significant investment for distribution networks, the proposed detection strategy is consistent with current trends and directives.

5.2 Space-Vector Modulated Multilevel Matrix Converter

5.2.1 Introduction

A MATRIX converter is a direct power-conversion topology that can convert energy from an ac source to an ac load without the need for bulky and limited-lifetime energy storage elements in the dc link. Even though this topology has some disadvantages, such as limited voltage transfer ratio (0.86) and a high number of power semiconductor device requirements, the matrix converter has received extensive research attention due to its significant advantages: adjustable input power factor, regeneration capability, and high-quality input current waveforms.

Matrix-converter topologies can be divided into two types: direct matrix converters (MCs) and indirect matrix converters (2MCs). Figure 5.8 (a) shows a conventional direct matrix converter, which has nine bidirectional switches. The typical bidirectional-switch configurations used in this converter are shown in Figure 5.8 (b). The 2MC comprises a four-quadrant current-source rectifier connected to a two-level voltage-source inverter (VSI) [Figure 5.8 (c)]. By applying appropriate modulation scheme, the direct converter and 2MC are able to generate input and output waveforms with the same quality.

However, in some applications, the 2MC may be preferred to the direct matrix converter due to simpler and safer commutation of switches, the possibility of further reducing the required number of the power semiconductor switches the possibility of constructing a direct converter topology with multiple input and output ports.

The multilevel matrix converter is a new topology from this family that incorporates the multilevel converter concept with a matrix converter. There are several types of multilevel matrix-converter topologies that have been proposed. Having the ability to generate multilevel output voltages, the multilevel matrix converter is able to generate better quality output waveforms in terms of harmonic content, but at the cost of a more complicated circuit configuration and modulation strategy.

A three-level matrix-converter topology with reduced number of switches has been proposed: indirect three-level sparse matrix converter (I3SMC). This topology is a combination of a three-level neutral-point-clamped VSI and a 2MC. Having two additional insulated-gate bipolar transistor switches connected as an additional inverter stage leg (neutralpoint commutator) in the dc link, as shown in Figure 5.9, the output voltage capability of I3SMC can be enhanced from two to three levels in the line-to-supply neutral voltages.

This addition results in an improved performance in terms of output-voltage harmonic

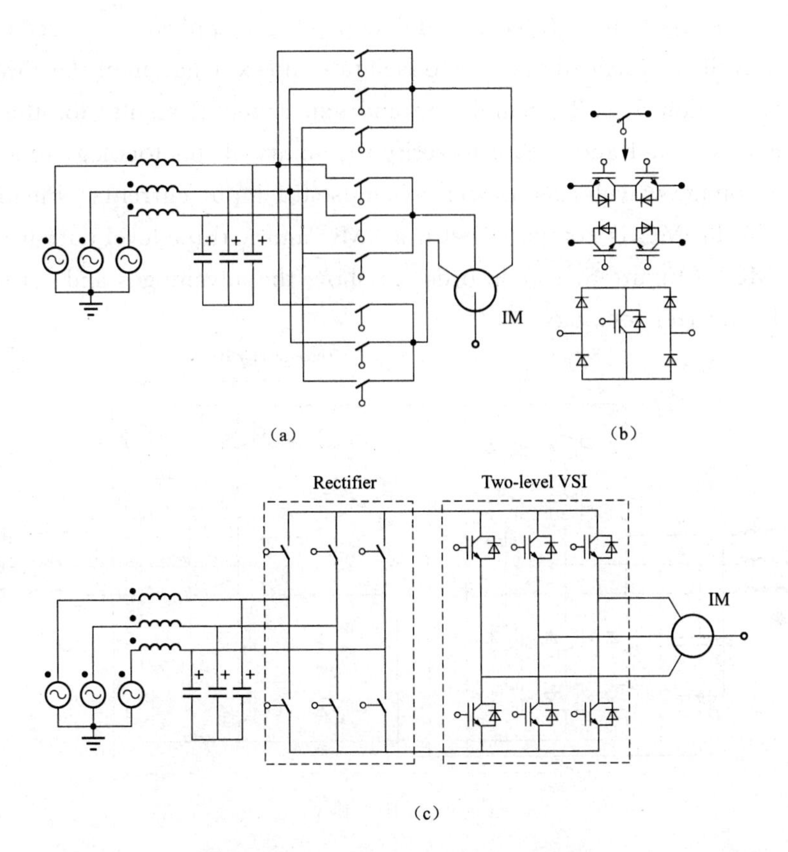

Figure 5.8 Matrix-converter topologies:
(a) Direct matrix converter; (b) Typical bidirectional switches; (c) 2MC.

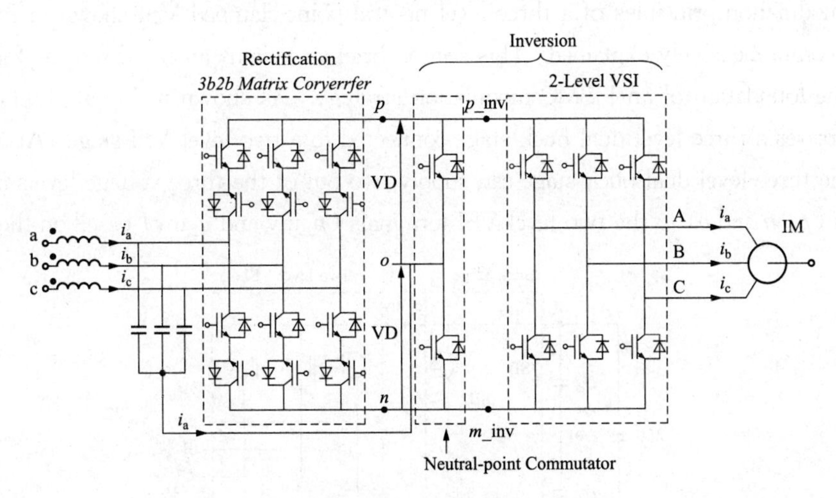

Figure 5.9 I3SMC.

contents. The additional voltage level is obtained by connecting the middle point (o) of the neutralpoint commutator to the neutral-point of the star-connected input filter capacitors.

This paper discusses the operating principles and a spacevector modulation (SVM) for the I3SMC. In Section 5.2.2, a modulation scheme applied to a three-level neutral-point-clamped VSI is explained in order to facilitate the explanation of the SVM applied to the I3SMC in Section 5.2.3. Simulation and experimental results for the I3SMC are shown in Sections 5.2.4 and 5.2.5 to verify the ability of this topology to generate multilevel output-voltage waveforms as well as sinusoidal input currents. Finally, the performance of the I3SMC is compared with a 2MC and a three-level-output-stage matrix converter (3MC) (Figure 5.10) in order to show the advantages and disadvantages of the proposed converter topology.

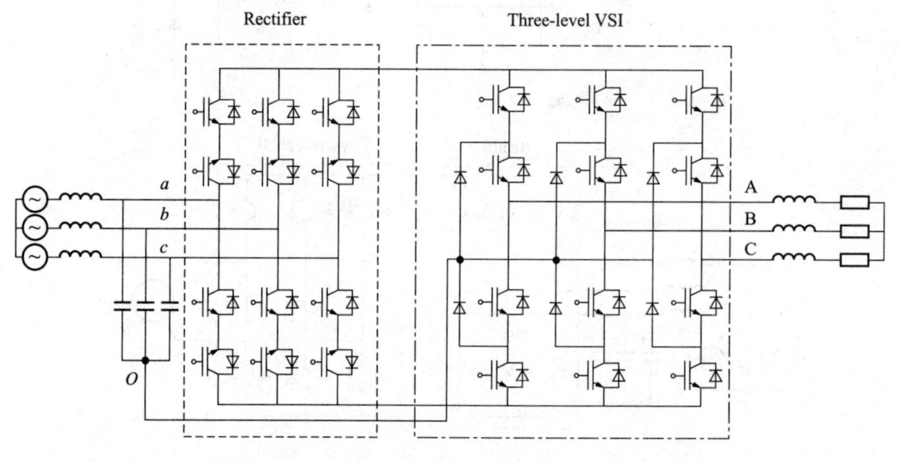

Figure 5.10 3MC.

5.2.2 SVM for the Three-Level Neutral-point-clamped VSI

The modulation principles of a three-level neutral-point-clamped VSI shown in Figure 5.11 have been comprehensively explained. This section briefly explains an SVM scheme for this converter as the foundation of an I3SMC modulation strategy. As shown in Figure 5.11, this converter comprises a three-level dual buck stage connected to a two-level VSI stage. At any instant in time, the three-level dual buck stage can supply two out of the three voltage levels available at the dc link (p, n and o) to the two-level VSI terminals (p_inv and n_inv) based on the switching

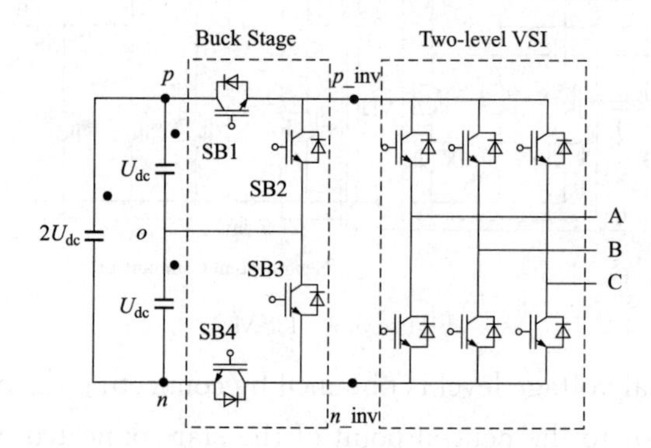

Figure 5.11 Three-level neutral-point-clamped VSI.

combinations presented in Table 5.1. Each voltage level is represented by a switching state: $P = U_{dc}$, $O = 0$, and $N = -U_{dc}$. Therefore, each output voltage, U_{Xo} ($x \in \{A, B, C\}$), of the two-level VSI has three possibilities: U_{dc}, 0 V, and $-U_{dc}$.

Table 5.1 Switching Combinations of the Buck Stage.

Switching Combination				Voltage Level Applied	
SB1	SB2	SB3	SB4	*p*_inv	*n*_inv
1	0	0	1	P	N
0	1	0	1	O	N
0	1	1	0	O	O
1	0	1	0	P	O

Figure 5.12 shows sector 1 of the space-vector diagram for the three-level neutral-point-clamped VSI. Three types of vectors are defined based on the switching-state combinations formed at the outputs: zero voltage vector (U_0), small voltage vectors (SVVs) (U_1 and U_3), and large voltage vectors (U_2 and U_4). Each sector consists of seven triangles ($T_1 - T_7$). To synthesize a reference output voltage vector U_{out}, three nearest voltage vectors are selected based on the triangle in which U_{out} is located. Table 5.2 presents the duty-cycle equation of each selected vector for each triangle, where m_U is the modulation index of the converter and θ_{out} is the angle of U_{out} within the respective sector.

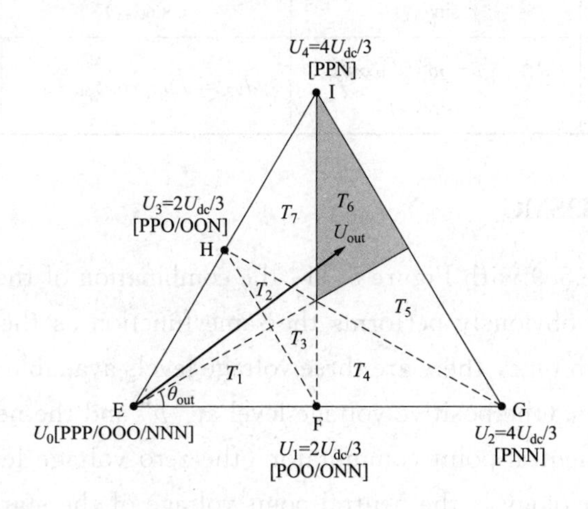

Figure 5.12 Definition of triangles within the sector 1 of the space-vector diagram of the three-level neutral-point-clamped VSI.

Similar to conventional multilevel neutral-point-clamped converter, the connection of the output to the neutral-point (o) of the three-level neutral-point-clamped VSI can cause uneven charging/discharging of the dc-link capacitors due to the neutral-point current.

Without proper control, the uneven charging voltage levels of the dc-link capacitors would impact on the ability of the converter to properly generate the three-level output waveform, causing output-voltage distortion. To maintain the dc-link capacitor voltages, the average neutral-point current over a switching period must be maintained at zero. For this three-level neutral-point-clamped VSI, only SVVs contribute neutral-point current. As discussed, the redundant switching states of each SVV connect the same output phase current to the neutral-point but with opposite sign. Therefore, in this SVM, both redundant switching states (e. g. , POO/ONN) of selected SVV (e. g. , U_1) have to be equally applied within a switching period in order to balance the neutralpoint current so that a zero-average neutral-point current can be obtained.

Table 5. 2　Duty-Cycle Equations for Selected Vectors of Each Triangle.

T_1 [ΔEFH]	$d_{V1}=m_U(\sqrt{3}\cos\theta_{out}-\sin\theta_{out})$	$d_{V3}=2m_U\sin\theta_{out}$	$d_0=1-m_U(\sqrt{3}\cos\theta_{out}+\sin\theta_{out})$
T_2/T_7 [ΔFIH]	$d_{V1}=m_U(\sqrt{3}\cos\theta_{out}-\sin\theta_{out})$	$d_{V3}=2-2\sqrt{3}m_U\cos\theta_{out}$	$d_{V4}=m_U(\sin\theta_{out}+\sqrt{3}\cos\theta_{out})-1$
T_3/T_4 [ΔFHG]	$d_{V1}=2-m_U(\sqrt{3}\cos\theta_{out}+3\sin\theta_{out})$	$d_{V3}=2m_U\sin\theta_{out}$	$d_{V2}=m_U(\sin\theta_{out}+\sqrt{3}\cos\theta_{out})-1$
T_6 [ΔGIH]	$d_{V3}=2-m_U(\sqrt{3}\cos\theta_{out}+\sin\theta_{out})$	$d_{V2}=0.5m_U(\sqrt{3}\cos\theta_{out}-\sin\theta_{out})$	$d_{V4}=0.5m_U(3\sin\theta_{out}+\sqrt{3}\cos\theta_{out})-1$
T_5 [ΔFGI]	$d_{V1}=2-m_U(\sqrt{3}\cos\theta_{out}+3\sin\theta_{out})$	$d_{V2}=\sqrt{3}m_U\cos\theta_{out}-1$	$d_{V4}=m_U\sin\theta_{out}$

5. 2. 3　SVM for the I3SMC

Comparing Figure 5. 9 with Figure 5. 11, the combination of the rectifier and the neutral-point commutator obviously performs the same function as the three-level dual buck stage. At any instant in time, there are three voltage levels available at the dc link; two are supplied by the rectifier (the positive voltage level at "*p*" and the negative voltage level at "*n*") and one by the neutral-point commutator (the zero voltage level at "*o*"). The zero voltage level of this topology is the neutral-point voltage of the star-connected input filter capacitors. By controlling the rectifier and neutral-point commutator to supply only two voltage levels to the two-level inversion stage at any instant, the I3SMC is obviously able to generate three-level output voltages in the same way as the three-level neutral-point-clamped VSI discussed in Section 5. 2. 2.

An SVM has been proposed to modulate the I3SMC to generate the desired input currents and output voltages. For each stage, SVM is used to produce a combination of vectors to synthesize the reference vector. The input current vector I_{in} is the reference vector for the rectification stage, while the output voltage vector U_{out} is the reference for the inversion stage. After determining the vectors and duty cycles, the switching pattern combines the switching states for the rectification and inversion stages uniformly in order to obtain the correct input currents and output voltages in each switching period.

A. Rectification Stage

For the I3SMC, the rectifier is modulated using SVM to generate maximum dc-link voltage and to maintain a set of balanced sinusoidal input currents. As shown in Figure 5.13, the space-vector diagram of the rectifier consists of six active current vectors with fixed directions and three zero current vectors, which are defined based on the valid switching combinations formed by the rectifier. Each current vector represents the connection of input phase voltages to the dc-link terminals. For example, vector I_1 (ac) represents the connection of the positive input phase voltage "a" to the p terminal and the negative input phase voltage "c" to the n terminal.

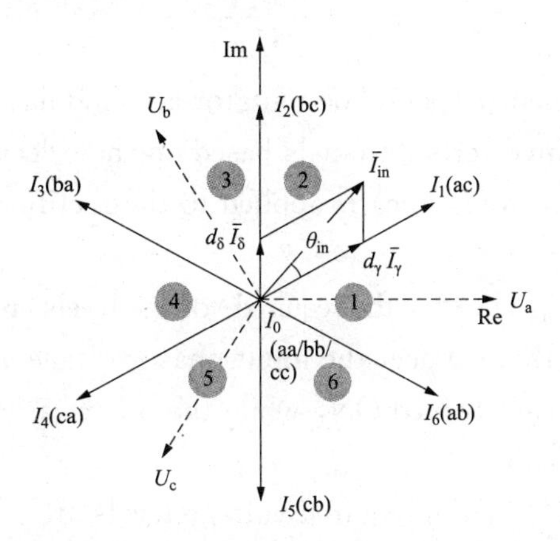

Figure 5.13 Generation of the reference input current vector I_{in} at the rectification stage.

To synthesize the reference vector I_{in} for the I3SMC, two adjacent active current vectors I_γ and I_δ are selected. The duty cycles of I_γ and I_δ are given by (5.8), where the rectifier's modulation index m_I is set to unity and θ_{in} is the angle of I_{in} within the respective sector, i.e..

$$d_\gamma = m_I \cdot \sin\left(\frac{\pi}{3} - \theta_{in}\right) d_\delta = m_I \cdot \sin(\theta_{in}) \tag{5.8}$$

The calculated duty cycles are then adjusted, using (5.9), so that they occupy the whole switching period. Over a switching period, the average dc-link voltage generated by the rectifier U_{pn}_avg is as given as

$$d_\gamma^R = \frac{d_\gamma}{d_\gamma + d_\delta} \quad d_\delta^R = \frac{d_\delta}{d_\gamma + d_\delta} \tag{5.9}$$

$$U_{pn\_avg} = d_\gamma^R U_{l-l\gamma} + d_\delta^R U_{l-l\delta} \tag{5.10}$$

For the I3SMC, the selected active current vectors define the voltage levels available at the dc link. For example, if the current vectorI_1 is applied, the voltage levels available at the dc link are the following: $P = U_{ao}$, $O =$ the neutral − point voltage, and $N = U_{co}$. At any time, only two voltage levels can be supplied to the inversion stage. Therefore, the modulation of the inversion stage determines how the rectifier and neutralpoint commutator supply the voltage levels to the inverter.

B. Inversion Stage

Due to the design similarity between the three-level neutralpoint-clamped VSI discussed in 5. 2. 2 and the I3SMC, the SVM for the three-level neutral-point-clamped VSI can be applied to the inversion stage of the I3SMC in order to generate multilevel output voltages. The reference vector U_{out} is synthesized using three nearest voltage vectors, and the duty cycle of each selected vector is determined using the equations shown in Table 5. 2, where the modulation index of the inversion stage $m_U$is given as

$$m_{\mathrm{U}} = \frac{\sqrt{3} \mid U_{\mathrm{out}} \mid}{U_{\mathrm{pn\_avg}}} \tag{5.11}$$

The rectifier and the neutral-point commutator are modulated so that only two voltage levels are supplied to the inverter's terminals based on the selected voltage vector. For example, when the current vectorI_1 (ac) is applied to the rectifier, the following can be derived:

1) For output vector V_1 (POO), the required voltage levels are $p\_inv = P$ and $n\_inv = O$. Hence, only the switches that connect the input phase voltage a to p_inv and the neutral-point voltage o to n_rminv are turned ON, while the others are OFF. The voltage applied across the inverter's terminals is $U_{p\_inv-n\_inv} = U_{ao}$.

2) For vector U_2 (PNN), the required voltage levels are $p\_inv = P$ and $n\_inv = N$. Only the switches that connect the input phase voltage a to p_inv and the input phase voltage c to n_inv are ON, while the others are OFF.

Therefore, $U_{p\_inv\_n\_inv} = U_{ac}$.

Based on the switching pattern, the voltage $U_p\_inv-n\_inv$ is applied to the outputs to generate the output line-to-line voltages for the I3SMC. By applying the input phase-to-neutral voltages to the outputs at low modulation indexes, the generated output voltage ripple is much lower than the voltage ripple generated by the use of the input line-to-line voltages. Hence, the output-waveform harmonic content is reduced, as will be shown in 5. 2. 5.

C. Switching Pattern for the I3SMC

The switching pattern shown in Figure 5. 14 clearly illustrates how the voltage levels are supplied to the inverter's terminals based on the selected current and voltage vectors within a switching period. Figure 5. 14 is based on an example where $I_{in}$is located in sector 2, and $U_{out}$is located in T_6 of sector 1. The current vectors I_1 ($= I_\gamma$) and I_2 ($= I_\delta$) are selected for the rectification stage, while the voltage vectors U_2, U_3, and U_4 are selected for the inversion stage. The voltage vectors are arranged in a double sided switching sequence: V3 - V4 - V2 - V3 - V3 - V2 - V4 - V3, but with unequal halves because each half has to be applied to each rectifier switching state: I_1 and I_2. As shown in Figure 5.14, both redundant switching states (PPO/OON) of U_3 are equally applied within the switching pattern. This arrangement ensures that zero-average neutral-point current is obtained within a switching period so that the voltage levels of the input filter capacitors as well as the output performance of the converter can be maintained, as discussed in Section 5.2.2.

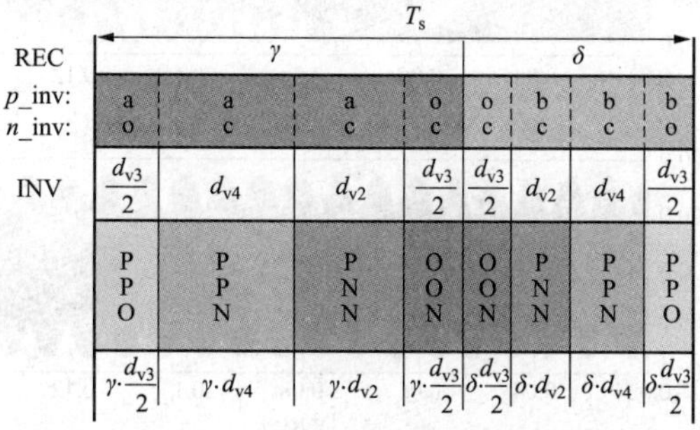

	T_s							
REC	γ				δ			
p_inv:	a	a	a	o	o	b	b	b
n_inv:	o	c	c	c	c	c	c	o
INV	$\frac{d_{v3}}{2}$	d_{v4}	d_{v2}	$\frac{d_{v3}}{2}$	$\frac{d_{v3}}{2}$	d_{v2}	d_{v4}	$\frac{d_{v3}}{2}$
	PPO	PPN	PNN	OON	OON	PNN	PPN	PPO
	$\gamma\cdot\frac{d_{v3}}{2}$	$\gamma\cdot d_{v4}$	$\gamma\cdot d_{v2}$	$\gamma\cdot\frac{d_{v3}}{2}$	$\delta\cdot\frac{d_{v3}}{2}$	$\delta\cdot d_{v2}$	$\delta\cdot d_{v4}$	$\delta\cdot\frac{d_{v3}}{2}$

Figure 5.14 Switching pattern of the I3SMC.

Table 5.3 Circuit Specifications

Circuit specifications	Value
Input	$U_{in\_rms}=240V$, $f_{in}=50Hz$
Filter	$L_f=0.6mH$, $C_f=10\ \mu F$
Load	$R_L=20\Omega$, $L_L=7mH$
Output frequency	$f_{out}=30Hz$
Switching frequency	$f_{sw}=5kHz$

5.2.4 Simulation Results

The I3SMC shown in Figure 5.9 has been simulated using SABER, based on the specifications shown in Table 5.3. The converter is implemented using ideal switches. The control strategy uses the measured input phase voltages (referenced to the neutral-point of the input filters) U_{ao}, U_{bo}, and U_{co}, to determine the reference angle of the input-current vector. This is because, according to the modulation strategy, the input currents generated by the I3SMC are synchronized with the input voltages. In addition, the magnitudes of the input phase voltages are used to determine the average dc-link voltage supplied to the inversion stage, as given in (5.10). In this simulation, the converter is evaluated using an RL load. The desired reference output voltage vector is provided for the control block of the inversion stage to determine the duty cycles for the switches.

Figure 5.15 shows the waveforms generated by the I3SMC as the voltage transfer ratio is stepped from 0.4 to 0.8. The waveforms shown in this figure consist of the dc-link voltage (U_{pn}), the potentials at the dc-link terminal referenced to the neutralpoint (U_{po} and U_{no}), the output terminal voltage (U_{Ao}), the output line-to-line voltage (U_{AB}), and the load currents (i_A, i_B, and i_C).

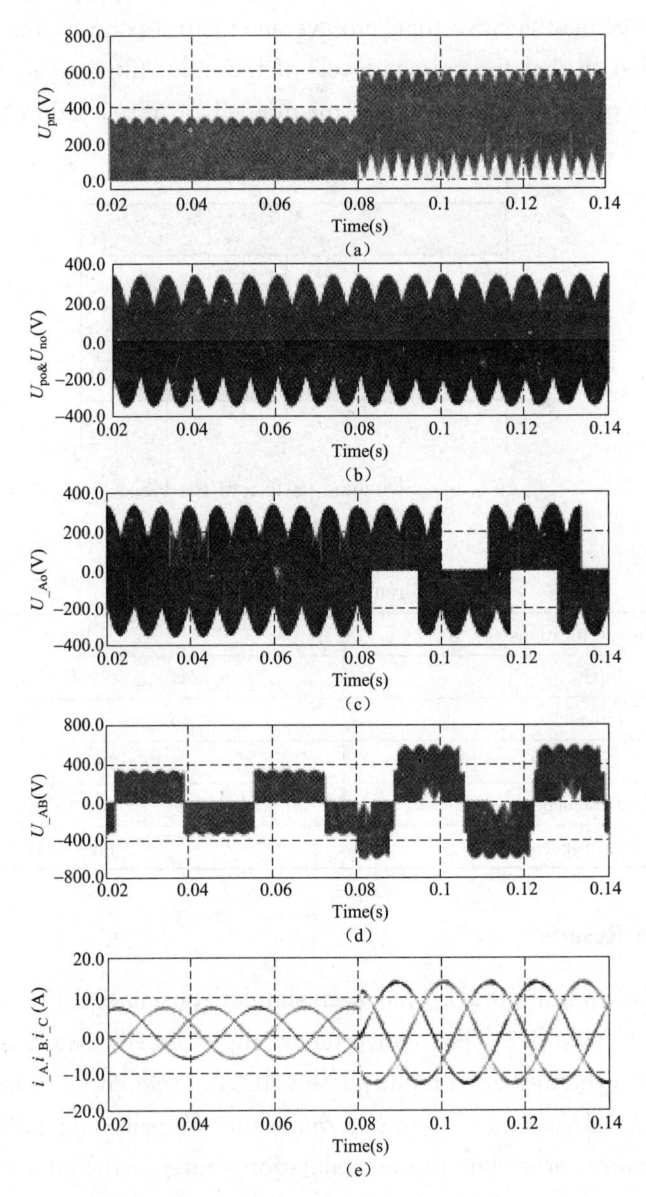

Figure 5.15　Output waveforms generated by the I3SMC with the voltage transfer ratio stepped from 0.4 to 0.8.

There are three voltage levels available at the dc links: U_{po}, 0 V, and U_{no}. However, at any instant, the inversion stage can only be operated with two voltage levels. Therefore, depending on the selected voltage vectors, the rectifier and neutral-point commutator provide the required voltage levels to the inversion stage. During the voltage-transfer-ratio transition, there is a noticeable increase in the dc-link voltage U_{pn}, as shown in Figure 5.15 (a). This is because, to generate higher output voltages at high modulation indexes, the rectifier and the neutral-point commutator constantly connect the input line-to-line voltages (e.g., U_{ab}) to the inversion stage's terminals *p*_inv and *n*_inv instead of the input line-to-neutral point voltages (e.g., U_{ao}), which are mostly used at low modulation indexes. The switching in U_{po} and U_{no}, shown in Figure 5.15 (b), shows the operation of the rectifier and neutral-point commutator in order to control the voltage levels supplied to the inversion stage.

The output terminal voltage, U_{Ao}, shown in Figure 5.15 (c), shows that the I3SMC generates three distinctive voltage levels at the output terminals. These levels are the positive and negative envelopes of the rectified input voltages and the zero voltage level. As shown in Figure 5.15 (d), a transition in the output line-toline voltage U_{AB} from three to five levels shows that the I3SMC is able to generate multilevel output voltages. To verify that the voltage levels are properly applied to generate the desired outputs, the load currents of the I3SMC are shown in Figure 5.15 (e). These currents are balanced and sinusoidal.

Table 5.4 Switching Frequency, Average Voltage Stress, and Current Rating for one Switch of the Neutral-Point Commutator.

Voltage transfer ratio	Switching frequency	Average voltage stress	Average DC-link current	Average current stress (p.u.)
0.2	5kHz	107V	1.9A	1.2A (0.63)
0.4	5kHz	125V	5.4A	2.9A (0.54)
0.6	5kHz	168V	9.1A	3.3A (0.36)
0.8	5kHz	220V	12.1A	1.9A (0.16)

Figure 5.16 shows the voltage levels of the star-connected input filter capacitors and the input-current waveforms. By applying the redundant switching states of selected SVVs equally, a zero average neutral-point current over a switching period can be obtained. As shown in Figure 5.16 (a), the voltage levels of the input filter capacitors remain balanced, proving that the modulation strategy is able to balance the neutral-point current. Then, the waveforms in Figure 5.16 (b) clearly show that the converter is able to generate sinusoidal and balanced input-current waveforms, even with the presence of neutral-point current.

Since the neutral-point commutator has to be constantly commutated to supply the required voltage levels to the inversion stage, it is important to investigate the voltage stress, switching frequency, and current rating for the switches of the neutralpoint commutator with respect to the load rating. Using the specifications given in Table 5.3, the converter

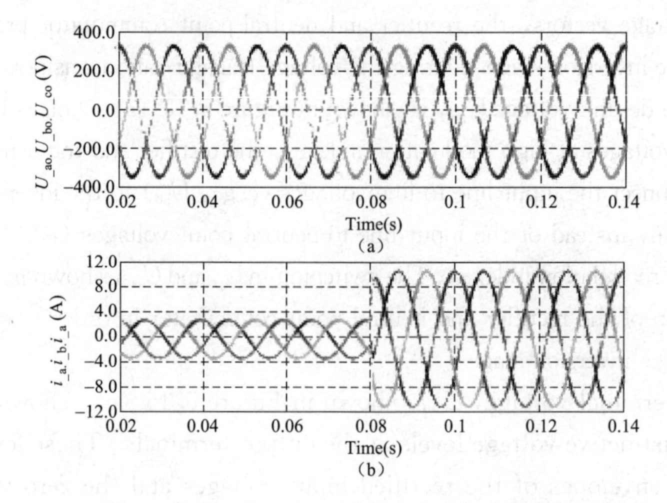

Figure 5.16 Input waveforms for the I3SMC:

(a) The voltage levels of the star-connected input filter capacitors; (b) The input-current waveforms.

is operated at different voltage transfer ratios, and one switch of the neutralpoint commutator is measured to determine the average voltage stress, switching frequency, and the average current stress, which are shown in Table 5.4.

Due to the connection of middle point o to the neutral-point of the input filter capacitors, the maximum voltage stresses across the switches of the neutral-point commutator are limited to the connected input phase voltages. As shown in Table IV, when the voltage transfer ratio is increased from 0.2 to 0.8, the average voltage stress across the switch is increased. On the other hand, the per unit (with the average dc-link current as the base unit) average current stress for the switch reduces. This is due to the dominance of large voltage vectors in synthesizing the reference vector when the voltage transfer ratio is increased, increasing the OFF-state time of the switches for the neutralpoint commutator.

5.2.5 Experimental Results

To validate the simulation results, a prototype of the I3SMC, shown in Figure 5.17, was developed. Using a 120-V (rms) supply, the converter was evaluated based on the specifications shown in Table 5.3, with different modulation indexes, using a balanced RL load.

The voltage transfer ratio is stepped from 0.4 to 0.8, and the waveforms from the prototype, shown in Figure 5.18, clearly correspond to the waveforms in Fig. 8. The increase in U_{pn} during the voltage-transfer-ratio transition proves that the rectifier and the neutral-point commutator connect the input line-to-line voltages to the inversion stage's terminals in order to generate higher output voltages at high modulation indexes. At low modulation indexes, only the input line-to-neutral point voltages are used, so the magnitude of U_{pn} is limited to the peak levels of the input phase voltages.

Referring to Figure 5.18, the input and output current waveforms of the prototype

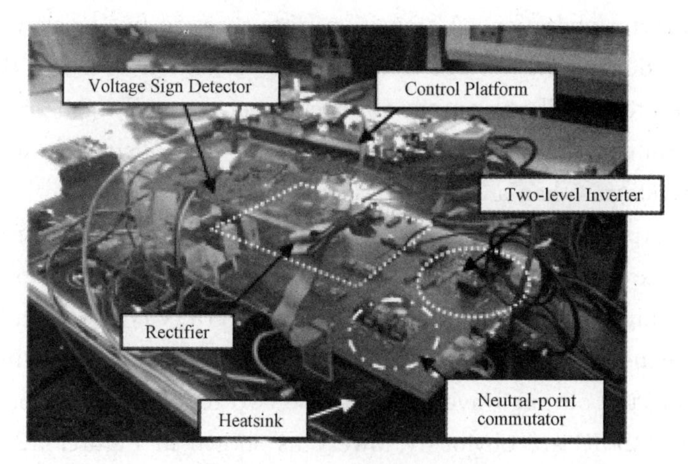

Figure 5.17　I3SMC prototype.

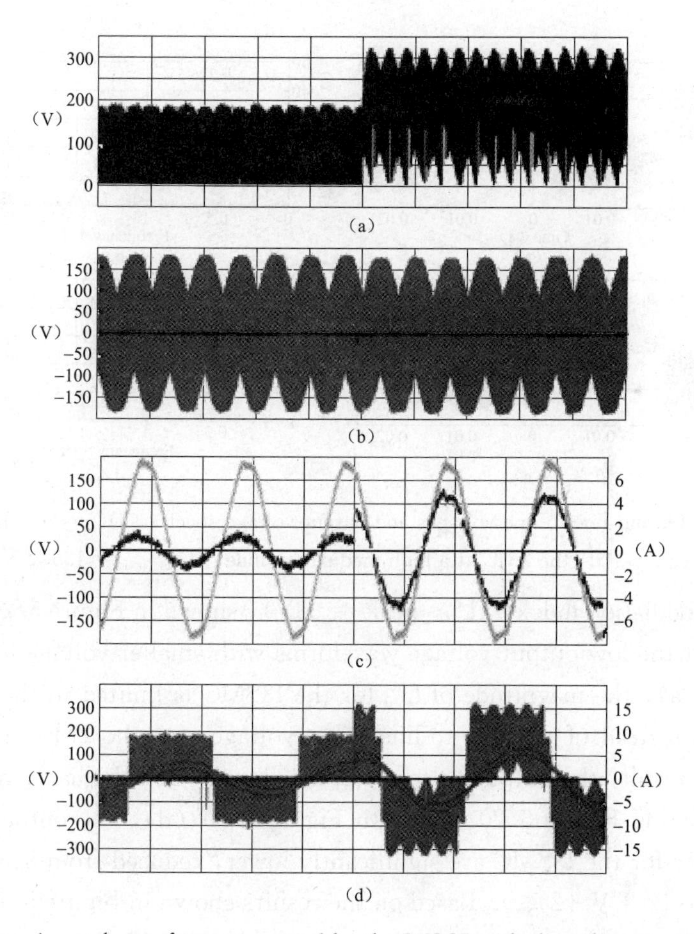

Figure 5.18　Experimental waveforms generated by the I3SMC with the voltage transfer ratio stepped from 0.4 to 0.8. (a) The dc-link voltage U_{pn}. (b) The potentials at the dc-link terminals referenced to the neutral-point U_{po} and U_{no}. (c) The input voltage U_{ao} and the input current i_a. (d) The output line-to-line voltage U_{AB} and the output current i_A.

show proper converter operation. A transition from three to five levels in the output line-to-line voltage U_{AB} during the voltage-transfer-ratio transition, verifies the ability of the converter to generate multilevel output-voltage waveforms. By constructing the output waveforms with multiple voltage levels, the I3SMC is obviously able to generate higher quality output waveforms than a 2MC with identical specifications.

To make an output performance comparison, the prototype of the I3SMC was operated as an indirect matrix-converter topology by disabling the gating signals to the neutral-point commutator. At a high modulation index (U_{out}_peak = 135V), shown in Figure 5.19, the I3SMC evidently generates five distinctive voltage levels for the output line-to-line voltage U_{AB}. By comparing the output waveforms for the two topologies, the out put wave form harmonics for the I3SMC are obviously lower, as shown in Figure. 12 (b). Compared with the 2MC, the out put voltage harmonics are reduced from 14.1 to 13.9 V (f_{sw}) and 38.1 to 25 V ($2f_{sw}$).

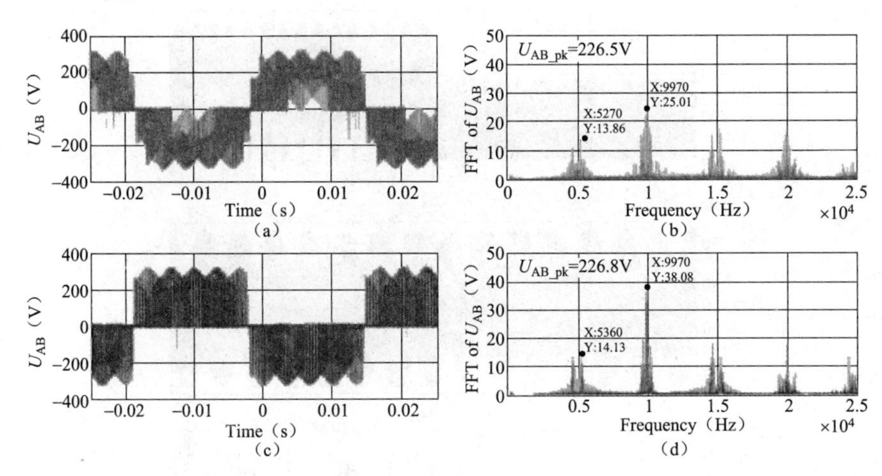

Figure 5.19 Output line-to-line voltages and output-voltage spectra (a) — (b) the I3SMC and (c) — (d) the 2MC at a high modulation index (U_{out_peak} = 135 V).

At a low modulation index, (U_{out}_peak = 68V), shown in Figure 5.20, the I3SMC is able to construct the low output voltage waveforms with smaller voltage levels. As shown in Figure 5.20 (a), the magnitude of U_{AB} for the I3SMC is limited to the input phase-to-neutral voltages instead of the line-to-line input voltages for the 2MC, shown in Figure 5.20 (c). As a result, the output-voltage ripple is lower, and the harmonic content is reduced. By comparing Figure 5.20 (b) with Figure 5.20 (d), the output switching frequency harmonics for the I3SMC are significantly lower, reduced from 8.5 to 5.6 V (f_{sw}) and from 40.5 to 12.1 V ($2f_{sw}$). Based on the results shown in Figure 5.19 and 5.20, the I3SMC evidently has a better output performance than the 2MC in terms of harmonic content of the output voltage.

In this paper, the output performance of the I3SMC is also compared with the 3MC. Both multilevel matrix converters integrate a three-level neutral-point-clamped converter

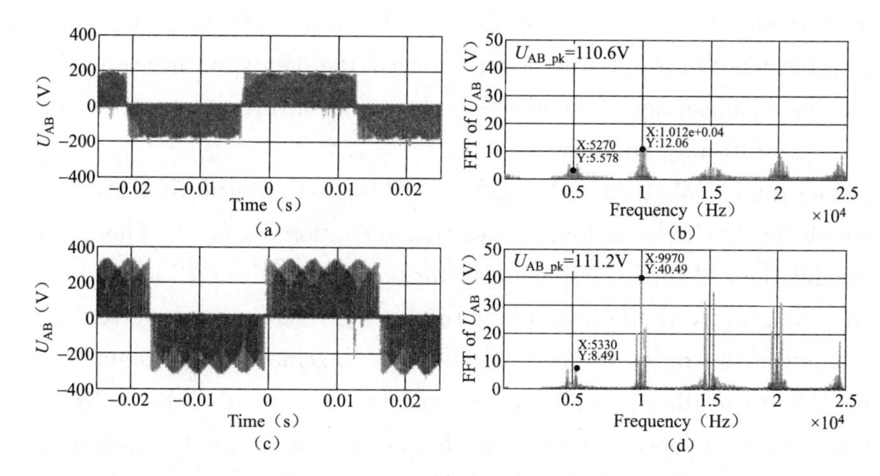

Figure 5. 20 Output line-to-line voltages and output-voltage spectra;

(a) - (b) the I3SMC and (c) - (d), the 2MC at a low modulation index (U_{out_peak} = 68V).

concept with an indirect matrix-converter topology. The only difference is that the I3SMC has a simpler circuit configuration, while the 3MC is more complicated and is able to achieve the switching states (medium voltage vectors) that are not achievable by the I3SMC. A prototype of the 3MC has been developed and was modulated using SVM based on the same specifications shown in Table 5. 3 with U_{in_rms} = 120 V. As discussed, the modulation on the 3MC ensures that a zero-average neutralpoint current is obtained for every switching period in order to maintain the voltage levels of the input filter capacitors as well as the performance of the converter.

To compare the output waveform quality for both topologies, the total harmonic distortion (THD) for the output line-to-line voltage U_{AB} are calculated, as shown in Figure 5. 21, where the fundamental output frequency is 30 Hz and the number of harmonics included in the THD calculation is 1000 (up to 30 kHz). In this figure, the THD for U_{AB} of the 2MC is also presented. The three-level matrix-converter topologies clearly have a better

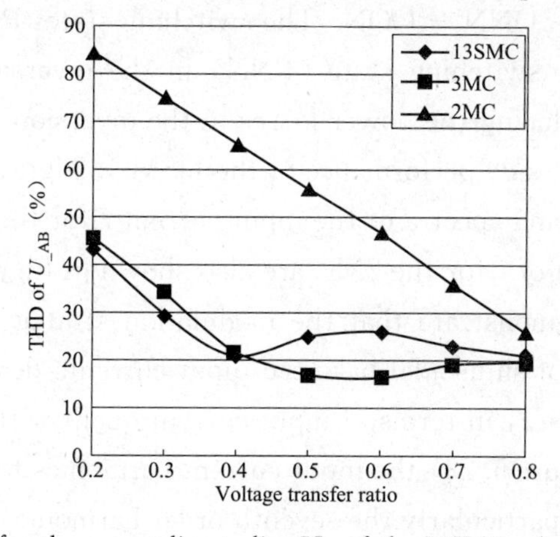

Figure 5. 21 THD for the output line-to-line U_{AB} of the I3SMC, the 3MC and the 2MC.

output performance than the 2MC in terms of the harmonic content in the output voltages. By having better quality output-voltage waveforms, the distortion in the load current will be lower, giving an advantage to the three-level matrix converters in applications where the load provides low filtering inductances.

Comparing the I3SMC to the 3MC, the THD for both multilevel matrix-converter topologies are obviously similar at low voltage transfer ratios (< 0.5). This similarity is because the modulations of the inversion stages for both topologies are identical; only the SVVs and zero voltage vectors are used to synthesize the reference output vector. However, at high voltage transfer ratios (> 0.5), the ability to achieve the medium voltage vectors enables the 3MC to synthesize the reference output vector with a better selection of the nearest three space vectors, reducing the harmonic content in the output waveforms. Therefore, as shown in Figure 5.21, the THD for the I3SMC is higher than for the 3MC. The decrease in the difference between the THDs, when the voltage transfer ratio approaches 0.8, is because the large voltage vectors of both multilevel matrix converters get more dominant in synthesizing the reference output vector. As a result, the THDs for both multilevel topologies, as well as the 2MC, converge when the voltage transfer ratio reaches 0.8.

Based on Figure 5.21, the 3MC appears superior to the I3SMC and the 2MC. However, the complicated circuit configuration of the 3MC is undoubtedly a disadvantage. On the other hand, the I3SMC has a simpler circuit configuration than the 3MC and is able to generate multilevel output voltages to improve the output waveforms quality as compared with the 2MC. This clearly gives an advantage to the I3SMC.

In addition, the inversion stage of the I3SMC has lower switching losses at high modulation indexes. This advantage is due to the fact that there is no switching in the inversion stage for certain change of switching states in the switching pattern. For example, if the reference vector U_{out} is located in triangle T_4, the switching pattern for the inversion stage is PPO - POO - PNN - ONN - OON. The switching states POO, PNN, and ONN actually represent the same switching state (PNN) in the inversion stage. Therefore, no switching is required, reducing the power losses in the inversion stage.

To evaluate the input-side performance of the I3SMC, Figure 5.22 and 5.23 show the input-current waveforms and spectra of the input current i_a for the converter. For comparison purposes, the spectra of i_a for the 2MC are also shown in Figure 5.23. The waveforms in Figure 5.22 clearly demonstrate that the modulation strategy is able to modulate the I3SMC to generate a set of sinusoidal balanced input currents despite the presence of neutral-point current. However, in terms of input-current quality, the 2MC is better than the I3SMC. As shown in Figure 5.23, the input-current harmonics for the I3SMC are relatively higher than the 2MC, particularly the seventh-order harmonic content at 350Hz. This is proven by the THD for the input current of each topology, shown in Figure 5.23, where the fundamental input frequency is 50 Hz and the number of harmonics included in the THD calculation is 40 (up to 2kHz).

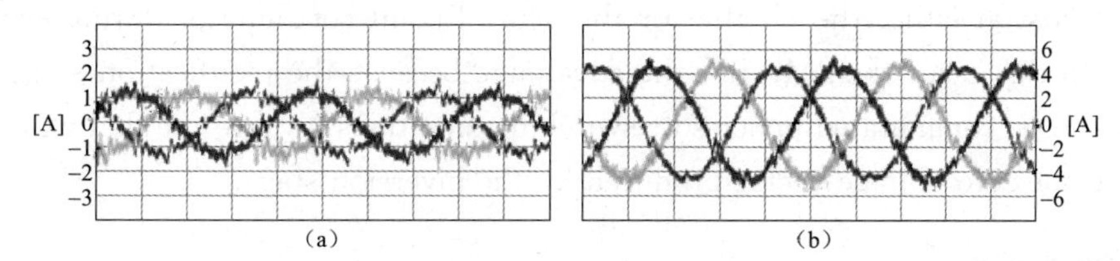

Figure 5.22 Input-current waveforms of the I3

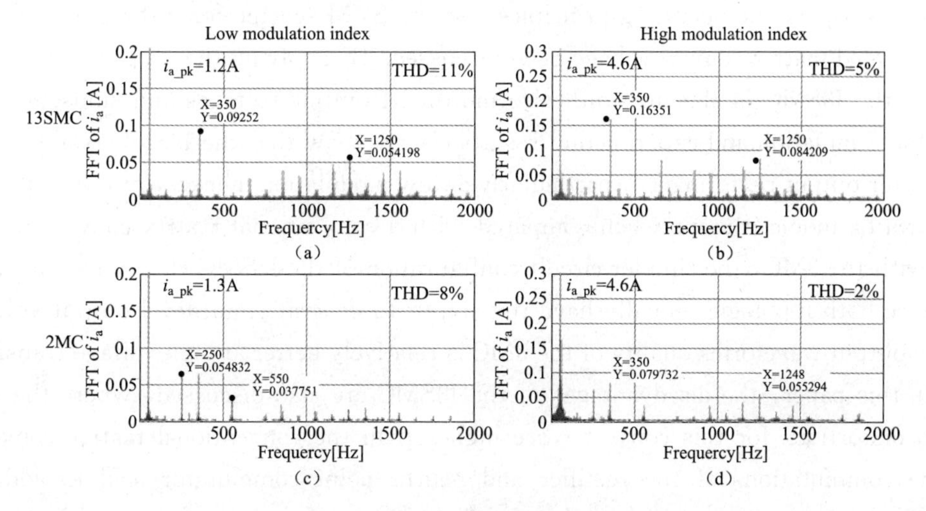

Figure 5.23 Input-side comparison between the I3SMC and the 2MC at a low modulation index (U_{out_peak} = 68V) and a high modulation index (U_{out_peak} = 135 V).

For the I3SMC, the operation of the rectifier and neutralpoint commutator to supply the voltage levels required by the inversion stage impacts on the ability of the rectifier to properly generate sinusoidal input currents. Based on the modulation on the rectifier, discussed in Section 5.2.2, the input currents are synthesized by distributing the impressed dc-link current accordingly to the input phases. However, the use of zero voltage vectors causes discontinuity in the dc-link current, and the SVVs are generated by disconnecting the rectifier from one of the dc link. This inevitably affects the rectifier to properly generate sinusoidal input currents. Therefore, the input-current distortions for the I3SMC are relatively higher than the 2MC, which is clearly a disadvantage.

In addition, the OFF-state switches for the inversion stage have to withstand high voltage stresses, which are proportional to the connected input line-to-line voltage, at high modulation indexes. In this case, the 3MC obviously has the advantage because the connections of clamping diodes always limit the voltage stresses across the OFF-state switches of the inversion stage to half the dc-link voltages (proportional to input phase voltages).

In addition, in order to supply the required voltage levels to the inversion stage, the rectifier and neutral-point commutator for the I3SMC have to be constantly commutated under hard switching situations, where neither the dc-link current nor the voltage across the devices are zero. Therefore, depending on the direction of the dc-link current, there is a switching loss at either the rectifier or the neutral-point commutator during every commutation. In this case, the 2MC has the advantage because the rectification stage for this converter can be commutated under soft-switching situations, where the dc-link current is zero due to the zero voltage vector produced by the inversion stage.

5.2.6 Conclusion

In this paper, the operating principles and the SVM scheme for a three-level neutral-point-clamped matrix converter have been discussed. By applying the proposed modulation scheme, the I3SMC is able to synthesize multilevel output voltages and sinusoidal input currents. Simulation and experimental results clearly show that the I3SMC is able to produce better output performance, particularly at low modulation indexes, in terms of output waveform harmonic content when compared with a conventional matrix converter. Compared with the 3MC, the simpler circuit configuration of the I3SMC clearly has the advantage since both topologies equally have the ability to generate multilevel output voltages, but the output waveforms quality of the 3MC is relatively better at high voltage transfer ratios. In this paper, the disadvantages of the I3SMC are also discussed, where the input-current distortions for this converter are higher than the conventional matrix converter, and the commutations of the rectifier and neutral-point commutator lead to additional switching losses.

5.3 A Compact nX DC-DC Converter for Photovoltaic Power Systems

5.3.1 Introduction

The proliferation of solar power requires more power electronic circuits operating on the grid than ever before. Efficiency, cost, reliability, and environmental tolerance present great importance for a photovoltaic (PV) power system. Power electronic circuits face many challenges to achieve high efficiency, low cost, high reliability, and environment friendly. First of all, the single PV panel voltage is too low to be of much use by itself. Although it may be tempting to suggest placing multiple panels in series to achieve a higher voltage, the reliability would be adversely affected and the current capabilities would be low. Multiple PV panel strings in parallel can extend current (power) capability, but series and parallel connections cause the low PV panels' efficiency. In situations involving mass production or the installation of a large number of facilities, the cost of power electronic circuits used must be minimized. With this in mind, the cost of employing one high power converter could easily outweigh the price of using multiple low power converters. Furthermore, the use of small DC-DC converters coupled to each PV panel would increase the system's overall reliability; even if one converter was to fail, the dc-bus voltage would still be useable. Cost and reliability make an excellent case for the use of low power DC-DC micro-converters fixed to each PV panel. Also, the inverter is necessary to get ac electricity for many situations such as grid-tie and ac load. As shown in Figure 5. 24, the task of regulating the voltage can fall to either the DC-DC converter, or to the inverter.

Figure 5. 25 and 5. 26 show the traditional DC-DC converter topologies. In theory, the voltage gain of Fig. 2 could even be infinity, but the truth is much less phenomenal. Losses in the switches and inductor would greatly limit the actual capabilities of the converter. As the voltage gain increases, the device stresses also increase, asymptotically approaching infinity as the duty cycle goes to one. This converter would not be practical for applications requiring very high voltage gains in PV power systems. Figure 5. 26 uses the transformer's turn ratio to achieve the high voltage boosting. Both an inductor and a transformer greatly increase the size and weight of the unit.

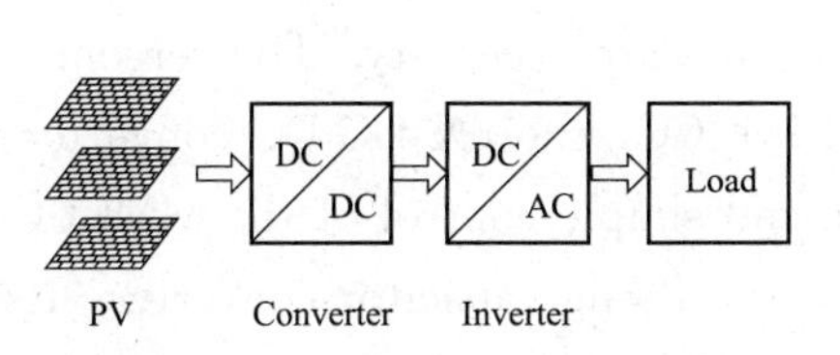

Figure 5. 24 System configuration.

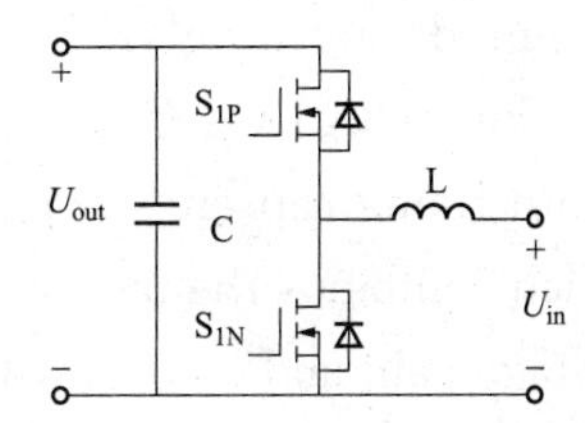

Figure 5. 25 DC-DC boost converter.

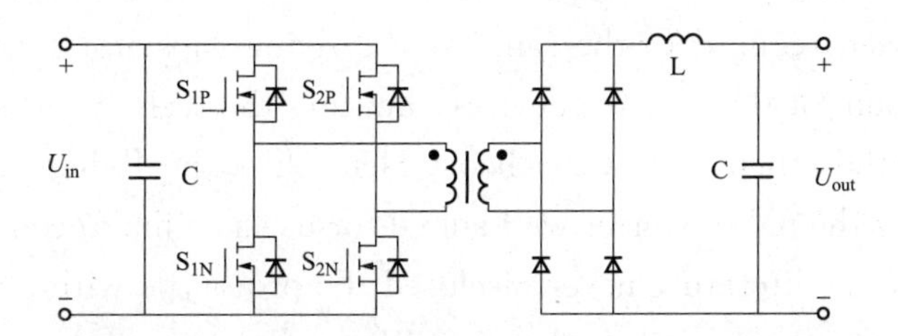

Figure 5. 26 Isolated full-bridge DC-DC converter.

Either Figure. 5. 25 or Figure 5. 26 employs magnetic cores (of inductors and transformer) that present a bottleneck for high temperature environments, because the permeability declines dramatically with increasing temperature; they are prone to saturation and instability as the temperature goes beyond a limitation, which leads to outputs nonlinearly related to the input, making effective control very difficult. Also, inductors and/or transformers, as some of the hottest components in the system, are bulky, lossy and are an obstacle to applying power converters in high temperatures and reducing their size and cost.

In order to reduce or eliminate bulky magnetic components from electronic converters, the flying capacitor converter was first proposed for 42-V/14-V automotive systems. As shown in Figure 5. 27, the larger the required boost is the more losses will be incurred from

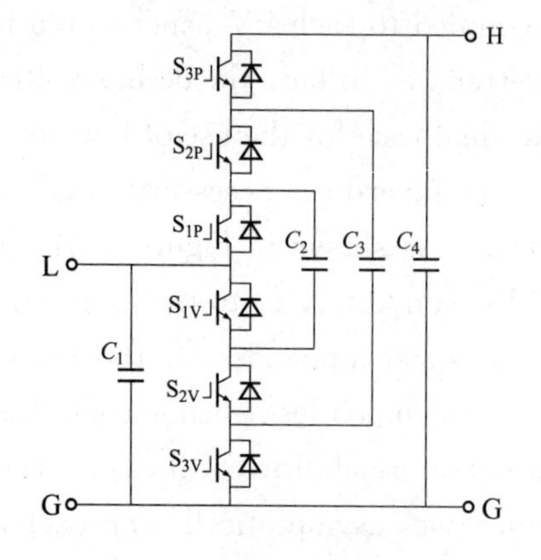

Figure 5.27　Flying capacitor converter.

the increased number of devices in the charge pump paths. The control becomes more complex with the boost ratio. Moreover, the flying capacitor structure is not modular, meaning that a fundamental, fully functioning unit of the circuit cannot be made. These issues offer the greatest argument against the use of the flying capacitor topology in situations involving high voltage gain. The multilevel modular capacitor clamped DC-DC converter (MMCCC) represents an attempt to surpass the flying capacitor converter's performance, for example, the control is very simple and the design is modular. However, it requires a large number of switches for a given voltage boost, and the n-2 switches need to handle higher voltage stresses. As shown in Figure 5.28, 10 switches are required to achieve 4X DC-DC converter, and 2 switches need to handle high voltage stress. For an 8X DC-DC converter, 22 switches are necessary. To overcome the drawbacks of both flying capacitor converter and MMCCC, the nX DC-DC converter was proposed, which combines the modular structure and simple control of the MMCCC with the greater voltage gain and lower switch count of the flying capacitor converter. Its modular design easily implements the high voltage gain. The control is also simple, just using a complementary 50% duty cycle to control all of the semiconductors involved. The overall switch count has been reduced to the same as the flying capacitor topology. The charge paths have a maximum of three active devices, limiting the resistive losses and eliminating any dependency of the efficiency on the boost factor like the flying capacitor converter. These features make the nX converter well-suited for use as a micro-converter in PV power systems. However, the literature never disclosed the prototype with a voltage gain larger than 6, and their power rating is less than 500 W.

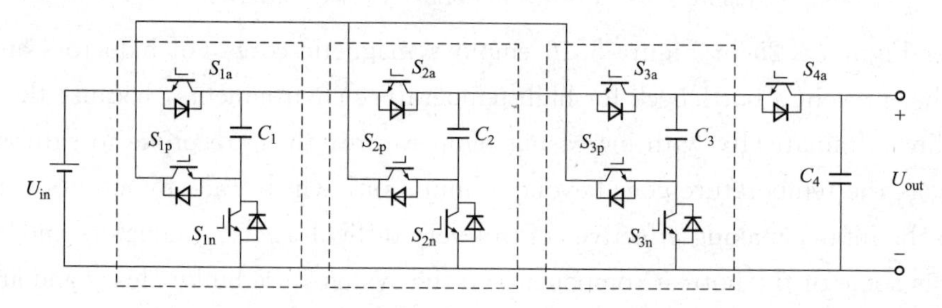

Figure 5.28　MMCCC.

This paper proposes a compact nX DC-DC converter-based PV power system. An inverter is cascaded to the nX DC-DC converter for achieving the desired AC voltage/power via adjusting the inverter's modulation index. The 1kW 8X dcdc converter is prototyped to investigate its performances, including size, weight, voltage gain, and efficiency. The results validate the 8X DC-DC converter, a best suitable for PV power systems, due to its small size (7 inches length by 3 inches width), light weight, and high efficiency (no need for special heat sink), and the magnetic-less design provides an excellent candidate in application to high temperature of solar power.

5.3.2 nX DC-DC Converter-Based PV Power System

Figure 5.29 shows the nX DC-DC converter based PV power system. For a PV panel voltage 25-40V, we can boost the voltage to 200 V- 320 V by using 8X ($n=8$) DC-DC converter to produce 120 V rms ac electricity. When the PV panel voltage changes, the inverter modulation index can be adjusted to keep a constant ac output voltage and achieve the PV panel's maximum power point tracking (MPPT). As shown in Figure 5.29, except the filter, whole PV power system does not include any magnetic components, which makes the system compact, high efficiency, and suitable for high temperature of solar power.

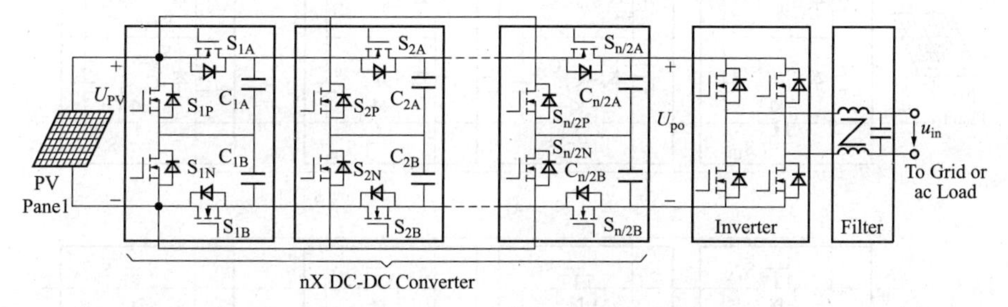

Figure 5.29 The nX DC-DC converter-based PV power system.

5.3.3 nX DC-DC Converter

The nX dc-dc converter, like the last two converters discussed in Figsure 5.27 and 5.28, is a type of switched capacitor or multilevel converter. It combines the modular structure and simple control of the MMCCC with the greater voltage gain and lower switch count of the flying capacitor converter. The operation relies on the transfer of charge from one capacitor to another, stepping up the voltage along the way. As shown in Figure 5.29, the converter does appear to be modular, and the basic cell with 2X boost factor is shown in Figure 5.30. The nX dc-dc converter can be fulfilled through cascading n/2 cells. The nX dc-dc converter has only two switching states, as shown in Figure 5.31. This greatly simplifies the control required to drive the converter. The other unique features include: 1) Two charge pump paths feed the load directly, leading to less power loss in the energy transfer; 2) Half of the capacitors reduce their voltages by $(n/2)\times U_{pv}$; 3) There is lower capacitance and ripple current requirement for the two output capacitors; 4) The nX converter employs fewer switches ($2n$ versus $3n$-2 in MMCCC) with no penalty of total device power ratings; 5) Each pair of complementary switch devices can be made truly capacitor clamped. These features allow further higher efficiency with more compact package and lighter weight for high voltage gain than the MMCCC and flying capacitor converter.

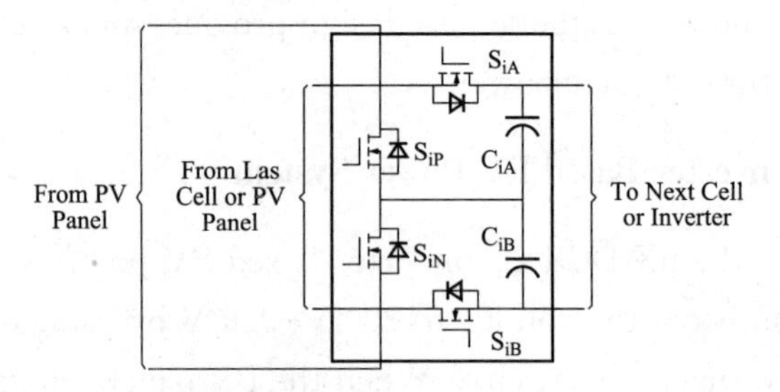

Figure 5.30 Basic cell.

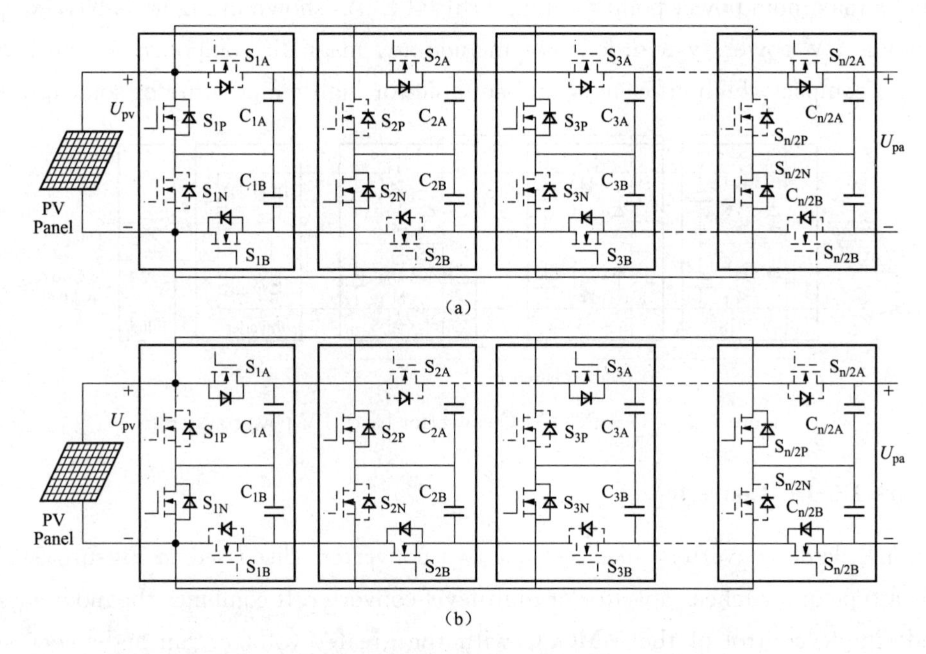

Figure 5.31 The nX converter's switching states: (a) state I; (b) state II.

5.3.4 Prototype of 8X DC-DC Converter

In order to test the actual capabilities of the nX converter topology, an 8X 1 kW converter was designed, built and tested. The operating input voltage range was between 20 V and 40 V. Since the input voltage can be as high as 40 V, MOSFETs capable of blocking at least one or two times that were needed. Using a multiplication factor of 1.5 for safety, half of the MOSFETs needed to be rated for 60 V and the other half for 120 V. The switches IPB017N06N3G and IPB036N12N3G were used to achieve low on-resistance and low gate charge. As for the stage capacitors, due to their small size, high-temperature capabilities, low ESR and ESL, multilayer ceramic capacitors (MLCC) are the optimal choice. Two types of capacitors were purchased: the 100-V/15-μF C5750X7S2A156M, and the 250-V/2.2-μF C5750X7T2E225K. The circuit was controlled by using a pulse width modulation (PWM) chip called the UC3525. Switching frequency is 165 kHz, with a deadtime of 500 ns.

The circuit was constructed on a four-layer printed circuit board (PCB) using the top layer for the power circuitry shown in Figure 5.32 (a), the bottom for the control electronics shown in Figure 5.32 (b), and the middle two layers as power and ground planes to the control layer. As shown in Figure 5.32 (a), the four pairs of MOSFETs on the board's left-hand side are the circuit's half-bridges, the light brown squares are the MLCCs, and the right-hand sets of MLCCs comprise the stage capacitors. The white wires are the AC nodes in Figure 5.32 (b) to allow the use of a current sensor. The dimensions of the board are about 3 inches wide by 7 inches long, making the converter quite small in size. All of this had the effect of increasing the power density.

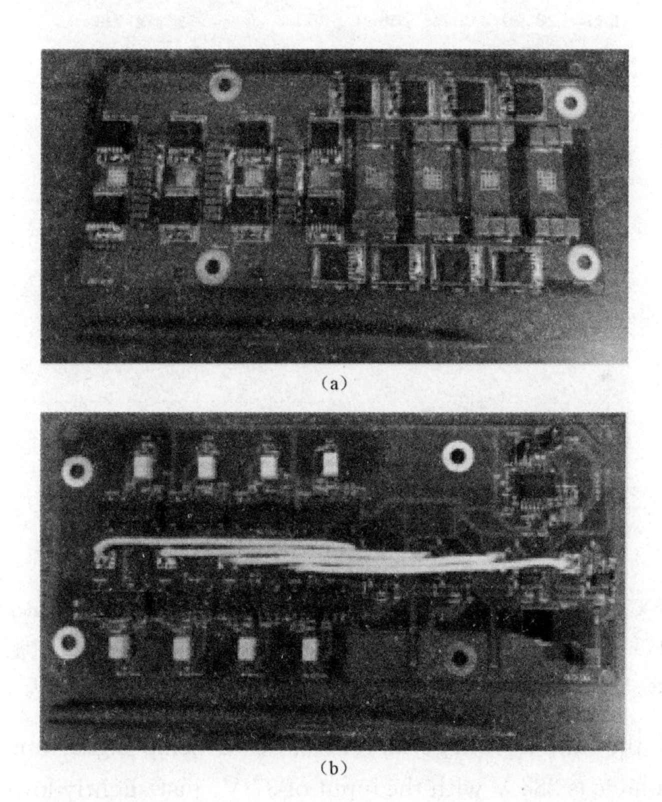

(a)

(b)

Figure 5.32 Prototype of 8X DC-DC converter: (a) top view; (b) bottom view.

5.3.5 Experimental Results

Figure 5.33-5.35 show the test results when the input voltage is 10 V, 20.4 V, and 37 V, respectively, where U_{gs} is the gate drive signal, U_{ds} is the MOSFET switching voltage, I_{ls} is the current passing through the so-called AC node between the half-bridges and the stage capacitors, U_{pn} and U_{pv} are the output and input voltages of 8X DC-DC converter, respectively.

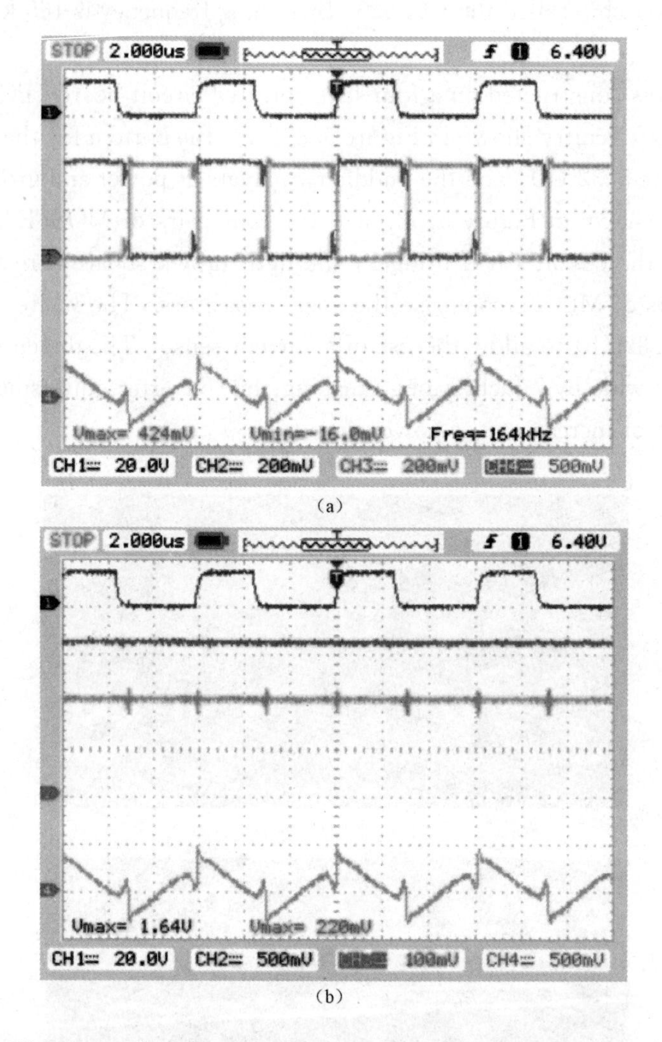

(a)

(b)

Figure 5.33　The 8X prototyped converter waveforms with $U_{pv}=10$ V. (a) Channel 1: U_{gs}, 20 V/div; Channels 2 and 3: two MOSFETs' switching voltages U_{ds}, 10 V/div; Channel 4: I_{ls}, 5 A/div; (b) Channel 1: U_{gs}, 20 V/div; Channel 2: U_{pn}, 25 V/div; Channel 3: U_{pv}, 5 V/div.

As expected, the output voltage is very close to being eight times the input voltage. The output voltage is 288 V with the input of 37 V, just slightly lower than the expected 296 V. The efficiency test is performed at both 156 kHz and 165 kHz, respectively, but both tests have the input voltage of 20 V, while the load resistance is changed to adjust the output power. Figure 5.36 shows fairly high efficiency over a wide output power range. Efficiencies greater than 98% were measured at several points during testing, confirming the converter's viability as a candidate for use in high temperature environments requiring both high efficiency and small size.

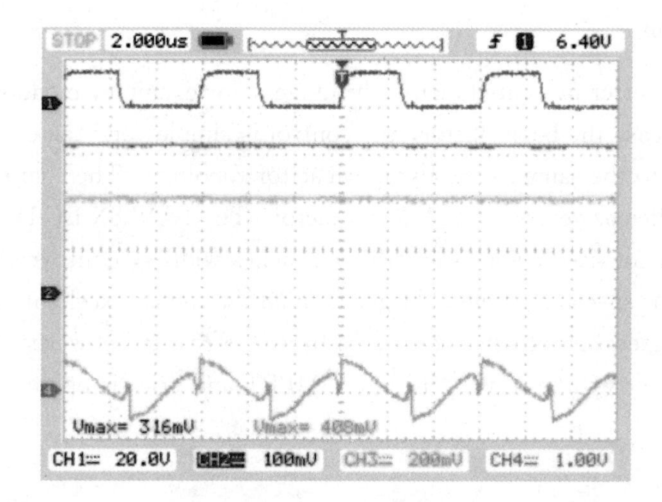

Figure 5.34　The 8X prototyped converter waveforms with $U_{pv}=20.4$ V (Channel 1: U_{gs}, 20 V/div; Channel 2: U_{pn}, 50 V/div; Channel 3: U_{pv}, 10 V/div; Channel 4: I_{ls}, 10 A/div).

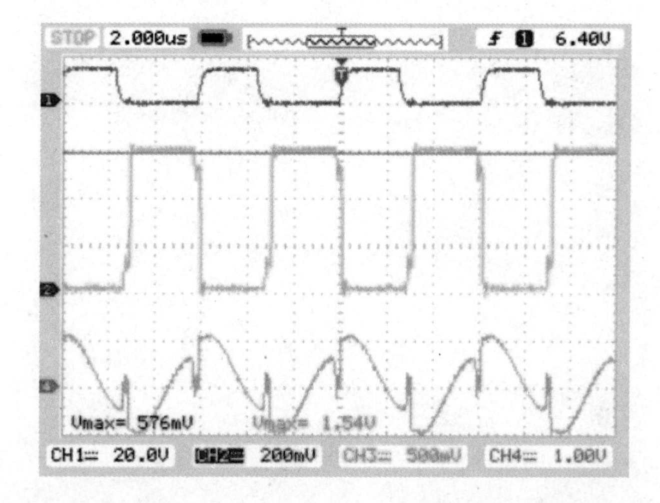

Figure 5.35　The 8X prototyped converter waveforms with $U_{pv}=37$ V (Channel 1: U_{gs}, 20 V/div; Channel 2: U_{pn}, 100 V/div; Channel 3: U_{ds}, 25 V/div; Channel 4: I_{ls}, 10 A/div).

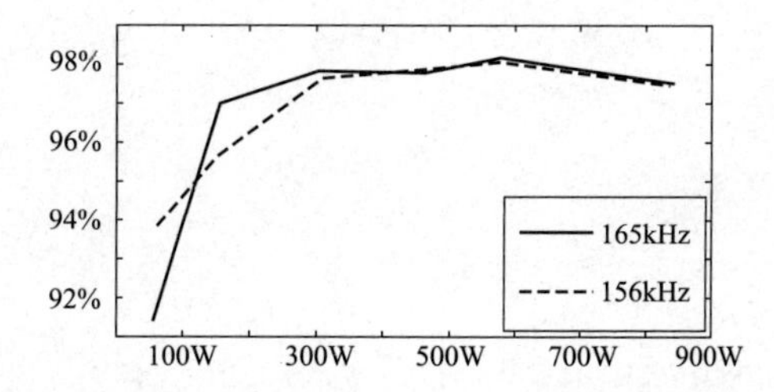

Figure 5.36　Efficiency vs. output power for the 8X prototype.

5.3.6 Conclusion

Then nX converter presented many advantages, for example, modular structure making it easy to increase the boost factor, the control is simple, and the overall switch count has been reduced to the same as the flying capacitor topology. The charge paths are minimized, so high efficiency even for high boost factor. The 1-kW 8X DC-DC converter prototype verified its small size (7 inches length by 3 inches width), light weight, and high efficiency (no need for special heat sink) larger than 98%, and an excellent candidate in application to high temperature of solar power. In future, a real PV power system based on the 8X DC-DC converter will be tested, including MPPT and grid-tie operations.

6

专业英语语法(3)

6.1 as 作关系代词引导的定语从句

as 作为关系代词时，通常用来引导定语从句。

6.1.1 as 用于 such … as … 或 such as … 结构中引导限制性定语从句

在这种限制性定语从句中，关系代词 as 与主句中的 such 相呼应，构成 such … as … 或 such as … 的结构。下面分别讨论之。

（1）as 用于 such … as … 结构中引导限制性定语从句这时主句中 such 为形容词，关系代词 as 在定语从句中可作主语、宾语或表语。汉译时，通常译成“……的这类……”，“（像）……之类的……”“（像）……这（那）样的……”，“那种……的……”。例如：

1）Such materials as can bear high temperature and pressure are urgently needed. 能耐高温高压的这类材料是极为需要的。(as 在从句中作主语，代表 materials)

2）Such light as is produced by an incandescent lamp is called incoherent light. 像白炽灯发出的那种光叫做非相干光。(as 代表 light，在从句中作主语)

3）This kind of transformer has such advantages as are described in the book. 这种变压器具有像本书所描述的那些优点。(as 在从句中作主语，代表 advantages)

4）Gravity tries to cause all bodies to move downward, and to accelerate such bodies as are already moving downward. 重力试图使一切物体朝下运动，并使已经朝下运动的物体加速。(as 在从句中作主语，代表 bodies)

5）The motion of ions is the motion of such atoms as have gained or lost electrons, which in most cases takes place in chemical solutions. 离子的运动也就是原子得到或失去电子时的运动，在大多数情况下，化学溶液中发生这种现象。(as 在从句中作主语，代表 atoms)

6）These are exactly such high voltage electricity installations as we have thought they are. 这些正是我们所想象的那样的高压电气设备。(as 代表 installations，在从句中作表语)

7）This is such a thing as we imagine it is. 这就是我们所想象的那样的东西。(as 在从句中作表语，代表 thing)

8）Such sophisticated systems as it are rare.（=Such sophisticated systems as it is are rare.）像这样精密的系统很少见。(as 在从句中作表语，代表 systems)

9）Such substances as tungsten emit electrons when heated in a vacuum. 像钨之类的物质，在真空中受热时便发射电子。（as 代表 substances，在从句中作表语，即定语从句的结构是 tungsten is as，将引导词 as 提前引导定语从句并将 is 省略，when 引导的状语从句中省略了 it is）

10）Jot down such points as the engineer has emphasized. 将那位工程师强调的那几点记下来。（as 在从句中作宾语，代表 points）

11）Such symbols as we often use in electric power engineering are English letters. 我们在电力工程中常用的那些符号是英语字母。（as 在从句中作宾语，代表 symbols）

在这种限制性定语从句中，as 可以和 such 呼应，并且 such 有时不放在所修饰的词前面，而和 as 放在一起。例如：

12）The principles which we shall study apply not only to cases such as have been mentioned but also wave-motion. 我们将研究的原理，不仅适用于已经讨论过的情况，而且也适用于波动。（as 在从句中作主语，其先行词为 cases）

13）We had hoped to give you a chance such as nobody else ever had. 我们希望给你一个其他人没有得到过的机会。（as 在从句中作宾语，其先行词为 chance）

14）The lab is full of facilities such as engineers want to use.（The lab is full of such facilities as engineers want to use.）实验室里堆满了工程师们想用的设备。（as 在从句中作宾语，其先行词为 facilities）

这时的 such as … 中 such 为代词，而在上面的 such as 结构中，such 不放在所修饰的词前面，而和 as 放在一起的情况中，这时 such 为形容词，它放在所修饰的名词之后。

（2）as 用于 such as … 结构中引导限制性定语从句 as 引导的定语从句有时跟 such 紧紧连在一起，而没有名词夹在中间，形成 such as … 结构，这时 as 的先行词就是代词 such 本身，它的意思是"像……这样一类的……"，汉译时往往需要把 such 所代的名词重复译出，译成"……的一种……"。例如：

1）The metric system is such as has a logical link between its units. 公制是各单位之间具有逻辑联系的一种度量衡制。（as 在从句中作主语，其先行词就是 such）

2）Ultrasonic sound is such as is inaudible to human ear. 超声波的声音是人耳听不见的一种声音。（as 在从句中作主语，代表不定代词 such）

在上述 such … as … 的结构中，关系代词 as 的意思是"of the（that）kind，character"。

6.1.2　as 用于 the same … as …（the same as …）结构中引导限制性定语从句

在这种限制性定语从句中，关系代词 as 与主句中的 same 相呼应，构成 the same … as … 或 same as … 的结构。下面分别论述之。

（1）as 用于 the same … as … 结构中引导限制性定语从句在这种限制性定语从句中，关系代词 as 与主句中 the same（same 为形容词）相呼应，构成 the same … as … 构。as 在定语从句中可作主语、宾语或表语。as 从句常常是省略句。汉译时，通常译成"和……同样的……""与……相同的……""和……一样的……"等。例如：

1）In the nuclear power station we used the same generator as is used in the common steam power station. 在核电站里，我们使用的发电机与普通热电站里所用的相同。（as 代表 generator，在从句中作主语）

2）Could you lend me the same devices as was used in your experiment in electric circuits? 你能把你做电路实验时用的那个仪器借给我吗？（as 在从句中作主语，代表 devices）

3）It is said that the moon is made of the same kind of matter as exists on the earth. 据说，月球是由与地球上存在的相同物质组成的。（as 在从句中作主语，代表 matter）

4）This force produces the same effect as is produced by the simultaneous action of the given forces.

这个力产生的效应与给定的几个力同时作用产生的效应相同。（as 在从句中作主语，代表 effect）

5）This conductor is the same length as that one.（＝This conductor is the same length as that one is.）这根导线与那根一样长。（as 在从句中作表语，代表 length）（as 在从句中作表语，代表 length）

6）He is using the same high voltage equipment as I used yesterday. 他在用的高压设备和我昨天用的是同一台。（as 代表 equipment，在从句中作宾语）

7）Although he tried another method，he arrived at the same conclusion as we had. 尽管他采用了另一种方法，但是他得出的结论和我们的一样。（as 代表 conclusion，在从句中作宾语）

8）Ordinary atoms always have the same number of protons as they do of electrons. 普通原子总是具有与电子数相等的质子。（as 在从句中作宾语，代表 number）

9）Have you used the same instrument as I referred to yesterday? 你已经用了我昨天提到的仪器吗？（as 在从句中作介词 to 的宾语，代表 instrument）

需要注意的是，as 引导的这种定语从句也常常构成省略句。例如：

10）This device has the same performances as that one.（＝This device has the same performances as that one has.）这台仪器与那台具有相同的性能。（as 在从句中作宾语，代表 performances）

11）They do the same work as I.（＝ They do the same work as I do）. 他们和我做同样的工作。（as 在从句中作宾语，代表 work）

12）X-rays have the same nature as visible light，but their wavelength is much less. X 射线与可见光具有相同的性质，但 X 射线的波长要短得多。（as 在从句中作宾语，代表 nature）

13）Heat radiation obeys the same laws as those of light. 热辐射与光所遵守的定律相同。（as 在从句中作宾语，代表 laws）

14）It was assumed that the proton would have the same mass as the electron. 人们以为质子与电子具有相同的质量。（as 在从句中作宾语，代表 mass）

15）Galileo believed if two bricks were put together，the two would fall at the same rate as one. 伽利略相信，假如两块砖捆在一起，那么它们落下时的速度和一块砖相同。（在 one 后面省略了谓语 would fall，故 as one 可看作定语从句；宾语从句里面的几个谓语用的都是虚拟语气。）

由上面的例句可见，the same＋名词在主句中作宾语，as 在从句中一般也作宾语。上面例句省略了惯常会省略的词 has、do、is。但是，若主从句的时态不同，动词就不能省略。例如：

16）They are doing the same work as I did last year. 他们正在做着我去年做过的那样的工作。

即使时态相同，若两个动词的形式不同，也以不省略为宜。例如：

17）They do the same work as he does. 他们和他做同样的工作。

（2）as 用于 the same as … 结构中引导限制性定语从句这一结构与上面常见的“the same＋名词＋as …”结构相同，关系代词 as 仍引导定语从句，两者的区别在于从句所修饰的先行词不同，这里 as 所引导的定语从句不是修饰 the same 后的名词，而是修饰 same 本身。此处的 same 不是形容词而是代词。as 从句常常是省略句。汉译时，一般要把 same 所代的名词重复译出，译成“和……的……一样”或“与……的……相同”。例如：

1）The tested conditions are the same as will be encountered in use. 测验条件和使用时要遇到的条件一样。（as 在从句中作主语，其先行词就是 same）

2）The weight of an object in space is not the same as its weight on the surface of the earth. 一个物体在太空中的重量与其在地面上的重量不同。（the same 在主句中作表语，as 在从句中也作表语，其先行词就是 same）

3）Have been read by many people，the handbook returned is no longer the same as it was. 这本手册经许多人阅读后，还来时已非原样。（the same 在主句中作表语，as 在从句中也作表语，其先行词就是 same）

4）The weight of the neutron is about the same as a proton，but it has no electric charge. 中子的重量和质子差不多相等，但中子不带电荷。（the same 在主句中作表语，as 在从句中也作表语，其先行词就是 same）

5）The forces acting on the moon are the same as those acting on other heavenly bodies. 作用在月球上的力与作用在其他天体上的力是一样的。（the same 在主句中作表语，as 在从句中也作表语，其先行词就是 same）

6）When water is stationary，the level of the water in each tube is the same as that in the reservoir. 水平静的时候，每根水管中的水位与储水容器中的水位是相同的。（the same 在主句中作表语，as 在从句中也作表语，其先行词就是 same，从句省略 is）

7）The voltage across each branch is the same as that across every other branch and is equal to the voltage of the battery. 每个支路上的电压与其他各支路上的电压是相同的，都等于电池电压。（the same 在主句中作表语，as 在从句中也作表语，其先行词就是 same）

由上面的例句可知，the same 在主句中作表语，as 在从句中通常也作表语。此外，as 引导的修饰某个名词的限制性定语从句还常与 such 连用，例如：

1）Such slight nonlinearities as are found In vacuum tubes may be neglected in the small-signal case. 像真空管所呈现的这种微弱的非线性，在小信号情况下可以忽略不计。(as 在从句中作主语，代表 nonlinearities)

2）Such meters as we use to measure current are called ammeters. 我们用来测量电流的这类仪表称为电流表。(as 在从句中作宾语，代表 meters)

6.1.3 用于 as many（或 much）… as … 或 as many（或 much）as 结构中引导限制性定语从句

在这种限制性定语从句中，关系代词 as 与主句中的 as many（或 much）相呼应，构成 as many（或 much）… as … 或 as many（或 much）as 的结构。下面分别进行讨论。

（1）用于 as many（或 much）… as … 结构中引导限制性定语从句。在这种结构中，第一个 as 为副词，第二个 as 为关系代词，引出定语从句，as 在从句中一般作主语或宾语，从句中常有省略。many 作形容词时，后跟可数名词复数，much 作形容词时，后跟不可数名词。只有在 much 用作形容词或代词时，as 才可用作关系代词，引导定语从句；汉译时，通常译成“的”即“……（那样多）的……”或“……的……都”以及“凡……的……都”或“如……一般多”。注意“as much (many) … as”结构不同于“as … as”结构。例如：

1）As many books as are on the shelf can be taken as teaching materials. 凡在书架上的书都可以用作教材。(as 在限制性定语从句中作主语)

2）You'd better get ready beforehand as many instruments as are necessary for the experiment. 你最好预先把实验所需要的仪器都准备好。(as 在限制性定语从句中作主语)

3）In processing petroleum as many fractions as are contained in it are to be separated from each other. 炼油过程中，要把油里所含的馏分部分馏出来。(as 在限制性定语从句中作主语)

4）The bullet must therefore possess as much energy as would be required to bore through the plank with a drill. 因此，枪弹一定具有用钻头在木板上钻同样一个孔所需要的能量。(as 在限制性定语从句中作主语)

5）It was not long until transistors were being manufactured in great quantities and in as many different configurations as were vacuum tubes. 不久，人们就生产出大批晶体管，而且型号同真空管一样多。(as 在限制性定语从句中作主语)

6）We should put as much emphasis on preventing AIDS as we do on caring it. 我们应当把预防艾滋病和治疗艾滋病放到同等重要的地位。[as 的先行词是 emphasis，as 在从句中作 do（=put）的宾语，先行词 emphasis 和关系代词被介词短语 on preventing 分隔 (split)]

7）A saturated solution contains as much solid as it can dissolve. 饱和溶液含有它所能溶解的最大量固体。(as 在限制性定语从句中作宾语)

8）Take as much paper as you need, but not more than is necessary. 按需要取纸，但勿拿多。(as 在从句中作宾语，先行词为 much)

(2) 用于 as many（或 much）as 结构中引导限制性定语从句。在这种结构中，many 或 much 不是形容词而是代词。例如：

1）There is plenty of sulfuric acid here. You may take as much as you want. 这里硫酸很多，你可以按需要尽量取用。（as 在从句中作宾语，先行词为 much）

需要注意的是，在 as much as 结构中，much 用作副词时，引出比较状语从句。例如：

2）Most power plants pollute the environment as much as anything else that consumes equivalent amounts of fuel. 大多数发电厂和消耗同等数量燃料的其他设施一样会污染环境。

3）Of course the glass expands too, but not as much the mercury, so the mercury moves up the scale as the temperature rises. 当然，玻璃也会膨胀，但其膨胀的程度不如水银，因此，随着温度的升高，水银就沿着刻度上升。

4）Physical changes depend upon energy changes just as much as do chemical changes. 物理变化正像化学变化一样，在很大程度上取决于能量变化。

6.1.4 as 单独引导限制性定语从句

as 单独可以引导限制性定语从句，例如：

1）He is the only one as can carry out the experiment now. 他是目前唯一能做这个实验的人。（as 在从句中作主语，先行词为 one，在这里 as=that）

2）Tell the story to them as know us. 把这件事告诉认识我们的那些人们。（as 在从句中作主语，先行词为 them，在这里 as=that）

3）We may achieve the equivalent switching function as may be required in a given application. 我们也许能达到相当于规定用途所需的开关功能。（as 在从句中作主语，先行词为 function，在这里 as=that）

6.1.5 省略 be 的限定性定语从句：as ＋ 过去分词

这种结构不是用来修饰整个句子，而是修饰单个词，不用逗号与句子隔开，它与被修饰的词关系较密切，可视为一种省略 be 的限定性定语从句。例如：

1）There are 107 elements, including all matter as found in nature, and the difference between one element and another is in the structure of its atoms. 包括自然界中所发现的一切物质在内，元素有 107 种，元素之间的区别在于其原子的结构。（as 在从句中作主语，先行词为 matter，在这里 as=that）

2）In fact, the materials as used in transistors do not exist in nature but man-made. 事实上，晶体管用的材料在自然界是不存在的，而是人造的。（as 在从句中作主语，先行词为 materials，在这里 as=that）

3）The first law of thermodynamics is really a restatement of the law of conservation of energy as applied to heat. 热力学第一定律实际上是热能的能量守恒定律的另一种说法。（as 在从句中作主语，先行词为 energy，在这里 as=that）

没有省略 be 的限定性定语从句举例如下：

4）Excess-three digits can be added by using a combinational full adder as was described earlier in this section. 其余三位的相加可以用本节早些时候讲到的组合式全加器来进行。（as 在从句中作主语，先行词为 adder）

6.1.6 as 单独引导非限制性定语从句

as 单独引导非限制性定语从句分两种情况，下面分别予以讨论。

（1）as 指的不是主句中的某一词，而是指整个主句所表达的内容。

这时，as 单独引导非限制性定语从句对主句所作的陈述作附加说明。这种从句在整个句中的位置相当于插入语，因此可位于主句之前、之中或之后，并且作为非限制性定语从句用逗号与主句分开。as 在从句中多用作主语或宾语，由于 as 指的是整个主句所表达的内容，所以 as 一定为单数。这种从句一般译成“这……”“如……”“正如……”“像……那样”等。例句：

1）Under certain conditions a body can be charged，as has been said before. 在一定的条件下，物体能够带电，这在前面已经讲到。（as 引导非限制性定语从句，as 代表整个主句所讲的内容并在从句中作主语）

2）As we have stated，resistance in the circuit produces losses which soon stop the oscillations. 如我们所提到的那样，电阻会在电路里产生使振荡很快就会停止的损耗。（as 引导非限制性定语从句，as 代表整个主句所讲的内容并在从句中作宾语）

3）The algorithm to be presented in this paper is significantly more efficient than algorithm in [1] when the total number of nonlinear resistors，inductors and capacitors is significantly less than the total number of linear inductors and capacitors in the circuit，as is often the case in practice. 当遇到实际中的常见情况即电路中所含的非线性电阻、电感和电容的总数大大少于线性电感和电容的总数时，本文所提出的算法比文献【1】所给出的要有效得多。（as 引导非限制性定语从句，as 代表 when 引导的整个从句所讲的内容并在自身所在的从句中作主语）

4）The red blood cells，as (is) shown in Figure. 14，are more numerous than the white blood cells. 如图 14 所示，红细胞比白细胞多。（as 引导非限制性定语从句，as 代表整个主句所讲的内容并在从句中作主语）

5）This material is elastic，as is shown in Figure. 2. 如图 2 所示，这种材料是弹性材料。

6）As this title indicates，this chapter will deal primarily with series of complex numbers. 如标题所示，本章主要讨论复数的级数。

7）As will be seen later。acids and bases play an important role in the functioning of the human organism. （我们）以后将会看到，酸和碱在人体组织的活动中起着十分重要的作用。

8）Are pure and applied sciences totally different activities，having little or no interconnection，as is often implied? 理论科学和应用科学是否就像大家常说的那样，是毫不相

干的两个方面？（as 引导非限制性定语从句，as 代表整个主句所讲的内容并在从句中作主语）

9）Water expands on freezing，as can be deduced from occasional breaks in water pipes in severe winter. 水结冰时会膨胀，这可由严冬季节水管偶然冻裂的现象中推论出来。（as 引导非限制性定语从句，as 代表整个主句所讲的内容并在从句中作主语）

10）As (has been) mentioned above，an object has positive charge when it has lost electrons. 如上所述，物体失去电子时带正电荷。（as 引导非限制性定语从句，as 代表整个主句所讲的内容并在从句中作主语）

11）As is mentioned above，a strengthened field is created where the lines of force in the two magnetic fields are parallel and have the same direction. 如上所述，两个磁场的磁力线相互平行且方向相同时，就会形成一个加强的磁场。（as 引导非限制性定语从句，as 代表整个主句所讲的内容并在从句中作主语）

12）If，as is usual，the electron-tube amplifier operates in a common-cathode configuration or if a transistor amplifier is used in a common-emitter configuration，there is an 180° phase difference between the input and output of the amplifier. 如果像通常那样，电子管放大器以共阴极接法工作，或者如果晶体管放大器以共射极接法使用，那么放大器的输入和输出之间便有一个 180°的相位差。或译：如果电子管放大器习惯地以共阴极接法工作，或者如果晶体管放大器习惯地以共射极接法使用，那么放大器的输入和输出之间便有一个 180°的相位差。（as 引导非限制性定语从句，as 代表整个主句所讲的内容并在从句中作主语）

13）As is known to us，inertia is an absolute quality possessed by all bodies. 正如我们所知，惯性是所有物体都具有的一种绝对属性。（as 引导非限制性定语从句，as 代表整个主句所讲的内容并在从句中作主语）

14）Transistors are small and efficient，as is well known to us. 晶体管体积小、效率高，这是我们大家所熟知的。

15）Friction，as you know，is resistance to motion. 人所共知，摩擦是阻碍运动的。（as 引导非限制性定语从句，as 代表整个主句所讲的内容并在从句中作宾语）

16）As we all know，matter exists in three physical states. 大家都知道，物质以三种物态存在。（as 引导非限制性定语从句，as 代表整个主句所讲的内容并在从句中作宾语）

17）As the name implies，a semiconductor is such a kind of matter as has a resistance between that of a conductor and that of an insulator. 顾名思义，半导体是电阻介于导体和绝缘体之间的一种物质。（第一个 as 引导非限制性定语从句，as 代表整个主句所讲的内容并在从句中作宾语；第二个 as 引导限制性定语从句并在从句中作主语，其先行词为 matter）

这里顺便将表示“顾名思义”的句型总结如下

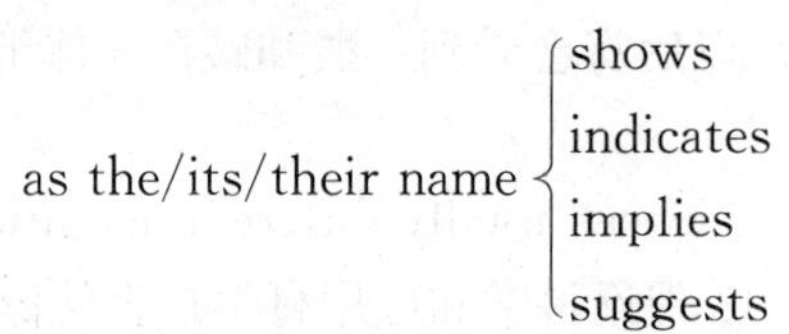

18）The decibel（dB），as its name shows，is just one-tenth of a bel. 顾名思义，分贝（dB）就是0.1dB。在这种非限制性定语从句中，as还可以和such呼应，并且such有时不放在所修饰的词前面，而和as放在一起。例如：

19）The combustion is simply a rapid burning process，such as was discussed in Chapter 3. 燃烧只不过是一种迅速氧化的过程，这一问题在第三章里讨论过。（as引导非限制性定语从句，as代表整个主句所讲的内容并在从句中作主语）

当as在所引导的非限制性定语从句中作主语，谓语为be＋过去分词时，be常可省略，只保留过去分词，其句式为as（省略be）＋过去分词，即代表整个主句所讲的内容且不含情态动词的as从句可以简化。其简化格式为：

as＋be＋动词-ed ⟶ as＋动词-ed

例如：

as has been stated ⟶ as stated（已经述及）

as has been proved ⟶ as proved（业已证明）

as is shown in Fig. 1 ⟶ as shown in Figure. 1（如图1所示）

20）A body at rest will never move without the influence of an outside force，as known from everyday observations. 正如日常观察所看到的，一个静止的物体没有外力影响绝不会移动。（as后省略is，as引导非限制性定语从句，as代表整个主句所讲的内容并在从句中作主语）

21）As shown in the Figure，current varies in the form of a sine wave. 如图所示，电流按正弦波的形状变化。（as后省略is，as引导非限制性定语从句，as代表整个主句所讲的内容并在从句中作主语）

22）As discussed in Section Ⅲ，in many cases，G_n can be obtained without first computing g_n. 正如第三节所述，在很多情况下，不用首先计算g_n就可以得到G_n。（as后省略was或has been，as引导非限制性定语从句，as代表整个主句所讲的内容并在从句中作主语）

23）As previously mentioned，certain applications require that the voltage applied to the load kept fairly constant. 如前所述，某些应用场合要求加到负载上的电压保持相对稳定。（as后省略was或has been，as引导非限制性定语从句，as代表整个主句所讲的内容并在从句中作主语）

24）The English system，as already referred to，uses foot，yard and mile as units of length. 正如已经谈到过的，英制把英尺、码和英里作为长度单位。（as后省略was或has been，as引导非限制性定语从句，as代表整个主句所讲的内容并在从句中作主语）

除了上述的谓语为be＋过去分词，be常可省略，只保留过去分词的情况外，还有在谓语为由介词短语和副词构成的系表结构中，省略be的情况，列陈如下：

as { 过去分词 / 介词短语 / 副词 }

as引导非限制性定语从句，主要作状语和定语，有时可作表语。例如：

25) As pointed out in the previous chapter, forces are not transmitted only by "direct contact."（作状语）正如前一节所指出的。力不仅仅是靠直接的接触来传递的。

26) As with the circular functions, we have the reciprocals of the above three hyperbolic functions.（作状语）如同圆函数的情况那样，我们可以获得上述三种双曲函数的倒数。

这种"as十介词短语"的结构，一般视为as引导的一个方式状语从句的省略句（省略了be)，但是也有人看作一个简单的状语。一般可译为"如""象"等。

27) The flow of current in semiconductor can be formed by a flow of negative charges, as in the case of conductors.（作状语）半导体中电流的流动，像导体一样是通过负电荷的流动面形成的。

28) Turns are made in a helicopter, as in an airplane, by banking.（作状语）直升机像飞机一样，是用倾斜的方法来转弯的。

29) These data are plotted on log—log paper, as shown in Fig. 2-6.（作状语）将这些数据如图2-6所示的那样画在双对数纸上。

30) The method as presented here will not work directly for such circuits as the Meacham bridge.（作限制性定语）这里介绍的方法不能直接用于像米契阿姆电桥这样的电路。

31) A sequence having a finite limit as described above is said to be convergent.（作限制性定语）像上面所讲的这种具有有限极限的数列被说成是收敛的。

32) The basic organization of such a system is as shown in Fig. 6.（作表语从句）这种系统的基本结构如图6所示。

但是，带有情态动词时则不能省略，这时一般表示表格、曲线、原理图。例如；

33) As can be seen from Table 9, the world production of transformers in 2006 was not far short of 2 million. 如表9所示，2006年世界变压器产量距2百万台不远。

34) As can be seen on page 8, this result is agreement with that obtained by them. 如第8页所示，该结果与他们所得结果一致。

还有一种省略情况是，当as不作主语时还可以省略主语，其格式为as（省略主语）＋过去式。例如：

35) As (I) stated in my last letter to you, I have begun to carry out the terms of the contract. 正如我上次给你的信中所述，我已开始执行合同的条款。

在这种非限制性定语从句中，有些省略或没省略的形式已构成习惯用法，科技书刊中常见的有：

as mentioned above　如上所述

as is well known to all　众所周知

as shown in the Figure　如图所示

as is often said　正如通常所说

as seen from the table　正如从表中看出

as explained before　如前面所解释的

as already discussed 正如已讨论过的

as is often the case 通常如此，通常就是这样

as has been pointed 正如所指出的那样

as has been said before 如上所述

as often happens 如同经常发生的那样

as might be expected 正如所料

这些 as 引导的从句一般位于先行词之后，但也可以位于先行词之前。其中还有一种从句中常见的动词为 call、know、refer to，从句意为“所说的”“所称的”“所谓的”，例如：

36）This revision resulted in the creation of “high-speed mechanics” or，as it is called，“relativistic mechanics”．这一修正导致了“高速力学”的形成，也就是现在人们所说的“相对论力学”。

37）These flaws，or “bugs”，as they are often called，must be found out and corrected. 这些毛病，也就是人们经常所说的“虫子”，必须要找出来加以纠正。

38）Unlike “snail mail”，as E-mailers derisively refer to it，a response can shoot back within hours. 不同于电子邮件发送者戏称的“蜗牛邮件”，（此电子邮件）能在几小时内收到回音。

（2）as 指的不是整个主句所表达的内容，而是指主句中的某一部分。

1）The electrons，as one kind of the small particles is called，revolve around the nucleus. 一种称为电子的小粒子绕原子核绕转。

2）The bare infinitive，as the infinitive without “to” is termed，is used with such verbs as “make” “let” and some others. 不带“to”的动词不定式又称“bare infinitive”，与“make”“let”和其他一些动词连用。

3）TIROS 1 was completely successful，as were other satellites of the same type. 像其他同类的卫星一样，电视红外线观察卫星一号也是完全成功的。

4）As was true of many Russian discoveries，this one was ignored by the West until one of their own scientists had repeated it. 这次发明如同俄国人以往的发明一样被忽视，直至西方的一位科学家也做出了同样的发明为止。

5）Carbon monoxide is poisonous，as are the gases we are going to discuss in the next chapter. 像下一章要讨论的气体一样，一氧化碳也是有毒的。

which 引导的非限制性定语从句说明整个主句。以 which 引导的定语从句，除了可以说明其先行词外，还可以用来对主句所述的事实或现象加以总结概括，补充说明或承上启下，其前都有逗号分开。所以翻译时，主句与从句分译，通常把“which”译成“这……”，但有时也可译成“从而……”“因而……”。例如：

1）Like charges repel，but opposite charges attract，which is one of the fundamental laws of electricity. 同性电荷相斥，异性电荷相吸，这就是电学的一个基本定律。

2）Energy can neither be created nor destroyed，which is the universally accepted law. 能量既不能创造也不能被消灭，这是一条普遍公认的规律。

3) All forces occur in pairs, which may conveniently be spoken of as action and reaction. 所有的力量都是成对出现的，这可以很方便地称为作用和反作用。

4) The properties of alloys are much better than those of pure metals, which makes them find wide application in industry. 合金的性能比纯金属好得多，这就使其在工业上得到广泛的运用。

5) The sun heats the earth, which makes it possible for plants to grow. 太阳晒热大地，这就使植物有可能生长。

6) To find the pressure we divide the force by the area on which it press, which gives us the force per unit area. 为了求得压强，则把力的大小除它所作用的面积，从而得出单位面积上的压力。

6.2 句子成分的分隔

英语中的所谓分隔结构（split structure）是指按照正常的语序和句法结构，本应紧密连续相邻的两个句子成分被另一句子成分所隔离，或者是词语的习惯搭配关系被拆分，从而使这两个部分被分隔开来。这种固定性原则的破坏使得两个相关的语言成分不再紧挨在一起，所以又称其为非连续成分（discontinuous constituents）。这种变异结构表现为主语与谓语的分隔、连系动词与表语的分隔、动词与宾语的分隔、宾语与宾语补足语的分隔、定语与中心词的分隔、同位语与中心词的分隔、谓语的分隔、介词与介词宾语的分隔、状语与其所修饰动词的分隔、固定词组内部的分隔等。下面进行简要讨论。

6.2.1 主谓分隔

这种分隔是为了保持句子平衡，在主语和谓语之间插入定语（多为介词短语、分词短语、不定式短语、关系从句）、状语（常为副词、介词短语或状语从句）、同位语、插入语等。

分隔主语与谓语的主要手段有 which、that 引导的定语从句；that 引导的同位语从句；if 引导的条件状语从句；破折号引出的句子；各种短语，如介词短语和分词短语等。

1. 主语谓语的分隔

(1) The longest electrical sparks that can commonly be generated in the laboratory measure 1 to 3 meters, with the maximum being 10 to 20m. 实验室里所产生的电火花的长度通常是1～3m，最长的可达10～20m。

(2) The force that pushes you toward the front of the bus when it stops is the inertia of your body. 汽车停止时，把你朝车前推动的力就是你身体的惯性。

(3) Observation plays an important part in high voltage engineering and testing, but experiment, which consists in modifying circumstance or events with a view to making more valuable observations, plays a more important part. 观察在高电压工程和测试中起着重要的作用，但实验起的作用更为重要，因为实验可以不断地改变各种环境条件或过程，以便做出更有价值的观察。

（4）The desire to prevent further OHLs from having a detrimental effect on landscape is consistent with the plan's wider emphasis on landscape and environmental protection. 进一步防止架空线路对景观造成有害影响的愿望与该计划广泛强调景观与环境保护是一致的。

（5）And this is exactly what a radar set，the equipment that sends out wireless signal and receives their echoes，based on. 而这正是雷达机—发射无线电信号并接受其回收波的设备——工作的基础。

（6）Ninety-nine people out of a hundred，if they were asked who presented the theory of relativity，would answer Albert Einstein. 如果要问提出相对论的是谁，100 人中有 99 个人会回答：是阿尔贝特·爱因斯坦。

（7）The increase in the number and variety of uses for electricity throughout the world has produced a wide range of standards of power quality that must be satisfied. 随着全世界用电量和用电类别的增加，已产生了各类必须满足的电能质量标准要求。

2. 复合谓语（谓语多个组成部分）的分隔

这种分隔通常是在复合谓语之间插入含有状语意义的状语、介词短语或状语从句，使复合谓语本身分隔开来。

（1）Environmental aspects can be significantly improved in many areas of the electrical power Engineering. 在电力工程的很多领域，环境方面都能够得到大大改善。

（2）We have never seen such a phenomenon in a laboratory. 我们从来没在实验室见里过这种现象。

（3）Regional Standardization Groups have also been formed to rationalize standardization within geographical areas . 建立多个地区标准化小组以实现区域内标准化的合理性。

（4）Most of plastics do not readily conduct heat or electricity. 大多数塑料都不易传热或导电。

（5）Clearly，deliberate release of SF_6 can now no longer be allowed. 显然，现在已不允许人为排放六氟化硫。

（6）All working parts are usually oiled so that friction may be greatly reduced. 全部活动部件通常都拥有润滑，以大大减少摩擦力。

6.2.2 动宾分隔

动宾分隔是指动词以及非谓语动词，包括动名词与其宾语的分隔，产生这种现象的主要原因在于作状语用的介词短语或状语从句直接置于动词、非谓语动词或动名词之后，用于修饰这些词，从而将它们与其支配的宾语分隔开来。例如：

（1）We cannot reach definite conclusions without considering in more detail the properties of the insulator. 如果不为更详细地研究这种绝缘子的特性，我们就不能得出明确的结论。

（2）A vacuum circuit-breaker has in greater or lesser degree such qualities as (a) safety；(b) reliability；(c) compactness；(d) low contact wear and (e) low maintenance. 真

空断路器在不同程度上具有这样一些性质：(a) 安全性好；(b) 性能可靠；(c) 体积小；(d) 耐磨损；(e) 维护费用低廉。

(3) The lack of air, the tremendous underwater pressure, the darkness, and the cold have all combined to prevent an oceanologist from penetrating, for a great length of time, the deep ocean to observe it first hand. 水下缺乏空气，巨大的水下压力，又暗又冷，这一切使海洋学家不能长时间地潜入深海直接进行观察。

(4) As helicopters get bigger and more efficient, more and more work will be found for them to do because they combine in one vehicle the lifting power of a crane with the carrying power of a lorry. 随着直升机型的增大和运载效率的提高，越来越多的工作可以让直升机来完成，因为直升机把起重机的起吊能力和载重卡车的运载能力结合起来体现在一种运载工具上。

(5) The voltage is great enough for an ion to acquire between collisions sufficient energy. 电压大得足以使离子在两次碰撞之间获得充分的能量。

6.2.3　宾语与宾语补足语的分隔

这种分隔一般是宾语后出现一个定语（后置定语），如分词短语、介词短语或从句等造成的，宾语在句首时，也会出现这种分隔。例如：

(1) We call materials which allow electrons to pass easily conductors. 我们将电子容易通过的那些材料称为导体。

(2) Any influence that causes a motion of matter we call a force. 使物体产生运动的影响称之为力。(宾语置于句首造成宾语 influence 及其补足语 a force 之间分隔)

(3) This substance we supposed to be a compound. 我们推测这种物质是一种化合物。(宾语置于句首造成宾语 substance 及其补足语 to be a compound 之间分隔)

6.2.4　定语（或定语从句）和其所修饰的词的分隔

1. 多个后置定语（从句）的分隔

当被修饰词后置有两个或两个以上的定语或定语从句时，其中一个或一个以上的定语只得和它所修饰的词分隔开，形成定语或多个定语连续后置的现，例如：

1) When ohm is too small a unit to show the resistance used, we may use the one a thousand times larger known as kiloohm. 在用欧姆作为表示所用电阻的单位过小时，我们可以用一个大 1000 倍的称为千欧的单位。

2) The element of an electronic computer that does the actual work of computation is called arithmetic element. 电子计算机进行实际计算工作的单元称为计算单元。

3) The widespread use of power electronics to control large items of power plant, and to replace conventional high voltage devices is revolutionizing the control of high-voltage networks. 广泛应用电力电子技术来控制发电厂的大型设备，取代传统的高压设备正在使高压网络的控制发生一场革命。

4) The element within a vacuum which is the source of electrons is known as the cath-

ode. 真空管中作为电子来源的那个元件称为阴极。

在被修饰词与其后置定语之间插入介词短语时，也会形成被修饰词与其后置定语或定语从句的分隔。例如：

5）In earlier years there was no restrictions，apart from cost，on the deliberate release of SF_6 to the atmosphere. 早些年，除了成本外，对于人为排放六氟化硫到大气没有明令限制。

2. 谓语分隔主语及其后置长定语（或定语从句或同位语从句）

若主句的结构比较简单而主句主语的后置定语较长，就会形成主谓两部分头重脚轻的现象，因此，为了使主谓结构紧凑，则需要将主语的后置定语置于谓语之后，从而使得句式成为定语（或定语从句）和它所修饰的主语被谓语（一般是被动语态或不及物动词）分隔的句式。但是，如果分隔后会引起句子产生歧义，则不可进行分隔，例如：

1）Attempts are being made of burning coal underground for getting gas. 人们正在做使煤在地下燃烧以获得煤气的尝试。（of burning…作 attempts 的定语）

2）Every effort is made to minimize current level or keep the power factors as close to unity as possible. 尽一切努力减小电流或使功率因数尽可能接近 1。（to minimize … possible 作 effort 的定语）

3）A new way has been shown of making use of this device to measure the amplitude of a direct voltage. 已有人提出了一种利用这种装置测量直流电压幅度的新方法。

4）Sealing problems have been widely encountered on silicon carbide type arresters. 使用碳化硅型避雷器经常会遇到密封问题。

5）A good understanding has been obtained of evaporation processes which are carried out at ordinary pressure. 对于常压下进行的蒸发过程已有了很好的理解。（of evaporation processes which … 作 understanding 的定语）

6）A list has been drawn up of the elements we have learned so far. 到目前为止，我们已知的元素表已经列出来了。（of the … so far 作 list 的定语）

7）A 16-candle-power lamp has a resistance when hot of 220 ohms. 炽热时一个 16 支烛光发光强度的电灯具有 220Ω 的电阻。（of 220 ohms 是 resistance 的定语）

8）Thus，a secondary current flows which increases when the primary current decreases. 在一次电流减小时，二次电流就增大，从而产生二次电流。

9）For permanent magnets，a material is desired which has a high residual magnetism. 最好是利用剩磁大的材料作永久磁铁。

10）No transformer exists which can not be made reasonably safe. 没什么变压器不可以制造得安全可靠。

11）For a long time special observations were made of these gases under different pressure and temperature. 人们对这些气体在不同压力和温度下的变化做了长期的专门观察。

12）Very wonderful changes in matter take place before our eyes to which we pay little attention. 我们很少注意到那些经常在我们眼前发生的物质的奇异变化。

13）A lecture was given by a professor of advances in high voltage engineering. 一位教授做了一次关于高电压工程进展的报告。

14）By using the mechanism called gear box，different speeds may be obtained suitable for different conditions of work. 使用称为齿轮箱的机械装置可以获得适合各种工作条件的不同速度。（suitable for … of work 作 speeds 的定语）

15）Scientists believe that the day will come when we can speak to or even talk with a computer in any language we like to use. 科学家们相信总有一天我们能用随便任何一种语言来对计算机说话，甚至和它交谈。

16）Many experiments can be done on the moon that are difficult to do on the earth. 许多在地球上难以完成的实验，在月球上却可以完成。（that 引导的定语从句作 experiments 的定语）

17）The nuclei of heavy elements are unstable，i. e. they break up，and that is why none occur in nature that are heavier than uranium. 重元素的原子核是不稳定的，即它们会分裂，这就是自然界没有任何比铀重的元素的原因。（定语从句 that are heavier than uranium 作 none 的定语）

18）But from time to time，out of this welter of uncoordinated activity，results emerge that we recognize as having an important bearing on whether some theory should be believed or not. 但有时从这种混乱而不协调的活动中，能出现一些我们认为对某种理论的可能性有极大关系的结果。（that … not 为 results 的定语从句）

19）This anyone can see who has tried to run on a highly polished floor or on ice. 这一点对于曾试图在极光滑的地板或冰上奔跑的人来说都很清楚。

20）In industry certain mixtures of metals or alloy are known to have been greatly developed that will resist the action of acids. 我们知道，在工业上已经研制出了能耐酸的某些金属混合物，即合金。

21）Here serious problems are encountered which have only been partly solved. 这里遇到一些尚未完全解决的重大问题。

3. 状语分隔先行词与其定语从句

当定语从句的先行词位于谓语动词之后时，修饰该动词的状语（或状语从句）可以插在先行词与定语从句之间从而使它们分隔开来。

（1）There are three quantities in a circuit which they need to measure. 电路中有三个量需要测量。

（2）Yet there exist complex computations in electrical engineering which people are unable to make. 然而，到目前为止在电气工程方面还存在许多人们无能为力的复杂计算。

（3）The molecules exert forces upon one another，which depend upon the distance between them. 分子相互之间都存在着力的作用，该力的大小取决于分子之间的距离。

（4）The Engineers heard the story from his own lips of how he had been hard at his design of these substations. 这些工程师听他亲口谈他是如何下功夫设计那些变电站的。

（5）A gas completely fills the volume of any container，however large，in which it may be placed. 不管盛放气体的容器体积有多大，气体都能把它完全充满。

(6) We have made a number of creative advances in theoretical research and applied sciences which are up to advanced world levels. 我们在理论研究与应用科学方面，获得了不少世界先进水平的创造性成就。

4. 定语分隔先行词与其定语从句

(1) These limitations do not apply to operations like boring which are performed with single point cutters. 这些限制不适用于像钻孔那样的操作，因为这些操作是使用单刃刀具进行的。

(2) Atomic cells, small and light, which have the advantage —— durability, can operate without being recharged for decades. 原子电池又小又轻，经久耐用，能够应用几十年而无须充电。

3. 和 4. 说明，当从句所修饰的词有介词短语、分词短语、形容词所表示的定语或状语时，从句往往放在这些成分之后，形成与它所修饰的词的分隔情况。

5. 谓语分隔同位语从句与其同位词

同位语从句一般都紧接在与其同位语的名词即先行词之后，但当先行词作主句的主语、而同位语较长、谓语很短（通常为被动语态或不及物动词）时，同位语从句常被置于谓语之后。这样就形成了同位语从句与其先行词之间的分隔。这种结构类似于谓语分隔主语和它的长定语。例如：

(1) In the first half of the last century the discovery was made that a magnet could be used to get an electric current. 上个世纪的前半叶，人们发现可用磁体产生电流。

(2) The assumption was made that the earth is a great magnet. 有人提出了地球是一个巨大的磁体的假设。

(3) As soon as man succeeded in launching satellites into orbit about earth, the question naturally arose: how would man fare in space? 人们成功地将卫星送入绕地球运行的轨道之后，问题自然而然地产生了：人们如何到太空去？

(4) The theory is of great importance that the hotter a body is, the more energy it radiates. 物体的温度越高，放射的能量就越多，这一理论非常重要。

6.2.5 某些词与所要求介词的分隔

动词、名词、形容词以及某些介词短语由于语法结构的需要，常常与其所要求的介词产生分隔现象。例如：

(1) One of the problems of modern power systems is the transportation of electrical energy from the place at which it is available to the place at which it can be used most effectively. 现代电力系统的问题之一，就是如何将电能从开发地输送到可以最有效地利用它的地方。

(2) This current, which is of the order of 100A, is associated to charges of around 10C and constitutes a direct transfer of charge from cloud to ground. 这一电流的数量级为100A，大约有 10C 的电荷，形成了电荷从云层到地面的直接传输。

(3) This book should therefore be useful in an advanced undergraduate course in power systems, to engineers interested in the faults of power systems. 因此，在电力系统方面的现代大学课程中，这本书想必对关心电力系统故障的工程师们是有用处的。

(4) Faced with such a situation it is obvious that we should search as widely as possible and with every available means for renewable energy sources that seem to be the least costly. 面对这样的形势，显然，我们必须尽可能广泛地并采取一切可能的手段来寻找那些费用不高的可再生能源。

(5) Thus far our analysis has been limited for the most part to dc circuits. 迄今为止，电路分析主要局限于直流电路。

(6) The electric resistance of a wire is the ratio of the potential difference between its two ends to the current in the wire. 导线的电阻等于该导线两端之间的电位差与导线中电流的比值。

(7) There will be atomic jet-plane and we will be able to fly right up through the atmosphere, through the stratosphere, and away out into space, to the Moon, and Venus and other planets . 将来会出现原子喷气飞机，我们能够径直穿过大气层、同温层，飞入太空，飞向月球、金星和其他星球。

(8) The propagation of a transverse wave through an anisotropic material such as a crystalline solid can be complicated by the response of the material to the impressed vibration. 横波通过晶体这样的各向异性材料的传播，假如一种结晶固体会由于材料对外加振动的反应而变得复杂起来。

(9) The development of the abrasive wheel and the grinding machine from being a simple accessory to the toolroom lathe, to a machine tool giving the highest precision, is due to many interacting factors. 砂轮和磨床从工具车床简单的辅助设备发展到精密度极高的机床，这是由许多相互作用的因素造成的。

(10) Work on the transmission of color TV signals over communication links (from camera to receiver), that is, on color TV systems, has been accompanied by studies into factors that control the quality of color reproduction in the receiver. 通过无线电微波接力线路（从照相机到接收机）传播彩色电视信号的工作，即彩色电视系统的工作，涉及接收机里彩色重现质量控制系统的研究。

(11) The attraction of the earth for other bodies is called the force of gravity. 地球对于其他物体的吸引力称为地心引力。

(12) The influence of temperature on the conductivity of metals is slight. 温度对金属导电性的影响不大。

(13) The steam turbine makes use of the direct impact of the molecules of steam on a large number of little curved blades attached to the circumference of a disk mounted on a long slender shaft. 蒸汽涡轮机利用蒸汽分子对数目众多的稍微弯曲的叶片产生直接冲击力，这些叶片装在一条细长轴的固定叶轮的圆周上。

(14) The introduction of hand and, later, machine welding to shipbuilding process in

the 1930's made increased use of sub-assembly practice desirable from both an economic and technological viewpoint. 20 世纪 30 年代先后把手工焊接和机器焊接技术引进到造船工序中，使合乎经济和技术观点要求的组件装配法应用越来越多。（desirable … viewpoint 为 practice 的定语）

（15）The planned proximity of housing and a primary school to the OHL generated particular local concerns. 所规划的住宅和小学临近架空线路，这在当地引起了特别的关注。

6.2.6 相同句子成分之间的分隔

若干个并列的相同（或称同等）句子成分（例如主语、谓语、宾语、定语或介词短语）之间由于某些句子成分（例如后置定语）的插入而会使它们形成分隔现象。例如：

（1）Experiments show that there is a definite relationship between the electrical pressure that makes a current flow, the rate at which the electricity flows and the resistance of the object or objects through which the current passes. 实验表明，使电流流动的电压、电流流动的速率与电流所通过的物体的电阻这三者之间有确定的关系。

（2）Solar energy seems to offer more hope than any other source of energy, particularly since those areas most in need of water lie rather close to the equator and have a relatively clear atmosphere. 太阳能似乎比任何其他能源更有希望，尤其是因为这些迫切需要水的地区离赤道很近，而且有较清洁的大气。

（3）The atoms differ from one another in the number of particles which they contain and in the arrangement of these particles. 各种原子在含有的粒子数量及这些粒子的排列情况等方面是彼此不同的。

6.3 "of ＋名词"引出不定式的逻辑主语

我们知道，不定式的逻辑主语一般由 for 引导，所形成的不定式短语结构为"for＋逻辑主语（名词或代词）＋ 动词不定式"，构成不定式的复合结构。但是，在科技英语中的某些抽象名词之后作定语的不定式之前，通常可用介词 of 引出不定式的逻辑主语，即用结构"of ＋ 逻辑主语（名词或代词）＋ 动词不定式"形成科技英语中一种特殊的但却极为常见的不定式复合结构。这些抽象名词中最为常见的是 ability（能力）、capability（能力）、其次为 tendency（趋势）；其他还有 failure（未能、无法）、intention（意图）、reluctance（不愿、反对）、capacity（能力）、desire（愿望）、right（权利）等。这种句型可以表示为"the ability/tendency, etc. ＋ of A to do B"。

例如：

（1）The ability of radar to measure velocity without contact with the moving objects is of considerable interest to industry. 雷达不接触运动物体而测量其速度的效能，对工业也有很大意义。

（2）The ability of a metal to be drawn into a wire is known as ductility. 金属被拉成金

属丝的性能称为展性。

（3）Energy is defined as the ability of a body to do work. 能量被定义为物体做功的能力。

（4）Hardness is the ability of a metal to resist being permanently deformed. 硬度是金属防止本身永久变形的性能。

（5）The next approach to the problem is that of improving the capacity of the river to carrying large volumes of water without overflowing its banks. 解决这一问题的另一方法是改善河流的过水能力，使河流能宣泄更多的水量而两岸并不漫溢。

（6）The ability of materials to resist currents is known as resistance. 材料阻止电流流动的能力称为电阻。

（7）The ability of a material to absorb energy from the sun and re-radiate only a small proportion of this heat is dependent on the nature and texture of the material. 一种材料吸收太阳能和再辐射出一小部分热量的能力，取决于该材料的性质和结构。

（8）Apart from the enormous outflow of funds from the oil-consuming countries, a more basic concern is the capability of the major exporters of crude oil to achieve a continuing production target roughly doubling output every ten years to meet the required oil demand if present trends continue. 除用油国的资金大量外流之外，一个更令人担心的重要问题是：按目前这种趋势发展，主要的原油出口国能否每十年大致将产量翻一番，以满足对石油的需要。(现在分词短语 roughly doubling output every ten years to meet the required oil demand if present trends continue 作 target 的定语，其中不定式短语 to meet the required oil demand 作目的状语)

（9）The apparent intention of the gas to shrink to zero volume at absolute zero is naturally never fulfilled. 气体在绝对零度时体积缩小到零是永远不可能实现的。

（10）The desire of man to control nature's force successfully has been the catalyst for progress throughout history. 人类想要成功地控制大自然的各种力的愿望，一直是推动整个历史进程的催化剂。

（11）The greater the tendency of an object to resist a change of velocity, the greater the inertia. 物体阻止速度变化的趋势越大，其惯性就越大。

（12）It may be hard for us today to understand the reluctance of our ancestors to shed their intuitive beliefs in the face of experimental evidence to the contrary. 今天我们可能难以理解，我们的祖先面对着与其直觉的信念截然相反的实验证据而不愿抛弃这些信念。

（13）There are wide differences in the ability of various substances to conduct heat. 各种物质的导热能力差异很大。

（14）Gravity is the tendency of all objects to attract and be attracted by each other. 万有引力就是所有物体相互吸引的趋势。

（15）A rise in temperature increases the ability of air to absorb water vapor. 温度的上升提高了空气吸收水蒸气的能力。

（16）A few factors affect the ability of a capacitor to store charge. 有好几个因素影响

电容器储存电荷的能力。

（17）The interference of light waves puts limit on the ability of any telescope to resolve the details of an object. 光波的干涉限制了望远镜分辨物体细节的能力。

（18）Elasticity may be defined as the tendency of a body to return to its original state after being deformed. 弹性可以被定义为物体在形变后恢复其原状的趋势。

（19）The deviations from the expected periodicity in Mendeleev's list were due to failure of contemporary chemistry to have discovered some of the elements existing in nature. 与门捷列夫周期表中所预料的周期性有出入的原因是当时化学界未能发现存在于自然界的某些元素。

此外，在个别情况下，以上结构中的 of 也可以用 for 来代替，特别多见于 tendency 之后，特别是当不定式的逻辑主语是人称代词 it 时。例如：

（20）Since the resultant force on the needless is zero，there is no tendency for it to move north or south as a whole. 由于指针上的合力为零，所以作为一个整体来说，该指针就不会趋于朝南或朝北运动。

（21）The tendency for an object to fall down proves that there is the force of gravity acting on it. 物体下落的趋势，证明有重力作用在它上面。

6.4 限定代词 one（ones）的用法

限定代词 one（ones）的语法功能是表示为避免重复而对前面已提及的可数名词进行替代，并且只能对已提及的可数名词作替代，它有单、复数形式，可有冠词、代词及形容词等修饰。one（ones）在句中作主语、宾语和表语。下面介绍限定代词 one（ones）的用法。

6.4.1 one（那个）用来代替句中已经出现过的某一可数名词单数

one 这样用时具有泛指性质（即 a/an），可指人或物。例如：

（1）He had not been bred an engineer，and had no inclination to become one. 他未曾受过工程师的教育，也不想成为工程师。（one＝an engineer）

（2）We haven't got an ammeter. We'll have to buy one. 我们没有电流表，得买一个。（one＝an ammeter）

（3）There are plenty of devices. Take one. 有很多的仪表。拿一台吧。（one＝a device）

6.4.2 one 在使用上的特点

（1）one 前面可带形容词或指示代词等所表示的定语说明。例如：

1）Their life in the substation was an interesting one. 他们在变电站过着有趣的生活。（one＝ life）

2）If you want a voltmeter. I'll give you a good one. 如果你想要一台电压表，我给你

一个好的。(one＝a voltmeter)

3) In short, machines are used because they can develop a large force using a rather small one or turn forces to advantage. 简单地说，人们使用机器是因为机器能利用一个相当小的力产生一个巨大的力，或者能有效地利用各种力。(one＝a force)

4) Among many problems of nonlinear wave propagation, the steady state one has a special interest. 在很多有关非线性波传播的问题中，稳态非线性波传播特别有趣味。

5) This device broke down. Would you bring me a new one, please? 这台仪器坏了，请拿一台新的给我。(one＝a device)

6) She had been to some substation, but what one she could not remember. 她曾到过某变电站，但她记不清是什么样的变电站。(one＝a substation)

注意，如果形容词前面有 a 或 an 时，仍保留 a 或 an。但是如果 one 前面的形容词带定冠词 the 时，one 可以省去。这种形容词一般是最高级形容词，或 next 和 last。例如：

7) That large transformer belongs to them, ours is the small (one). 那台大变压器是他们的，我们的是小的。

8) Of all these installations, this is the most expensive (one). 在所有这些设备中，这台是最贵的。

9) This book is the most difficult (one) I've ever read. 这本书是我读过的书里头最难的一本。

10) Let's finish this experiment so we can go on to the next (one). 咱们做完这个实验，好继续做下一个。

11) Is this experiment as important as the last one? 这个实验是否像上次的那样重要？(one ＝an experiment)

12) I'm looking for an outlet for my measuring instrument. Is there one near the window? 我在找测量仪器的插座，窗口附近有吗？(one＝an outlet)

(2) one 后面可以跟介词短语、分词短语或定语从句作后置定语。例如：

1) An elastic body is one that tends to return to its original shape and size when the deforming force is removed. 弹性体，就是移去变性力之后能恢复原来形状和大小的物体。(one ＝a body)

2) A multi-purpose machine tool is one that is capable of doing a number of different types of operation. 万能机床是一种进行多种不同类型操作的机床。(one ＝a machine tool)

3) The normal state for a body to be in is one of rest or of uniform motion in a straight line. 物体所处的正常状态，就是静止状态或作匀速直线运动的状态。(one ＝a state，其后有介词短语"of rest or of …"作为定语)

4) Once the heavier body is moving at a given speed, it is harder to stop than the lighter one moving at the same speed. 较重的物体一旦以一定的速度运动，要停下来就比以相同的速度而运动的较轻的物体来得困难。(one ＝ a body，其前面有形容词 lighter 作定语，其后有分词短语 moving at … 作定语)

6.4.3 ones（那些）代替可数名词的复数

one 的复数形式 ones 代替可数名词的复数，这样用时也具有泛指性质，前面一般带定语。例如：

（1）Are synthetic materials as good as natural ones? 合成材料和天然材料一样好吗？（ones＝materials）

（2）The attractions between gas molecules are slight ones. 气体分子之间的引力很小。（ones＝ attractions）

（3）These transformers are better than those old-fashioned ones. 这些变压器比那些老式的变压器要好。（ones＝ transformers）

（4）For nearly two thousand years it was believed that all heavy objects fell faster than light ones. This is a mistaken idea. 将近两千年来，人们认为一切重物比轻物落得快些。这是一个错误的概念。（ones＝objects，其前有形容词"light"作为定语，译成"轻物"）

6.4.4 ones 在使用上的特点：ones 前面紧接有 these 或 those 时，ones 可以省去。

例如：

I like these devices better than those（ones）. 我喜欢（用）这些仪器，胜过那些仪器。

但是，one 用在 this 和 that 之后，one 可以省略也可以不省略。例如：

I like this device better than that one. 或 I like this device better than that.

Few persons are so careful as this one（＝a person）. 或 Few persons are so careful as this.

很少有人像这个人那么仔细。

跟 one 一样，ones 前面有最高级形容词时，ones 可以省去。

6.4.5 one 和 ones 在使用上的共同特点

（1）one（ones）前可带定冠词 the，one（ones）这样用时，特指某（些）人或某（些）物，后面要跟修饰语。修饰语可以是从句、介词短语，或分词短语等。例如：

1）This device is not so good as the one we purchased last week. 这台仪器没有我们上周购买的那台好。（定语从句修饰 the one）

2）Three measuring instruments are on the working table. Please pass me the one on the right. 工桌台上有三台测量仪器，请把右边的那台递给我。（介词短语修饰 the one）

3）Hand me my measuring instrument，please. It is the one hanging on the third hook. 请把我的测试仪表拿给我。挂在第三个钩上的那一台。（分词短语修饰 the one）

4）He is the one to be trusted. 他是可信赖的人。（动词不定式修饰 the one）

5）Of all his books，I like the ones that were concerned with power system stability. 在所有他写的书中，我喜欢有关电力系统稳定的书。（定语从句修饰 the ones）

如果从上下文看可以看出修饰语是什么时，可以省去修饰语，但要保存定冠词 the。例如：

Someone borrowed my ammeter. I think Joe was the one (who borrowed it). 某人借了我的电流表。我想是乔借的。

(2) 所有格形容词 my、your、our、her 和 their 被 mine、yours、ours、hers 和 theirs（his 和 its 没有变化）所代替时，不用 one 和 ones。例如：

This is his AC voltmeter，not mine. 这是他的交流电压表，不是我的。（不能说 not my one）

在这种形式中，单复数不分。例如：

I brought my book，but I forgot yours.（单数）

I brought my books，but I forgot yours.（复数）

我带来我的书，但是忘带你的书了。

own 单独使用时，前面总是带所有格形容词。例如：

Thank you for offering me your instrument，but I prefer to use my own. 谢谢你把你的仪表给我，但是我倒愿意用我自己的。（不能说 my own one）

whose 之后也不跟 one 或 ones。例如：

That is a new book on high voltage insulation. Whose is it? 那是本关于高电压绝缘的新书。是谁的（书）？

用所有符号“，”构成的所有格之后也不用 one 或 ones。例如：

Bill's book is newer than John's. 比尔的书比约翰的书新。

(3) either、former、latter、neither 和 which 可以后接 one 或 ones，也可以不接。例如：

1) Some of your answers were correct，but I do not remember which. 或 Some of your answers were correct，but I do not remember which ones. 你有些回答是正确的，但我记不起是哪些。

2) Neither one is satisfactory. 或 Neither is satisfactory. 两个都不令人满意。

3) Of these alternatives I prefer the former one. 或 Of these alternatives I prefer the former. 在这两者之间我愿选择前者。

(4) 作为单数的 another 和 other，可以单独用，也可以后接 one；作为复数可用 others 或 other ones。例如：

1) If that device does not work，try another. 或 If that device does not work，try another one. 如果那台仪器不能用，试试另外一台。

2) You can take this book. I will keep the other. 或　You can take this book. I will keep the other one. 你可以拿这一本书。我保留另外一本。

3) You can take this book. I will keep the others. 或　You can take this book. I will keep the other ones. 你可以拿这一本书。我保留另外几本书。

4) As soon as you learn these methods，I will show you some others. 或 As soon as you learn these methods，I will show you some other ones. 你一学会这些方法，我就教你

一些其他的。

（5）基数词如 one、two、three 等和序数词如 first、second 等，通常不和 one 和 ones 连用。例如：

1）You have three devices，but I have only two.（不说 two ones）你有三台仪器，我只有两台。

2）Henry carried six chairs，but Jones carried seven. 亨利搬了六把椅子，琼斯搬了七把椅子。

He stands in the second row，and I stand in the eighth.（不说 the eighth one）他站在第二排，我站在第八排。

6.5 that 和 those 作为代替词的用法

指示代词 that 和 those 可以用来代替句中已出现过的某一相同的名词或名词词组以避免重复。下面分别介绍。

（1）that 代替可数名词（或名词词组）单数或不可数名词（或名词词组）。that 只能指物，不能指人，具有特指性质（即 the），这是它与 one 的基本不同点。此外，that 后面一定要跟修饰语，最常见的是 of 引起的介词短语，也可以是分词短语或定语从句等。例如：

1）The strength of the magnetic field increase with the increase of that of the current. 磁场强度随着电流强度的增加而增加。（that＝the strength）

2）The weight of a neutron is a little larger than that of a proton. 中子的重量比质子的质量稍大一些。（that＝the weight）

3）The term "work" has been given by science a somewhat more limited meaning than that which we have been accustomed to. 科学给予"work"（功、工作）这个术语的含义比我们习惯所使用的范围多少要小一些。（that＝the meaning）

4）The air gap of an induction motor is always much less than that in dc machines. 感应电动机的空气间隙总是比直流机的（空气间隙）小很多。（that＝the air gap）

5）It's a different kind of measuring technique from that I am used to. 这种测量方法跟我所习惯的那种不一样。（that＝the kind of measuring technique）

that 代替可数名词单数时，在口语中通常用 the one，而不太用 that。例如：

I'll take the seat next to the one（＝that）by the window. 我就坐在靠窗口的那个座位。

（2）those 代替句中已出现的某一相同的可数名词的复数（或名词词组）。that 的复数形式 those 作为指示代词，既可指人也可指物，也具有特指性质（即 the），这是它与 ones 的基本不同点，一般后跟介词短语、分词短语或定语从句等修饰语。例如：

1）If the forces causing motion are greater than those opposing it，an increase in speed takes place. 如果引起运动的力大于反抗运动的力，速率就会增加。（those ＝the forces）

2）Parallel currents flowing in the same direction attract each other，and those that flow in opposite direction repel. 在同一方向上流动的平行电流相吸，在相反方向上流动的平行电流相斥。(those ＝the parallel currents)

3) Today's engines are of much greater difference from those used in the past. 现在的发动机与过去所使用的（发电机）有很大的不同。(those ＝the engines)

4) The results obtained agreed approximately with those expected. 所得到的结果几乎和所预料的一致。(those ＝the results)

6.6　科技英语中的一些特殊结构

6.6.1　不定冠词置于修饰词后

当形容词修饰一单数可数名词时，若该形容词又被 so、as how、however、too 等修饰时，不定冠词要置于这一形容词之后；遇到 quite 和 rather 时，多半置于其后；遇到 such、many 和感叹句中的 what 时，也只能置于其后。

1) We obtained so precise a result in the experiment. 我们在试验中得到了如此精确的结果。

2) He is as good a worker as he was. 他和从前一样，是一个好工人。

3) How fine a sight it is! 多美丽的景色啊!

4) That is too difficult a problem. 那是个过分困难的问题。

5) It is rather a long time since the installation broke. 该设备坏了较长一段时间了。

6) They did the experiment in quite a different way. 他们用不同的方法做了这个实验。

7) I have not seen such a device before. 我从未见过这样的仪表。

8) Many an engineer are good at math. 很多工程师数学很好。

9) What a beautiful substation (it is)! 多么漂亮的变电站!

6.6.2　inasmuch as

这是一个短语连词，用以引出原因状语从句，其意义与 because 和 since 相同，但比较正式，现只用于书面语，故而也会用于科技文章中。通常译为“因为”“由于”或“既然”。例如：

(1) Machine is not an economical method of producing a shape, inasmuch as good raw materials is converted into scrap chips. 机械加工不是一种经济的成型方法，因为宝贵的原材料变成了废屑。

(2) Inasmuch as the pressure increases with depth, there is a greater pressure at the lower surface of the submerged body than at the upper surface. 由于压力随深度而增加，所以浸入（水中）的物体的底面所受到的压力比顶面大。

6.6.3　as … so …

这一结构不作“因为……，所以……”解，而做“正像……，也……”解。要表达前一种意思，可用 as（后面不用 so），或用 so（前面不用 as）。这里的连词 as 和 so 用来连接两个概念，指出程度上或关系上相似的地方。as 引出的是方式状语从句，为着加强语气，还可以在 as 之前加 just，so 引出的是主句，主句有时可以部分倒装。一般译为“正如……一样，……也……”。例如：

1）As water is the most important of liquids，so air is the most important of gas. 正如水是最重要的液体一样，空气也是最重要的气体。

2）As the sun is the central body of the solar system，so the nucleus is the core of the atom. 正如太阳是太阳系的中心体一样，原子核也是原子的中心。

6.6.4 such … as ＋ 不定式

这一结构与“so … as ＋ 不定式”结构相似，都起结果状语作用，不同点在于副词 so 后面接形容词或副词，而形容词 such 后面接名词。这一结构相当于“such … that”引出的结果状语从句，表示“这样……以致”的意义。汉译时，可加“使”“得”“就”“得使”等词，有时还可以译成名词的定语，即“……的……”。例如：

1）These two compounds react in such a way as to liberate oxygen.（＝… in such a way that oxygen is liberated.）这两种化合物按这样的方式起反应，就能放出氧气。

2）The computer is programmed in such a way as to gather a wide range of information from outer space. 这台计算机的程序设计得使它能从外层空间广泛地收集信息。

3）Modern forms of transportation are designed and constructed of such shape as to give little resistance to the air. 现代交通工具应设计和制造得使其形状对空气呈现很小的阻力。

4）The induced current flows in such a direction as to oppose the change of flux. 感应电流以阻碍磁通变化的方向流动。

6.6.5 as ＋ 介词短语

这是科技英语的一种特殊结构，可位于句首、句末或句子中间，常用逗号分开，在句中起状语作用，一般译为“像……一样”“如同……一样”“如……等”。例如：

1）As in the case of silicon，phosphorus remains dissolved in solid iron. 象硅一样，磷在固态铁中处于溶解状态。

2）An increased angle of attack gives more lift，as with the kite，but also increases the drag. 如同风筝一样，攻角增大，升力增大，阻力也增大。

3）A particular source of radiant energy may emit a single wave length，as in the light from a laser，or may contain many different wave lengths，as in the radiations from a light bulb or a star. 一种特殊的辐射能源可发出单一波长，如激光，或含有许多波长，如灯泡或星星发出的辐射。

4）As in the first method，the gas must be preheated to 150℃. 如同第一种方法，该气体必须预热到 150℃。

6.6.6　…as＋形容词或副词原级＋as＋形容词或副词原级＋…

这种是“as … as”结构的一种特殊用法，它不作“和……一样”解，而是“又……又……”，“not only … but also”。汉译时注意原文与译文的语序，“as A as B”译成“既B又A”，或“又B又A”。例如：

(1) This method of measurement is as simple as practical.（＝This method of measurement is not only practical but also simple.）这种测量方法既实用又简单。（不可译为：这种测量方法和实用一样简单。）

(2) The wheel turns as fast as stably. 这轮子转动得又稳又快。

(3) Insulators are as important as useful. 绝缘子既有用又重要。

6.6.7　A as well as B

这个词组有“和……一样好”等多种含义，但在这里 as 和 well 都失去原来的词义，一起构成一个短语连词，连接两个并列的成分。它有两个意思：一是相当于“both … and”，一是相当于“not only … but also”。作第一意时，译成“A 和 B 都”、“A 以及 B”或“A 和 B 一样都”。作第二意时，这个词组强调的是前面的词（A），所以翻译时要注意原文与译文语序不同，译为“不但 B 而且 A”或“既 B 又 A”。例如：

(1) Freezing points as well as boiling points are affected by external pressure. 冰点和沸点都受外界压力的影响。(Both freezing points and boiling points …)（或译：冰点和沸点一样都要受外界压力的影响。/冰点和沸点一样同样都要受外界压力的影响。）

(2) An engineer has practical experience as well as book knowledge. 工程师不但要有书本知识，而且还要有实际经验。(as well as ＝ “not only … but also”)

(3) Electric energy can be changed into light energy as well as into sound energy. 电能既可以变为声能，又可以变为光能。(as well as ＝ “not only … but also”)

(4) In many instances it is necessary to consider the rate at which work is done as well as the total amount of work done. 在许多情况下，必须考虑做功的速率以及所做功的总量。(as well as ＝ “not only … but also”)

(5) Notice that a force has direction as well as magnitude. 要注意，力不但有大小，而且有方向。(direction as well as magnitude ＝ not only magnitude but also direction)

第4篇

新能源与专业英语语法

7 新能源

7.1 A Review of Short-Term Wind Power Forecasting Approaches

7.1.1 Introduction

During the last few years, with the new paradigm shift in the electricity environment and the gradual reduction in greenhouse gases emissions, producers have been faced with the need to delivery electricity with clean energy resources and tackle tough competition in deregulated electricity markets. In this context, wind power resources have had the greatest jump in exploration and implementation in the electric grid, in comparison with other clean energy technologies. This large jump mainly occurred due to the ratio between production and implementation costs, maintenance costs of a wind farm, the maturity of the technology, and the increasing production capacity of wind machines. However, due to the stochastic nature of wind power resources, its integration is responsible for the introduction of a greater variability, volatility and uncertainty in the operation system, which complicates the proper management of all production resources.

The aforementioned behaviour in wind farms and in the electricity markets depends on the quality and variation in wind speed, the weather conditions, total wind power capacity connected in the electrical grid, the maintenance scheduling of the wind farm, and the acceptance of wind power in the network when it is available.

The non-dispatchable feature of wind power may affect the power balance and quality requirements of the system, compromising the profits of wind farm owners. In this context, Portugal is not an exception since it is one of the countries with the fastest growth in wind power resources, aiming to attain a significant installed capacity by 2020. Thus, great endeavours are being carried out to minimise the impacts of the aforementioned issues, which are mainly being undertaken by the scientific community in the presentation of new ideas to predict wind power behaviour and its corresponding electricity production, thus helping to reduce the fluctuating power and optimise the wind power resources installed.

Wind power forecasting can be organised by the time-scales, which means it can be divided into very-short-term, short term, and long-term horizons, which can be a few minutes (very-short-term) to multiple days (long-term) or more of forecasting.

In this paper, only the short-term horizon is considered (between 24 to 168 hours). Several wind power forecasting methodologies have been developed and reported in the technical literature in the last few years, which can divided into physical and statistical approaches.

The physical methodologies are constituted by an extended number of physical specifications, and their inputs are also physical variables, such as orography, pressure, atmospheric temperature, machine specifications, and others types of physical data, which can have advantages in long-term forecasting. The statistical methodologies try to find the relationships in inherent structures within the measured data, which can have advantages in short-term forecasting. Statistical forecasting methodologies can surpass other methodologies in terms of accuracy in the very-short-term due the stable efficiency in forecasting.

Moreover, new intelligent and hybrid approaches have been reported in the technical literature with interesting results, such as neural networks, neural networks with wavelet transform, adaptive wavelet transform with neural networks, neural networks with fuzzy systems, evolutionary approaches and hybrid methods, which can be more efficient and accurate if the proper inputs are selected from the predicted system.

The present paper aims to present a valuable state-of-art on the topic of short-term wind power forecasting, taking into account the journal papers published over the last few years. The review work will help the reader understand the pros and cons of all major forecasting approaches, using tabular forms to aggregate important information. Comparative studies will also be highlighted.

7.1.2 State-of-the-art review

A method was proposed to forecast wind power in the short-term, based on the application of an Evolutionary Algorithm optimisation for the automated specification of neural networks (NN) and nearest neighbour search. In the same work, the forecasted results were compared with two others algorithms based on particle swarm optimisation (PSO) and differential evolution. The proposed method used weather data combined with historical wind power data from several wind farms in Germany. Also, the system was tested with data from 2004 to 2007 with a time step of 1 hour.

A forecasting method was presented to predict the wind power in two wind farms in Portugal for the subsequent 72 hours, combining feed forward NN and Entropy and Correntropy Theories to help reduce the forecasted error distribution. The proposed system was tested in online and offline frameworks for the years 2005 and 2006.

A forecasting method was proposed to predict the wind speed for the next 24 and 48 hours using a Fractional Auto Regressive Integrated Moving Average (ARIMA), or fractional ARIMA model. The presented results were collected for 4 wind farms in North of Dakota, USA. After wind speed forecasting, the obtained results were combined with the mechanical characteristics of wind driven data to determine the wind power output. Furthermore, the final results were compared with a persistence model.

A forecasting method was proposed for the very-short term, combining an exponential sweetening method and data mining. The proposed method combined the collected data of a SCADA system with weather, physical and mechanical wind-driven data. In addition, the forecasting system was compared with other systems such as NN and support vector machines (SVM). The system predicted, with different time steps, the results for more than 168 hours ahead. In summary, model 1 predicts wind-driven function coefficients, model 2 uses mechanical wind-driven data and wind speed to predict the wind power output, and model 3 uses data mining parameters combined with previous models to predict wind power data.

A forecasting method was proposed using a differential evolutionary algorithm with a new crossover operator and selection mechanism to train the Ridgelet NN and wavelet transform (WT) for the next 24 hours ahead without exogenous variables. The case study used historical wind power data from Ireland from 2010 to predict wind speed, and wind power in Spain from 2010 to predict wind power.

A wind power forecasting method was proposed to predict 24 and 48 hours ahead, composed of feature selection components which perform irrelevance and redundancy filtering of historical data. This method also used a forecasting engine based on a NN cascaded structure with enhanced PSO. The system was tested at two wind farms located in Alberta, Canada, and Oklahoma, USA.

A wind power forecasting method was proposed based on WT and NN to predict the next 3 hours ahead up to 24 hours ahead with a time step of 15 minutes. The system used historical data of wind power provided by the SCADA system in Portugal between 2006 and 2007 without exogenous or weather data.

A forecasting method was proposed based on an Adaptive Neuro Fuzzy Inference System (ANFIS) to predict the next 3 hours ahead up to 24 hours ahead with a time step of 15 minutes. The system used historical data of wind power results provided by the SCADA system in Portugal between 2006 and 2007 without exogenous data. The proposed system was compared with ARIMA and NN forecasting systems.

A hybrid forecasting methodology was proposed based on ANFIS and PSO to predict the wind power in Portugal without exogenous or weather data, forecasting with only historical wind power data from the SCADA Portuguese system from 2006 to 2007.

A new hybrid and evolutionary forecasting method was presented, based on a combination of evolutionary PSO and ANFIS algorithms to predict the next 24 hours ahead, with a time step of 15 minutes for wind power production in Portugal, without exogenous or weather data. The proposed forecasting system was compared with other forecasting approaches, such as ARIMA, NN, Data Mining, and others.

A forecasting model was proposed based on multi observation points divided into 2 stages, to predict the speed and direction of wind in stage 1, and stage 2 uses the obtained data from stage 1 to predict the wind power output of the wind farm utilising dependent power curves. The study is performed with physical data from a wind farm at an Australian island. The proposed method was also compared with the grey model and persistence model.

A forecasting method was presented with a switching regime based on artificial intelligence to predict wind power, specifically the extreme events associated with the uncertainty of numerical weather prediction (NWP). The NN used was based on resonance theory and probabilistic methods, and was tested at two different wind farms, namely one in Denmark with historical data from 2000 to 2002 and one at Crete, Greece, with historical data from 2006 to 2008.

The problem regarding the large penetration of new wind farms into the electric grid was tackled, reviewing the pros and cons, and the advances in wind power forecasting approaches.

A NN was proposed to predict the active and reactive power in the electrical grid using the study case of a wind farm in Germany. The time step of this approach is 1 hour to predict from 24 to 48 hours ahead. The predicted results will help in wind farm management and also in controlling the power transmission system.

A probabilistic model forecasting of wind power was proposed, which uses prediction points and uncertainty data from deterministic models. These results come from the quality of NWP data, daily wind power forecasting, and weather stability (speed and direction of wind). Also, this forecasting approach used a combination of a multiple NN with PSO algorithm. The historical data used comes from the wind farms located in Denmark and Greece. Furthermore, this method predicts the wind power for the next 60 hours ahead.

A wind power forecasting approach was proposed based on 3 models of WT and SVM to predict, with a time step of 1 to 3 hours ahead, the wind power output of a wind farm located in the State of Texas, USA. Model 1 is ensemble accordingly with the wind-drives characteristics and WT principles. Model 2 combines the wind-driven characteristic with a substitution of Kernel RBF functions. While model 3 is a combination of the two previous models and the output is the wind power forecast.

A wind speed and wind power forecasting method was proposed for the next 30 hours ahead using in a first stage a combination of WT and NN to predict the wind speed, and in the second stage a feed forward NN to create a non-linear mapping between the wind speed and wind power results. These results were obtained without weather variables and performed for a wind farm located in Denver, USA.

7.1.3 Reported results

This section shows the results summary of the state-of-art, namely the study of several recent wind power forecasting techniques, considering their location, time horizon, time of historical data, time ahead forecasting, and also the type of data used to predict the wind power.

Table 7.1 show the countries where wind power forecasting approaches were tested in this state-of-art, i.e. Australia (AUS), Canada (CAN), Denmark (DEN), Germany (GER), Greece (GRE), Ireland (IRE), Spain (SPA), Portugal (POR), and the USA. Table 7.2 shows the period of time of the historical data used in most approaches. Note that the year 2009 was not found in any reported publication used. Table 7.3 presents the type of data used in each forecasting approach. Also the time step used and the horizon time of forecasting are described. Table 7.4 shows the three types of horizon most commonly used and their applications. As stated, in the scientific community there is still no consensus on the forecasting time ahead to be used. It should be noted that this depends on the market policy for where the forecasting approach is to be used, the purposes of forecasting and others.

Table 7.1 Countries where forecasting approaches were applied.

	AUS	CAN	DEN	GER	GRE	IRE	SPA	POR	USA
Ref [3]						X	X		
Ref [4]	X								
Ref [6]			X		X				
Ref [9]				X					
Ref [10]			X		X				
Ref [11]									X
Ref [15]		X							X
Ref [19]									X
Ref [22]								X	
Ref [23]									X
Ref [24]								X	
Ref [25]				X					
Ref [26]								X	
Ref [27]								X	
Ref [28]								X	

Table 7.2　　Historical data year used in forecasting approaches.

Year	2000	01	02	03	04	05	06	07	08	2010
Ref [3]										X
Ref [6]	X	X	X				X	X	X	
Ref [10]	X	X	X				X	X	X	
Ref [22]							X	X		
Ref [24]							X	X		
Ref [25]				X	X	X	X			
Ref [26]						X	X			
Ref [27]							X	X		
Ref [28]							X	X		

Table 7.3　　General characteristics of forecasting approaches.

	Physical	Wind Power	NWP	Time Step	Time-Ahead
Ref [3]		X		—	24h
Ref [4]		X	X	50min	7h
Ref [5]	X	X	X	10-60s	168h
Ref [6]		X	X	6h	48h
Ref [9]	X			1h	24-48h
Ref [10]		X	X	6h	48h
Ref [11]	X			1-3h	3h
Ref [15]		X	X	1h	1-48h
Ref [19]	X	X	X	1h	24-48h
Ref [22]		X		15min	24h
Ref [23]		X		—	30h
Ref [24]		X		15min	24h
Ref [25]		X	X	1h	—
Ref [26]		X		0.5h	72h
Ref [27]		X		15min	24h
Ref [28]		X		15min	24h

Table 7.4　　Application of forecasting approaches.

Horizon Type	Time Ahead	Application
Very-Short-Term	10 sec till 30min	-Electricity market -Wind forecasting control
Short-Term	30 min till 168h	-Strategy planning -Commitment decisions
Long-Term	More than 168h	-Maintenance planning -Wind farm study

7.1.4 Conclusions

This paper presents an overview of the most recent wind power forecasting approaches. The information provided here is aimed to represent a valuable database and a good starting point for future development in the field.

7.2 Economic and Efficient Voltage Management Using Customer-Owned Energy Storage Systems in a Distribution Network with High Penetration of Photovoltaic Systems

7.2.1 Introduction

IT is important to extensively adopt renewable energy systems worldwide in order to significantly reduce greenhouse gas emissions. Photovoltaic (PV) systems and wind power generators have been utilized as renewable energy sources for generating electric power. However, connecting such systems to existing electrical power systems increases the number of connections, which could affect the power quality. In particular, in distribution networks, locally concentrated connections from PV systems can cause unexpected voltage rises in distribution lines due to reverse power flows. As a result, overvoltages could occur depending on the simultaneous occurrence of high PV output and low consumer consumption; the chances of such a simultaneous occurrence are often extremely low throughout the year because the electricity demand in distribution networks changes daily and seasonally, and PV output has probabilistic characteristics. Distribution network operators (DNOs) can manage the bus voltages in a distribution network by installing only voltage regulators, but a DNO primarily determines the capacity of voltage regulators after considering worst-case scenarios (i. e. , the highest PV output and the lowest daytime load); consequently, the cost is significantly expensive in networks with high PV penetration.

In a closely related study that researched the actual fluctuation data of wind power generation, various voltage regulation methods were examined from an economic viewpoint by comparing the power factor control of existing inverters, reinforcement of distribution feeders, and customers' load control. The use of load control to regulate voltage is required only infrequently, and such a voltage regulation scheme by load control can be financially attractive. If DNOs need to use customer owned equipment, it is important to establish a cooperative operation framework between the DNOs and the customers so that the customers can earn substantial profits from the cooperation. Also, there are some existing services for spinning reserve or standby reserve in the electricity market. For example, standing reserve services in the U. K. are bought by the network operator at a fixed price. Voltage control in distribution networks is evaluated from a technical point of view assuming that customer-owned distributed generators are used in a power factor control mode under normal conditions and a voltage control mode under emergency conditions of distribution network voltage constraints. However, it has not been evaluated in terms of economics on the basis of data monitored throughout the year. There is no existing service for voltage regulation in a distribution network that uses customer-owned equipment.

In recent years, electrical energy storage systems have been practically applied; these are expected to improve the economic efficiency and reliability of energy supplies with fewer voltage sags, but the equipment costs are high currently. The authors have already analyzed customer-owned energy storage systems in terms of economics. It is concluded that the economic benefit achieved by only load leveling for each customer is not enough for installing the energy-storage systems from the viewpoint of cost and benefit analysis. As a multi-objective evaluation of customer-owned energy storage systems, this paper focuses on voltage regulation in a distributed network with intermittent renewable energy resources.

In this paper, the authors propose a concept in which a DNO controls the output of the energy storage systems of commercial customers during a specific time period in exchange for providing a subsidy covering a part of the initial cost of the storage system. Further, a model that allows customers to optimally install energy storage systems is developed. The customers receive a subsidy fromDNOs, and hence, they can install energy storage systems with more capacity with optimal planning. In order to make the voltage regulation of a distribution network using customers' energy storage systems more realistic, this paper utilizes the output of PV systems based on minute-by-minute irradiation data and focuses on 1) the voltage variations in distribution lines caused by output fluctuations and 2) voltage regulation using energy storage systems. The authors also conduct cost comparisons with a reference case wherein only a voltage regulator is installed by the DNO.

7.2.2 Proposed Concept for Voltage Regulation in Cooperation with Customer-Side Energy Storage Systems

In general, a DNO is primarily concerned with economic and stable operations of distribution networks without any violations of voltage constraints. Therefore, the capacity of voltage regulators such as static var compensators (SVCs) is determined by the worst-cases cenario, i.e., high PV output and less demand.

On the other hand, it is very important for commercial customers to reduce energy procurement costs such as energy charge per kWh and demand charge per kW (for purchasing electric energy) and equipment costs. In particular, the economic performance of a customer's energy system under normal operating conditions of distribution networks is more important than that under emergency operation conditions of power systems because emergency conditions last for very short durations in a year.

This paper proposes a novel concept of cooperative operation between DNO and commercial customers with energy storage systems, as shown in Figure 7.1. The DNO is allowed to control the output of the energy storage systems of customers during a specific time period in exchange for supplying a subsidy covering a part of the initial cost of the storage system. Since the capacity of inverter subsystems is important for voltage regulation in distribution networks, it is assumed that the subsidy is paid by the DNO to customers who install energy storage systems in proportion to the inverter capacity. This reduces the required capacity of voltage regulators for the secure operation of distribution networks by utilizing customers' energy storage systems only for a pre-specified time period. Moreover, from the customer's point of view, the energy procurement cost in a year is reduced since a subsidy [JPY/kVA] in proportion to the inverter capacity is received; the capacity of energy storage systems is also larger. Consequently, commercial customers can reduce their energy procurement costs by using a larger energy storage system under the normal operating conditions.

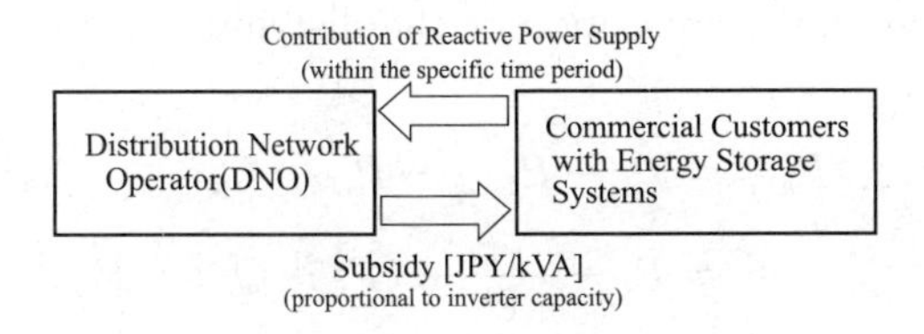

Figure 7.1 Proposed concept of cooperative operation.

Some important assumptions in this paper are summarized as follows.

1) A DNO can give a subsidy to commercial customers with storage systems in proportion to the inverter capacity. After receiving the subsidy, the commercial customers can change the capacity of their storage systems according to their economic rationale by load leveling.

2) As some different types of commercial customers, five types of commercial customer—offices, hotels, hospitals, stores, and restaurants—are assumed to install energy storage systems. Those customers are assumed to occupy medium-sized building areas (2000 m).

3) On the network, the sending voltage at a substation can be adjusted at maximum of three times a day. The installation of PV systems does not change those sending voltages.

4) Neglecting the concrete low-voltage (100/200 V) network, this paper focuses on a medium-voltage network with narrower voltage constraints by considering a voltage drop at low voltage.

5) The economic evaluation is based on minute-by-minute solar irradiation data. More detailed data fluctuations on time scales of less than 1 min are not considered. The smoothing effect of solar irradiation in the area covered by the distribution feeder is also ignored.

6) When voltage violations cannot be fully eliminated only by the customer's coopera-

tion, a DNO installs an SVC on the same bus as the commercial customers. The SVC installation cost is proportional to the capacity.

7) A DNO can detect violations of the voltage constraints. After detecting a violation, the DNO can use the advanced metering infrastructure to send a control signal to commercial customers' inverters immediately.

7.2.3 Distribution Network Model

This section evaluates the influence of highly penetrated PV systems on the voltage variations in a distribution network without considering customer-owned energy storage systems.

A. Power Flow Calculation

Since our analysis assumes a radial distribution network, the authors have calculated the active and reactive power flows from upstream to downstream of a distribution line. Each bus voltage is also calculated based on the given sending end voltage and bus load. The power flow equations for each section of the distribution line (Figure 7.2) are as follows:

$$P_{k+1} = P_k - P_{loss,k} - P_{Lk+1} = P_k - \frac{r_k}{|U_k|^2}(P_k^2 + Q_k^2) - P_{Lk+1} \quad (7.1)$$

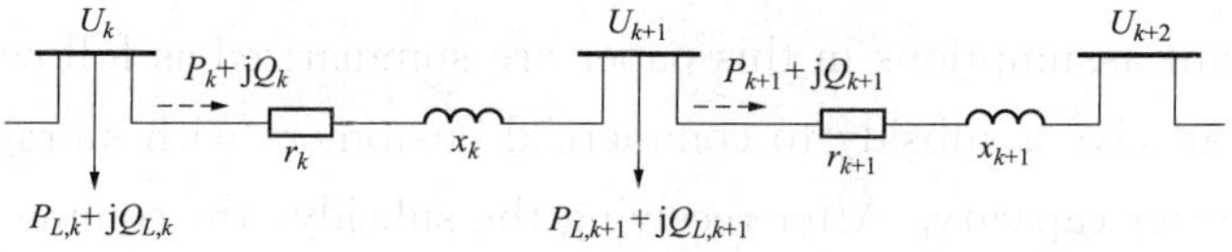

Figure 7.2　One line network.

$$Q_{k+1} = Q_k - Q_{loss,k} - Q_{Lk+1} = Q_k - \frac{x_k}{|U_k|^2}(P_k^2 + Q_k^2) - Q_{Lk+1} \quad (7.2)$$

$$|U_{k+1}|^2 = |U_k|^2 + \frac{r_k^2 + x_k^2}{|U_k|^2}(P_k^2 + Q_k^2) - 2(r_k P_k + x_k Q_k). \quad (7.3)$$

For the evaluation of the effectiveness of the proposed concept, a detailed model of low voltage lines is neglected. Although voltage variations for low-voltage customers must be maintained within [V] in Japan, this study maintains the bus voltage in a high voltage distribution network within

[p. u.] in this numerical simulation because voltage drop in a low voltage line is considered.

The power balance is satisfied in the load flow calculations in (7.1) - (7.3). Further, the rated capacity of conductors can be considered in load flow calculations. However, this paper assumes that the conductors in distribution lines have sufficient capacity.

B. Determination of Sending Voltage at Substation

The sending voltages at a distribution substation are adjusted to the daily and seasonal load profiles such that the distribution line voltage is maintained within a specified range. To evaluate the effectiveness of the proposed concept as a reasonable method of determining the sending voltage, the sending end voltage is determined such that the sum of the squared differences between the bus voltage and voltage limits over all buses is minimized, as in (7.4). Moreover, the authors have also assumed that the sending voltage can be adjusted a maximum of thrice a day, and it is determined for three intervals—0：00 - 8：00, 8：00 - 16：00, and 16：00 - 24：00—on the basis of the PV output profile:

$$Min \quad J(U_{0t}) = \int_{t}^{T} \sum_{k=1}^{M} \{(U_{kt} - U_{max})^2 + (U_{kt} - U_{min})^2\} dt. \qquad (7.4)$$

Here

T	*length of tine interval, h;*
M	*total number of buses;*
U_{kt}	*voltage of bus k at time t;*
U_{0t}	*sending - end voltage at time t;*
U_{max}, U_{min}	*upper and lower limits for bus voltage* ($\pm 0.03 p.u.$ *from the reference voltage*).

In this paper, even when PV and energy storage systems are introduced in the distribution system model, they are not considered when the sending voltage is determined.

Table 7.5　Parameters of Distribution Network Model.

Reference voltage	600 [V]
Reference capacity	5 [MVA]
Distance between buses	1 [km]
line impedance ($r_k + jx_k$)	0.305+j0.405 [Ω/km] ($Cu\ \varphi 60mm^2$)
SVC cost	4821.5 [JPY/ (kVA · year)] (300kVA, 15Million JPY)[17]

C. Simulations of Voltage Fluctuations in Distribution Networks With PV Systems

The radial distribution network model and parameters are shown in Figure 7.3 and Table 7.5, respectively. In this model, 500 households are connected at both buses #1 and #2. The households connected to bus #1 have a 2.7kW PV system with reverse power flow to the grids. The PV generation data used in this study are obtained by converting

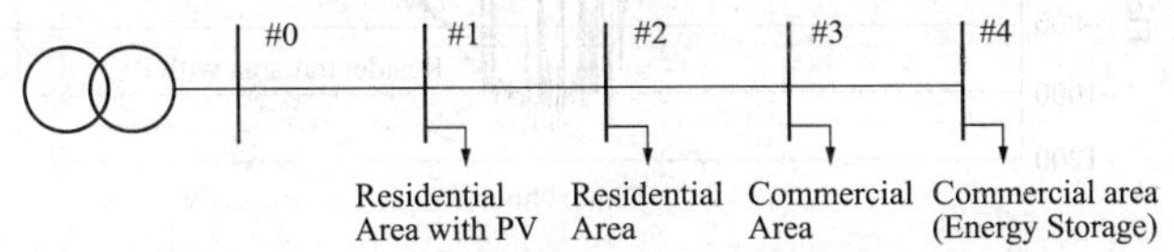

Figure 7.3　Distribution network model (test system 1).

minute-by-minute data of the global solar irradiation observed in Tokyo in 2005 using the International Energy Agency (IEA) conversion method (tilt angle: 30; azimuth: 180). Buses #3 and #4 have loads comprising the five commercial customer types discussed in Section 7. 2. 4-A.

The authors have also analyzed the effectiveness of the proposed concept in regulating the voltage for different types of commercial customers after assuming that among the customers connected to bus #4, only one type of commercial customer installs energy storage systems. Moreover, when the proposed voltage management concept does not lead to sufficient effectiveness in voltage regulation, the study assumes that the DNO eliminates violations of the voltage constraint by installing an SVC on bus #4. For an SVC costing 15 [million JPY] per 300 [kVA] and having a useful life of 15 years, its installation cost is 4821. 5 [JPY/kVA-year]. An inverter for energy-storage systems can be controlled within a second for suppressing any voltage rise. SVC is a voltage regulator with such fast response ability; thus, it has been chosen from the viewpoint of a fair comparison. Figure 7. 4 shows the energy demand of the residential area (buses #1 and #2) and that of the commercial area (buses #3 and #4). As an example, the energy demand for bus #1 is obtained by subtracting the PV output (1-min intervals) on August 21 from August's daily load curve (monthly average; 1-h intervals); negative demand implies reverse power flows from the customer to the grids. Figure 7. 5 shows the corresponding changes in the bus voltage when neither subsidy nor voltage regulator are present. The voltage constraint was not observed to be violated when both PV and storage systems were not introduced into the network model. The maximum bus #1 voltage (V1), which has locally concentrated connections from PV systems, is 1. 03 [p. u.] due to reverse power flows from the PV systems during the day. The voltage violation in the test system (Figure 7. 3) is not large because all PV systems are assumed to be installed at bus 1, as explained in Appendix B.

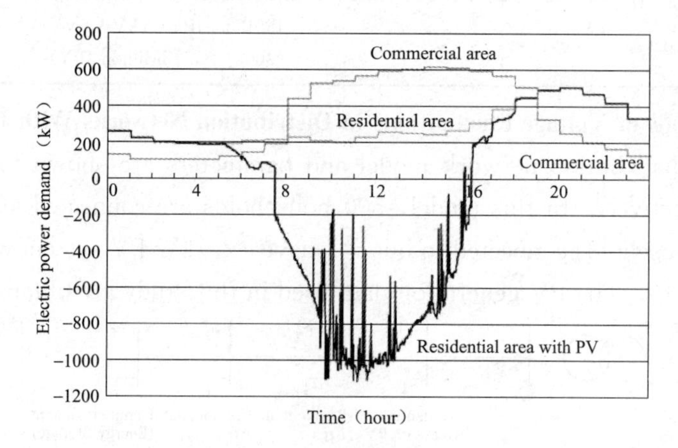

Figure 7. 4 Example of electric power demand in the commercial-area and residential-area bus (PV output: August 21).

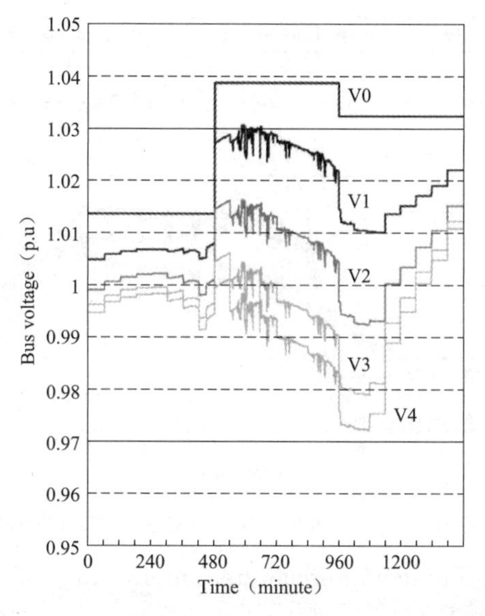

Figure 7. 5 Example of bus voltage (August 21).

7. 2. 4 Simulation of Energy Storage System Installation by Customers

This section describes the optimal planning model for installing energy storage systems at the customer side by considering subsidy payment. Further, the influence of subsidy payment on the capacity of storage systems and inverters, which can be installed by customers, is evaluated by performing numerical simulations.

A. Simulation of Charge and Discharge Patterns of Energy Storage Systems Considering a Subsidy

The authors have considered five types of commercial customers—hospitals and hotels, which have high load factors on the daily load profile, and offices, restaurants, and retail stores having low load factors—and assumed that each has a floor space of 2000 [m], as given in Figure 7. 6. The energy load curve data are typical load data when designing a cogeneration system in Japan.

Table 7. 6 Time-of-Use Pricing [JPY/KWH] for Commercial Customers

Time (h)	0 22	8 24	10		17
Summer	7. 12	11. 28	16. 36	11. 28	7. 12
Others	7. 12		11. 28		7. 12

Figure 7. 7 shows the configuration of the energy system of the commercial customers. A heat pump air conditioner was used for space cooling and heating. Electric power was supplied through the receiving equipment, and the load power factor was assumed to

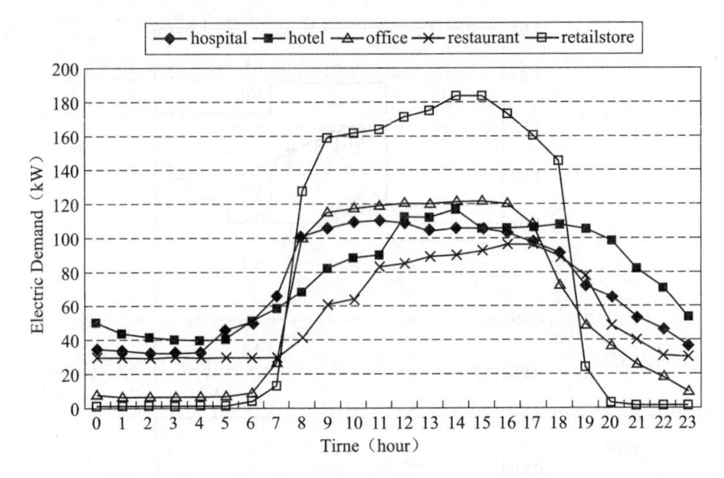

Figure 7.6　Load curve for all customers (August).

be 0.9 for all five types. The commercial customers must install an energy storage system to flatten the daily load curve, while the energy system is considered flexible in the sense that the storage system's inverter, output, and storage component capacities can be freely set. Further, the demand charge for the contracted power is assumed to be 1609.5 [JPY/kW/month], and the energy charge is shown in Table 7.6, which is based on the electricity tariff in a Japanese electric power company. The parameters associated with each equipment component are shown in Table 7.7. The cost of energy storage systems is assumed to be 250000 yen per kW for a 7.5h system used for load leveling. For the cost parameters, it is assumed that a redox-flow battery or natrium sulfur battery is used. For other equipment, the cost parameters are based on, which were estimated by linear regression analysis proportional to the rated capacity. The coefficients of performance (COP) and efficiencies are typical averages described in the datasheets of various products. Further, in Table 7.7, the efficiency due to the charge and discharge is assumed to be 87% for energy storage. Assuming that the efficiency of the inverter is 95%, the total efficiency of the energy storage system becomes 78.5% (= 0.95×0.87×0.95).

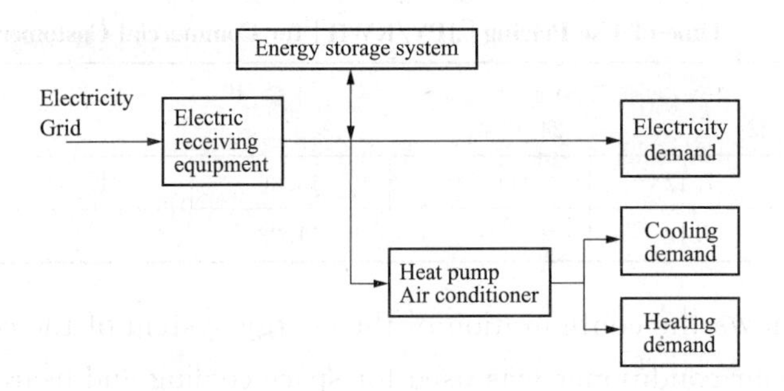

Figure 7.7　Energy system of commercial customers.

Table 7.7 **System Parameters for Commercial Customers.**

Equipment	Initial cost	Efficiency& COP	Lifetime [year]
Heat pump Air conditioner	6000 [JPY/kW]	Cooling: 3.42 Heating: 3.84	15
Hot water boiler	8 [JPY/kcal/h]	0.85	15
Receiving electricity equipment	56000 [JPY/kVA]	—	15
Inverter	50000 [JPY/kVA]	0.95	10
Energy storage System (kW)	110000 [JPY/kW]	—	10
Energy storage System (kWh)	12000 [JPY/kWh]	0.87	10

The formulation of optimal planning and operation for energy storage systems is described below. This study focuses on the economic benefits of customer's load leveling by installing energy storage systems under time-of-use pricing. The objective function is to minimize the annual total cost comprising purchased energy cost (energy charge [kWh], demandcharge [kW]) and annualized equipment cost. The authors assume that the financial subsidies provided by DNOs for energy storage systems are for the cost of inverters that supply reactive power. The parameter "d" in (7.7) implies subsidy cost for inverter [JPY/kVA]. Given "d", the optimal capacity of energy storage systems for each customer can be determined using this optimization model (7.5) - (7.12). The other symbols are listed in the Appendix.

$$Min\ E_{cost} + \delta_{es} \cdot ES_{cost} + \delta \cdot EQ_{cost} \tag{7.5}$$

Subj to.

$$E_{cost} = \sum_{m=Jan}^{D_{ec}} [E_{k,m} \cdot \overline{E_k(m)} + \sum_{t=t1}^{124} \{E_{Z,mt} \cdot E(m,t)\} \cdot S_m] \tag{7.6}$$

$$ES_{cost} = ES_{kW} \cdot C_{kW} + ES_{kWh} \cdot C_{kWh} + ES_{inv} \cdot C_{inv} \cdot \left(1 - \frac{d}{100}\right) \tag{7.7}$$

$$ES_{cost}(m,t+1) = ES_{sto}(m,t) + \eta_{inv} \cdot \eta_{es} \cdot ES_{in}(s,t) - ES_{out}(m,t) \tag{7.8}$$

$$v_{in}(m,t) \geqslant u_{in}(m,t) - u_{in}(m,t-1) \tag{7.9}$$

$$v_{out}(m,t) \geqslant u_{out}(m,t) - u_{out}(m,t-1) \tag{7.10}$$

$$\sum_{t \in t1}^{124} v_{in}(m,t) \leqslant 1 \tag{7.11}$$

$$\sum_{t \in t1}^{124} v_{out}(m,t) \leq 1 \tag{7.12}$$

The annual cost for commercial customers consists of electricity charges as well as annualized costs for their initial equipment. The energy-storage-system capacity and its charge and discharge patterns are determined assuming the most economically efficient operations, i. e. , such that the annual costs for all types of commercial customers are minimized. The charge and discharge patterns of the storage system for a hotel in August, i. e. , a customer with high load factor, and an office, i. e. , a customer with low load factor (Figure 7. 6), are shown in Figure 7. 8 and 7. 9, respectively. Given a subsidy, both hotel and offices increase their energy storage system output.

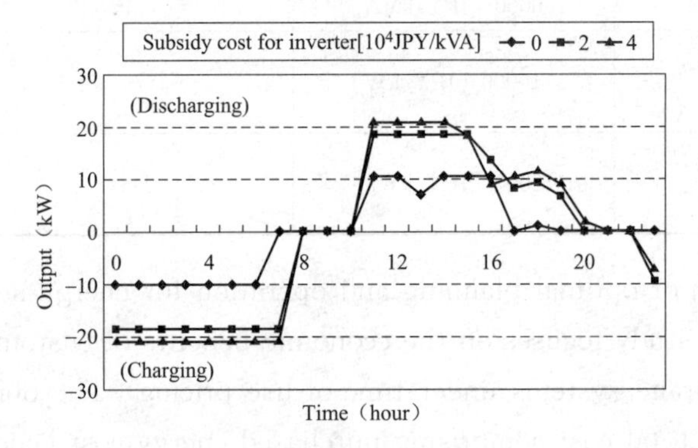

Figure 7. 8　Charging and discharging patterns of energy storage system (hotel, August).

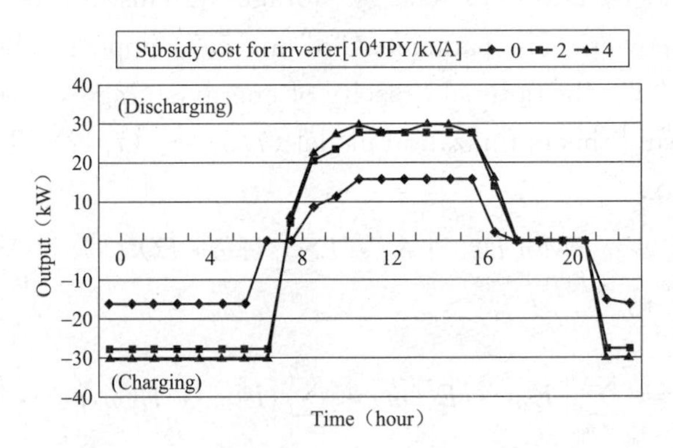

Figure 7. 9　Charging and discharging patterns of energy storage system (office, August).

Figure 7. 10 and 7. 11 show the changes in the inverter capacity and the annual cost for the five customer types, respectively, which decrease the initial costs involving the inverter installation. The inverter capacity varies with the customer type because the amount of energy demanded by each customer is different. However, Figure 7. 10 and 7. 11 show that for all five types, a subsidy for a storage system inverter increases the capacity [kVA] of the installed inverter.

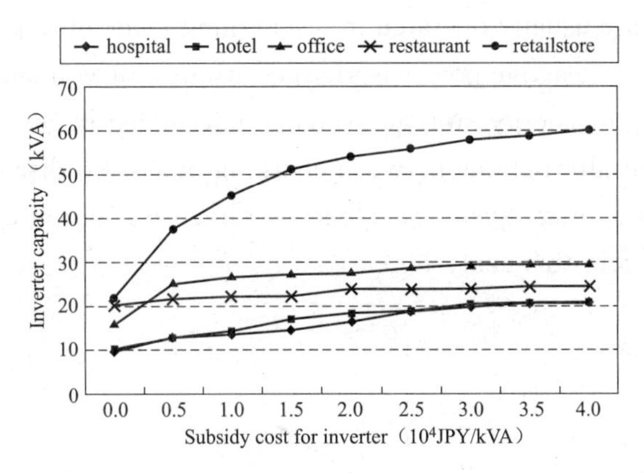

Figure 7. 10 Inverter capacity and subsidy cost.

Among the customer types, customers with low load factors (office and retail store) tend to increase the inverter capacity installed with a subsidy due to decreasing expected costs by peak shaving; this leads to an increased capacity of the storage system and eventually to larger changes in costs.

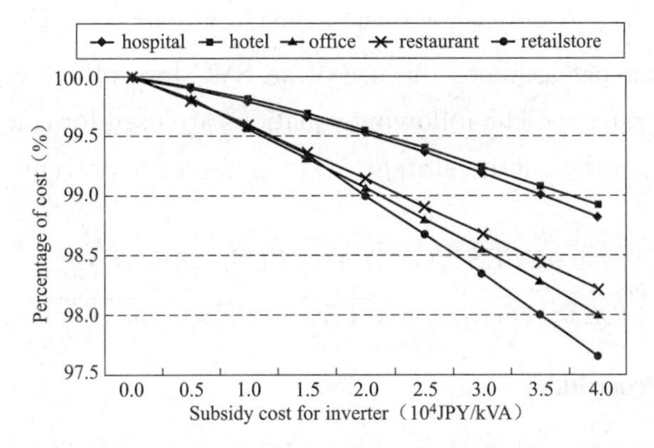

Figure 7. 11 Cost per year of commercial customers to subsidy.

B. Cooperative Operation of Distribution Network and Energy Storage Systems

In this paper, voltage regulations are performed such that the distribution network voltage is maintained within its upper and lower limits by cooperating with the customer-side energy storage systems during violations. To prevent any voltage constraint violations, the cooperative operation takes advantage of the reactive power compensation by the inverter of the storage systems. Figure 7. 12 shows the concept of the control by the inverter of the storage systems. However, when violations still occur, despite the contributions from the energy storage systems, the DNO finally eliminates them by using its own voltage regulator (i. e. , SVC). To prevent the voltage constraint violation by using only energy storage systems, the active power output of energy storage systems must increase until it reaches the output level with the most economically efficient operation (based on the charge and discharge patterns determined in Section 7. 2. 4-A). However, because energy storage systems are usually operated in discharging mode during the day, as shown in Figure 7.8 and 7.9, if the active power is charged to prevent voltage constraint violation, the equipment capacity of energy storage systems determined in Section IV-A may be exceeded. Also, frequent alternations between charging mode and discharging mode may affect the lifetime of the storage systems. Hence, voltage regulation by charging the active power is not performed in this analysis.

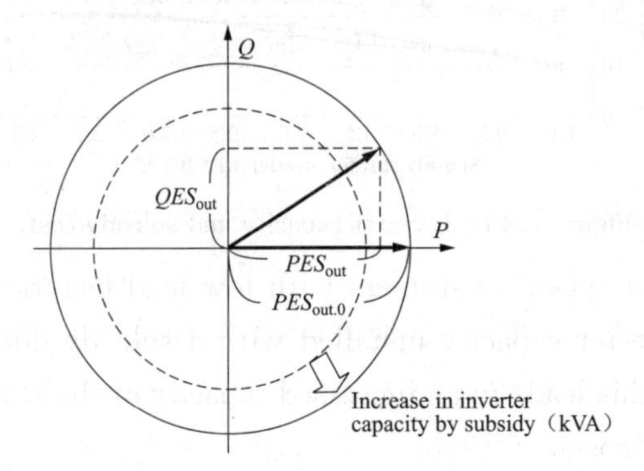

Figure 7.12 Reactive power compensation by inverter of energy storage.

In addition, this paper assumes the use of an SVC for voltage regulation cooperating with energy storage systems. The following equations are used for calculating the output of the reactive power from the energy storage systems used for preventing voltage constraint violations:

$$0 \leqslant PES_{out}(m_c, t_c) \leqslant PES_{out,0}(m_c, t_c) \tag{7.13}$$

$$QES_{out}(m_c, t_c) = \sqrt{ES_{inv}^2 - PES_{out}(m_t, t_c)^2} \tag{7.14}$$

7.2.5 Evaluation Procedure

The evaluation procedure has three steps. First, an optimal planning model for customer-owned energy storage systems is developed considering the subsidy payment for covering initial energy storage costs. The sending voltage at the substation is then determined by (7.4) for the original network without either the PV or the energy storage systems. Thirdly, load flow calculations are performed using (7.1) - (7.3) for checking voltage constraint violations considering charge/discharge patterns of customer-owned storage systems.

When a voltage violation occurs, the reactive power output from inverter subsystems of energy storage systems are calculated by (7.13) and (7.14). If the voltage violation cannot be eliminated by only customer's reactive support, the necessary reactive power requirements for SVC can be evaluated. Depending on the annual simulation results for load flow, the SVC capacity can be determined using the maximum reactive power requirement expected during the course of a year. The maximum requirement is obtained by identifying the maximum reactive power required to avoid voltage violations based on the minute-by-minute solar irradiation data for one year.

7.2.6 Economic Evaluation from Perspective of Customer and Network Operator

Using the distribution network and customer models (as explained above), the proposed cooperative operation is evaluated by performing numerical simulations on the basis of minute-by-minute solar irradiation data for one year. Further, another test system is evaluated in general. Finally, the proposed method is examined from the viewpoint of practical implementation.

A. Duration Time of Voltage Constraint Violation

Figure 7.13 shows the annual duration time of the voltage constraint violations per year when reactive power compensation is performed by energy storage systems installed at commercial customer sides, following the previously discussed cooperative operation scheme. A subsidy of 0 [JPY/kVA] implies that no cooperative operation with energy storage systems is performed; the duration time of the voltage constraint violations varies with the type of customer. When storage systems are installed at any of the5 commercial customers, the duration time of voltage constraint violations decreases as the amount of subsidy increases, but the reduction flattens gradually. This is because the reactive power output plateaus as the inverter capacity plateaus (Figure 7.10). The effect of cooperative

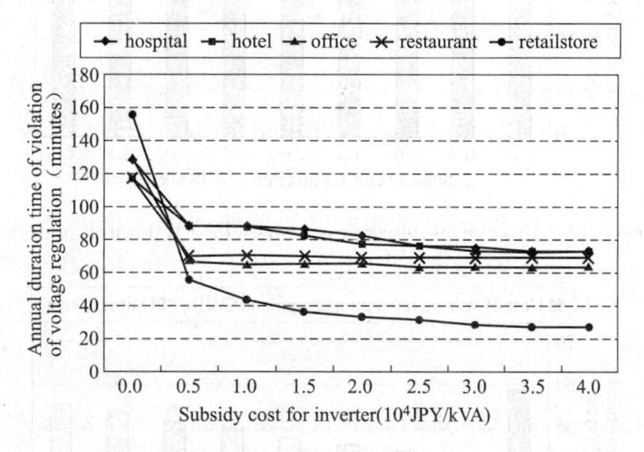

Figure 7.13 Duration time of violations of voltage regulation per year.

operation is especially remarkable in retail stores because of the large inverter capacity. The sending voltage at the substation is determined when no commercial customer has installed a storage system; hence, once storage systems with different capacities are installed, the electric load of distribution system changes. This explains the difference in the number of violations.

B. Costs for Network Operator

The annual cost of preventing voltage constraint violations to the DNO is defined as the sum of the inverter subsidy and the annualized cost of the voltage regulator required to compensate for insufficient reactive power in the cooperative operation. The load data is an annual load profile (monthly averaged daily load profile) for each customer. The charge/discharge patterns of energy storage systems are calculated using the monthly averaged daily load profile. Consequently, taking into consideration inverter capacity margins, the additional capability of reactive power output from energy storage systems can be evaluated on an hourly basis in a year. Figure 7.14 and 7.15 show the DNO's cost when the proposed concept is applied to hotels (high load factors) and offices (low load factors), respectively. In both cases, the DNO's cost is lowest when the inverter subsidy is 5000 [JPY/kVA] and increases with the subsidy amount. Moreover, since the network operator's cost is minimized when the subsidy is 5000 [JPY/kVA], and even with a small subsidy, the effect of cooperative operation with customer-side storage systems is remarkably large. Further, installing energy storage systems at the customer side with low load factors is more effective in regulating voltages and reducing costs because the customers install inverters with a large capacity.

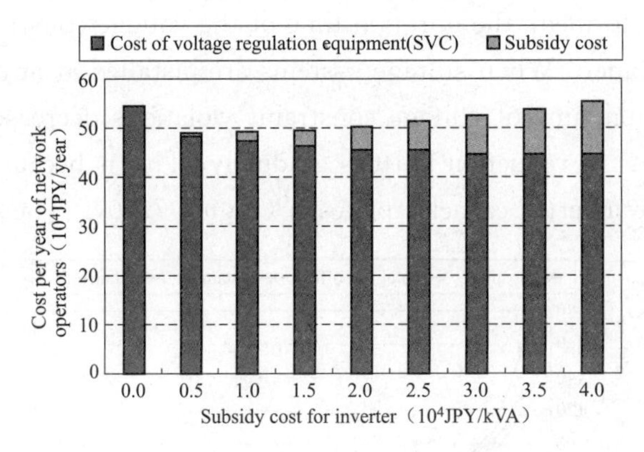

Figure 7.14　Cost of network operators (DNO) (hotel customer).

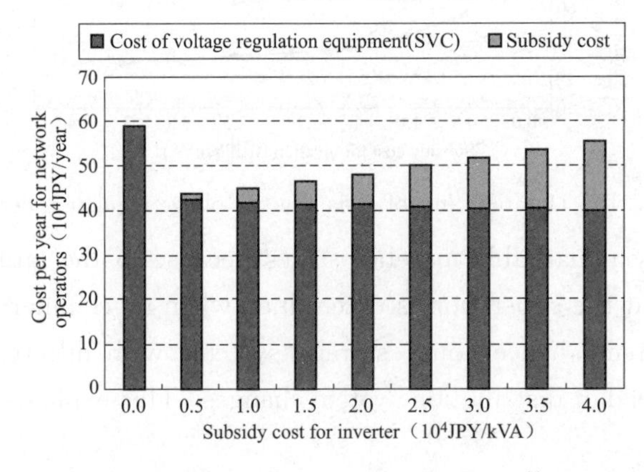

Figure 7.15　Cost of network operators (DNO) (office customer).

C. Cost Increases Due to Cooperative Operation at Customer Side

The proposed cooperative operation system requires commercial customers to increase their purchases of electricity from grids because when a voltage constraint violation occurs, the active power output from the energy storage systems must be reduced in order to increase the reactive power output, as shown in Figure 7.12. For all types of commercial customers, the cost for the increased electricity purchases is extremely low at approximately several hundreds [JPY/year]. Considering that the annual costs for commercial customers and the DNO are in millions [JPY/year] and hundreds of thousands, respectively, the increased cost due to the cooperative operation is relatively small.

D. Cost for Network Operator With Another Test System

In order to evaluate the proposed concept from a more general viewpoint, the DNO cost is evaluated using a different test system shown in Figure 7.16. In comparison to the test system for the base case shown in Figure 7.3, only the locations of the customers in the distribution line are different from those of the base case test system. Since the voltage profile along the distribution line in Figure 7.16 is different from that in original test system, the sending voltage at the substation in Figure 7.16 is determined by (7.4) again. In Figure 7.16, a voltage constraint violation occurs in bus #4, and the customer's own energy storage system is located in bus #1. As the voltage fluctuations for the test system shown in Figure 7.16 are larger than those for the original test system, the capacity of the SVC in the former system should be larger than that of the SVC in the latter system (see Appendix B). Also, customer-owned energy storage systems are connected to bus #1, and they can adjust power flow only between buses #0 and #1.

From the simulation results for the different test system, the DNO cost for two customer's cases—hotel and office—are shown in Figure 7.18 and Figure 7.19, respectively. By comparing with Figure 7.14 and Figure 7.15, the cost of SVC in the different test system is much larger than in the original test system because the location of high penetration of PV is changed to the end of the feeder. In addition, Figure 7.17 shows the annual duration of the voltage constraint violations per year for another test system. As compared to the annual duration in Figure 7.13, the voltage fluctuations for this test system are much larger than those for the original test system. On the other hand, the voltage regulation effect of customer-owned energy storage systems, shown in Figure 7.18 and Figure 7.19 are almost identical to those in Figure 7.14 and Figure 7.15, respectively. Figure 7.17 confirms that the proposed method using another test system has almost the same effect on the annual duration of voltage violations as the original test system. This is because the effect of suppressing the voltage rise at bus #4 (Figure 7.16) is determined by the line current between bus #0 and bus #1, while the effect of suppressing the voltage rise at bus #1 (Figure 7.3) is also determined by the line current between bus #0 and bus #1. Consequently, another test system will result in an increase in voltage violation and reduction in the effect (percentage-wise) of voltage regulation due to the proposed management concept.

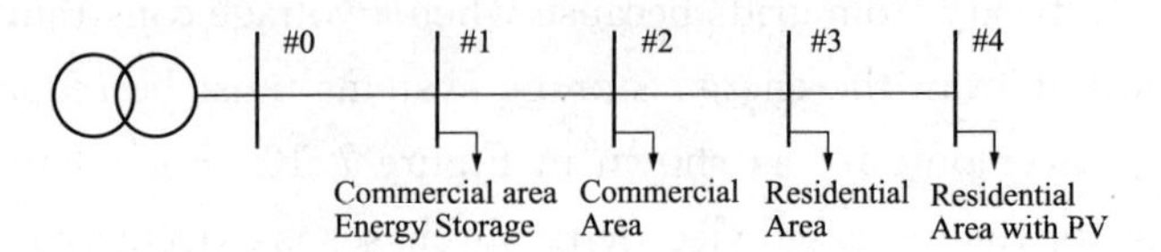

Figure 7.16　Different test system for sensitivity analysis (test system 2).

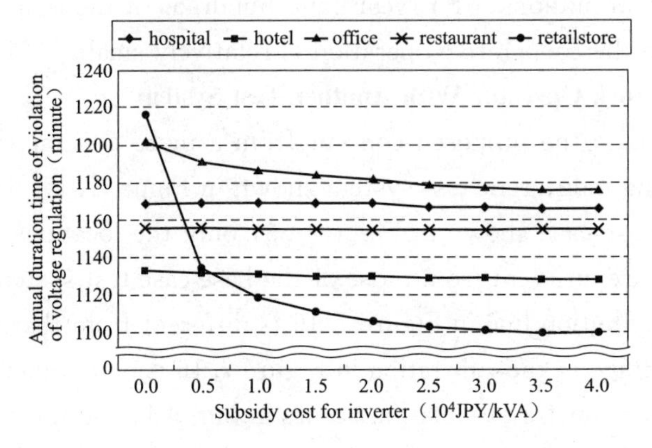

Figure 7.17　Duration time of violations of voltage regulation per year in another test system.

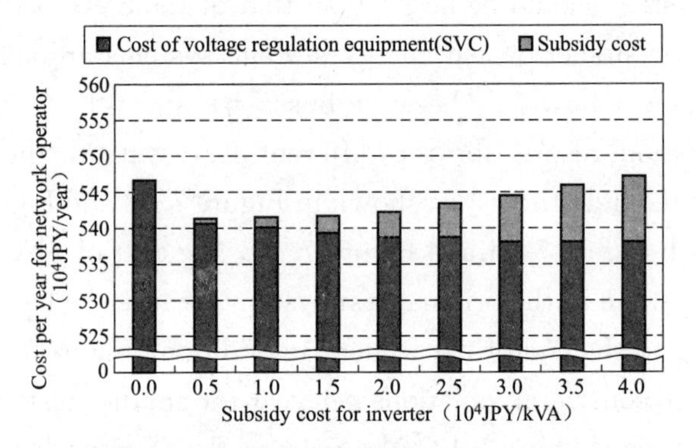

Figure 7.18　Cost of DNO in test system B (hotel).

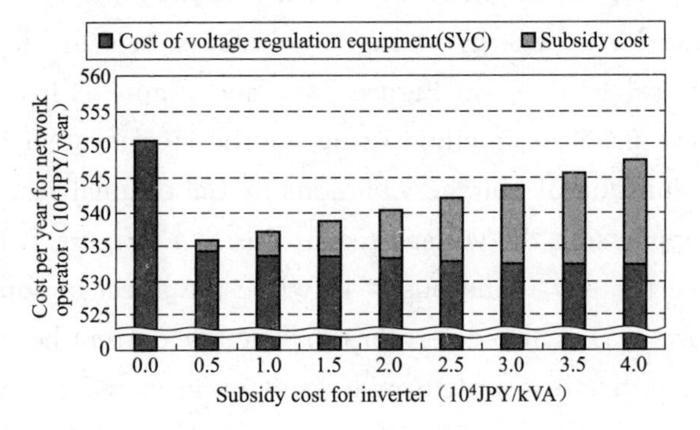

Figure 7.19　Cost of DNO in test system B (office).

In other words, in the first test systems as Figure 7. 3, power flow between bus #0 and #1 can be changed by installing PV at bus #1. Therefore, by referring to Appendix B, the authors can infer that capacity of SVC is not very large because the voltage fluctuations are relatively small. However, in another test system as Figure 7. 16, power flow between bus #0 and #4 is changed by installing PV at bus #4. Consequently, initial cost of SVC is larger than that in Figure 7. 3. On the other hand, energy storage systems at bus # 1 in Figure 7. 16, can control the power flow between #0 and #1. Therefore, the reduction effect of SVC cost in Figure 7. 18 and Figure 7. 19 is almost the same as that in Figure 7. 14 and 7. 15.

In both test systems (Figure 7. 3 and Figure 7. 16), the locations of the energy storage system and SVC relative to PV are important for evaluating the impact of the proposed method. In general, as explained in Appendix B, when PV is located toward the end of the feeders, the increase in voltage becomes higher. Moreover, when the storage system (and SVC) is also located toward the end of the feeders, the effect of voltage increase is suppressed. Therefore, the conditions under which the proposed method would be most effective are 1) when PV and storage systems are located at the end of the feeders and 2) when SVC is installed at only the buses near the substation (beginning of the feeders). However, because these conditions are somewhat impractical, the storage systems in test system 2 area lsoinstalledat the beginning of the feeders on the basis of assumption 6 given in Section 7. 2. 2. Consequently, the simulation results show the certain effect of the proposed method in test system 2.

As a further development, assuming a large-scale distribution network, the proposed method will be evaluated by changing the location of SVC as an independent parameter.

E. Practical Implementation of Proposed Concept

In order to implement the dispatch of reactive power in practice, the operating point of the inverter must be recorded and communicated by the DNO to estimate the contribution of the reactive and active power output from the distributed generators and storage systems. The cost of implementation including inverter, customer's meters and communication networks between DNO and customer-side energy storage, will have significant impact on the economic merits of the proposed cooperation system. However, Advanced Metering Infrastructure (AMI) is expected to be available in future power systems, which will reduce additional costs for realizing the cooperation system. It would be important to reduce the investment of voltage regulators appropriately in distribution network planning by utilizing the customer-owned reactive power sources using communication networks.

In addition, this paper focuses on the evaluation of the effectiveness of the proposed method under the assumption that it is not possible to forecast PV output with a 1min interval. Therefore, the charge and dischargc planning for storage systems is carried out by day-ahead scheduling without considering the PV output. After detecting or estimating a voltage violation in a distribution network, the DNO sends a control signal to the commercial customer with the energy storage system. However, from a practical point of view, if the PV output can be forecasted with adequate accuracy, it is possible to use this data for charge and discharge planning for storage systems in advance. This possibility will be considered in a future study. Readers might think that the capacity of the installed PV system can be gradually increased with load as a way to reduce the voltage rise. However, moving clouds may cause the output of PV systems to reduce suddenly. Therefore, the DNO should be equipped with control systems for both increase and decrease in the output of PV systems.

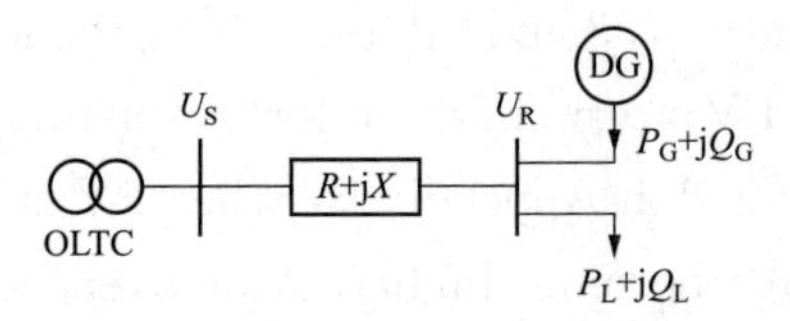

Figure 7. 20　Two-bus distribution system with DG.

7. 2. 7　Conclusion

This paper examined the voltage regulation problem of a distribution network with high penetration of PV systems and proposed a novel concept utilizing customer-side energy storage systems. Further, its cost effectiveness from the viewpoints of both commercial customers installing storage systems as well as the network operator was examined. The proposed concept involves cooperative operation systems between distribution system operators and customer-owned energy storage systems. Numerical simulations based on minute-by-minute irradiation data from actual observations revealed that energy storage systems are effective in regulating voltages even when the subsidy for inverters is small. Our results from the evaluation of the cooperative operation concept utilizing customer-side energy storage systems from an economic point of view clarified the possibilities of making voltage management more economical in distribution networks.

The main points are summarized as follows.

1) From the commercial customer's point of view, the subsidy can reduce the annual cost for normal operation, particularly when the load factor is low. The cost increase for emergency cooperative operation is relatively small in terms of annual operational cost.

2) From the DNO's point of view, both the duration of voltage violations and the required SVC capacity can be reduced by giving subsidies to commercial customers with storage systems, but the effect flattens gradually.

3) The proposed concept has the potential to reduce the total subsidy cost and SVC cost in a distribution network with high penetration of renewable resources.

To demonstrate the effect of the novel voltage management concept based on actual solar irradiation data throughout a year, two types of simple test system are considered in this paper. In further development, the proposed concept will be tested using a larger test system in which data are monitored.

Besides the proposed concept, other approaches (multi-agent based dispatching scheme and distributed automatic control method) have also been proposed. In future studies, such alternative voltage regulation approaches must be examined, evaluated, and integrated. As a further development of the appendix.

A. List of Symbols

Set $t \in t_1, t_2, \cdots, t_{23}, t_{24}$
$m \in Jan, Feb, \cdots, Nov, Dec$

Discrete Variables

(0 or 1) $u_{in}(t)$, $u_{out}(t)$: *Index for output of energy storage system (output=1; no output=0)*; $v_{in}(t)$, $v_{out}(t)$: *Index for starting output of energy storage (Starting=1; other = 0)*

Continuous Variables $\overline{E_k(m)}$: *Maximum Contracted Power [kW]*; $E(m, t)$: *Purchased Electric Energy [kWh]*; ES_{kW}, ES_{kWh}, ES_{inv}: *Energy Storage Output Capacity [kW]; Storage Capacity [kWh] and Inverter Capacity [VA]*; $ES_{sto}(m, t)$: *Other Equipment Cost (Excep for Energy Storage)*

Constants $E_{k,m}$: *Demand Charge [JPY/kW]*; $E_{z,mt}$: *Electricity Price [JPY/kW]*; S_m: *Number of Days*; δ_{es}, δ: *Annual Expenditure Rate (ES, other)*; C_{kW}, C_{kWh}, C_{inv}: *ES's Cost Parameters of Output Part [JPY/kW]; Energy Part [JPY kW]; Inverter Part [JPY/VA]*; d: *Subsidy Rate* (d=0, 10, 20, ⋯, 80); η_{inv}: *Inverter Efficiency*; η_{es}: *Efficiency of ES*

B. Voltage Rise in Distribution Network With DG

To understand the voltage rise phenomena in distribution networks with DG, a simplified two-bus distribution system is considered as shown in Figure 7.20. The effect of voltage rise by DG is approximately expressed as follows:

$$\Delta U = U_R - U_S \approx R(P_G - P_L) + X(Q_G - Q_L)$$

When the same active power output of DG is connected to the end of feeders, the effect of voltage rise at the connecting bus is the largest because the resistance R is generally maximum. Further, when the same reactive power output of DG or voltage regulators are connected to the end of feeders, in general, the effect of voltage regulation is the largest because the reactance X is also maximum at the end of the feeders.

7.3 Virtual Power Plant (VPP), Definition, Concept, Components and Types

7.3.1 Introduction

The penetration of Distributed Energy Resources (DER) is rising fast worldwide, which is mainly associated with the requirement of a sustainable energy system with less environmental problems, more diversified energy resources and enhanced energy efficiency. In the meanwhile, the continuing process of liberalization of the electricity market, i. e. the changeover from the monopoly system to competitive market structures, also attracts more and more attention. In the context of these two tendencies, running a great number of DER units under market conditions is inevitable, which yet poses new challenges that have to be addressed including:

• Taking part in market: Considered as small, modular power sources, storage technologies and controllable loads, DER is generally prohibited from entering the current electricity market.

• Intermittent nature: As many DER technologies like solar cells and wind turbines are weather dependent, their fluctuating output is therefore considered non-dispatchable which not only limits their contribution to grid operation, but also causes economic penalties associated with unexpected unbalances.

• Standing alone: Many DER units are working isolated due to their different ownerships. Cooperation and communication often lack between neighboring DER units, thus the capability of DER is restricted to satisfy the local needs rather than the whole grid.

One way to address these issues is to aggregate a number of DER units in a so-called Virtual Power Plant (VPP). In this construction, the group of DER units will have the same visibility, controllability and market functionality as the conventional transmission-connected power plants.

The remainder of paper is as follow: in section7. 3. 2 various VPP definitions introduced in literature and general concept of it has presented. Section 7. 3. 3 states the main VPP components. Then, section 7. 3. 4 explains the terms technical VPP (TVPP) and commercial VPP (CVPP). Finally, section V mentions the conclusion remarks.

7.3.2 VPP Definition and Concept

A. Definition (Literature Review)

Actually, VPP is still in hypothesis stage and there is not a unique definition for the framework of VPP in the literature. VPP is defined the same as an autonomous micro-grid. VPP is defined as an aggregation of different type of distributed resources which may be dispersed in different points of medium voltage distribution network. In VPP is composed of a number of various technologies with various operating patterns and availability which they can connect to different point of distribution network. VPP defines as a multi-technology and multi-site heterogeneous entity. In the FENIX project concept of VPP is defined as: "A Virtual Power Plant (VPP) aggregates the capacity of many diverse DERs, it creates a single operating profile from a composite of the parameters characterizing each DERs and can incorporate the impact of the network on aggregate DERs output. A VPP is a flexible representation of a portfolio of DERs that can be used to make contracts in the wholesale market and to offer services to the system operator."

B. General Concept

A virtual power plant is a cluster of dispersed generator units, controllable loads and storages systems, aggregated in order to operate as a unique power plant. The generators can use both fossil and renewable energy source. The heart of a VPP is an energy management system (EMS) which coordinates the power flows coming from the generators, controllable loads and storages. The communication is bidirectional, so that the VPP can not only receive information about the current status of each unit, but it can also send the signals to control the objects.

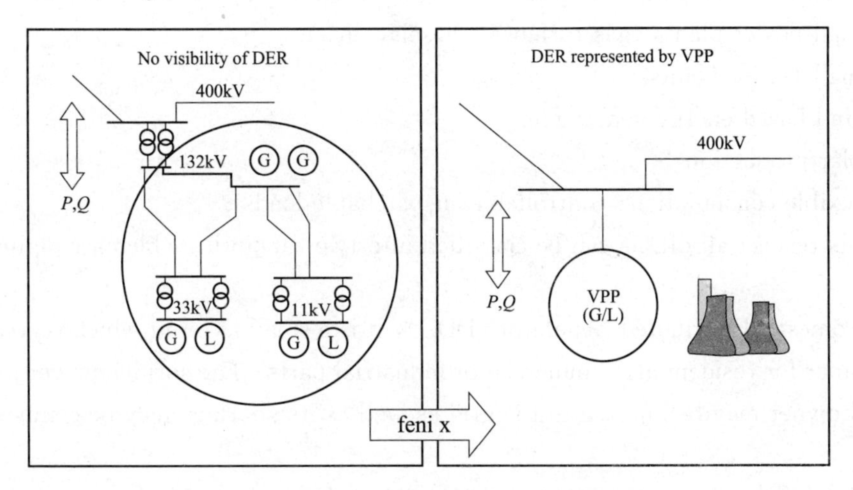

Figure 7.21 VPP Concept in FENIX.

The mentioned Energy Management System (EMS) can operate according to its targets which can be, for example, the minimization of the generation costs, minimization of production of green house gasses (GHG) and maximization of the profits. In order to achieve such targets the EMS needs to receive information about the status of each unit on the one hand, and on the other hand forecast - especially for renewable units like wind and photovoltaic (PV). Furthermore, the information about the possible bottlenecks in the grid plays a relevant role in the optimization process of the VPP operation. In this way the EMS can choose the optimal "modus operandi" . Due to the fluctuating nature of renewable energy sources, the prediction of the energy production is not an easy procedure. Actually, for Wind Park, the day ahead forecasting errors are between 9% and 19%. Due to such errors, power networks with a high penetration of renewable energy sources, can easily have bottleneck and balancing problems. These problems can be faced either by using ESS or NSM. Although ESS and NSM are the most used instruments, in some cases alternative solutions can be also used. For example, in regions poor of fresh water, desalination plants that are driven by the electricity surplus can be an optimal solution to face bottleneck situations.

7.3.3 VPP Components

The ideal Virtual Power Plant consists of three main parts including:

A. Generation Technology

The DG specification is useful to broadly mention the range of capabilities for various technologies, generally falling under the distributed generation category. DER considered for integration in VPP:

- CHP (Combined Heat and Power).
- Biomass and biogas.
- Small power plants (gas turbines, diesels, etc.).
- Small Hydro-plants.
- Wind based energy generation.
- Solar production.
- Flexible consumption (controllable/dispatchable loads).

In this respect all ofDGs can be classified into two categories which are defined in following.

1) Domestic Distributed Generator (DDG), it is a small DG unit which serves individual consumer for residential, commercial or industrial parts. The surplus power production of a DDG owner may be injected to the grid as well as its shortage may be compensated by the grid.

2) Public Distributed Generator (PDG), it is a DG unit which does not belong to an individual consumer and its primary aim is to inject its power production to the grid. Generally, both DDG and PDG can be equipped with energy storage. DDG is referred to a generator with a load and probably an energy storage which is usually connected to low voltage

distribution network. On the other hand, PDG is referred to a generator and probably an energy storage which can only be connected to the medium voltage distribution network. The distinctions of DDGs and PDGs are as follow:

1) The aim of the owners of DDGs is to provide economically their electrical and probably heating needs as well as to promote probably their services reliability. They are uninformed of the power business rules. On the other hand, the aim of the owners of PDGs is to sell their power production to the network customers.

2) Generally, the capacities of DDGs generation are small in comparison with PDGs. So, a DDG is never being able to participate in the power market independently as an individual participant, but a PDG may test its chance in the power market.

SomePDGs or DDGs have stochastic nature, e. g. wind and photovoltaic units which are not equipped by energy storages. But some others, e. g. fuel cells and micro turbines, are dispatchable that is; they are capable to vary their regime of operation almost quickly. In this respect, both PDGs and DDGs can be subdivided into two categories, i. e. Dispatchable PDGs (DPDGs) and Stochastic PDGs (SPDGs) for the former and also Dispatchable DDGs (DDDGs) and Stochastic DDGs (SDDGs) for the later.

B. Energy Storage Technologies

Energy storage systems can be considered today as a new mean to adapt the variations of the power demand to the given level of power generation. In context of use renewable generation, can be used also as additional sources or as energy buffers in the case of non-dispatchable or stochastic generation, e. g. wind turbines or PV technologies especially in weak networks. ESS considered for integration in VPP:

- hydraulic Pumped Energy Storage (HPES).
- compressed air energy storage (CAES).
- flywheel energy storage (FWES).
- super conductor magnetic energy storage (SMES).
- battery energy storage system (BESS).
- supercapacitor energy storage (SCES).
- hydrogen along with fuel cell (FC).

C. Information Communication Technology (ICT)

The important requirement for VPP is communication technologies and infrastructure. In many different communications, media technologies can be considered for communications in Energy Management Systems (EMS), Supervisory Control and Data Acquisition (SCADA) and Distribution Dispatching Center (DCC).

7.3.4 Technical VPP (TVPP) & Comercial VPP (CVPP)

A. Technical VPP (TVPP)

The TVPP consists of DER from the same geographic location. The TVPP includes the real-time influence of the local network on DER aggregated profile as well as representing the cost and operating characteristics of the portfolio. Services and functions from a TVPP include local system management for Distribution System Operator (DSO), as well as providing Transmission System Operator (TSO) system balancing and ancillary services. The operator of a TVPP requires detailed information on the local network; typically this will be the DSO. The TVPP enables:

- Visibility of DER units to the system operator (s).
- Contribution of DER units to system management.
- Optimal use of the capacity of DER units to provide ancillary services incorporating local network constraints.

This allows small units to provide ancillary services and reduces unavailability risks by diversifying portfolios and capacity compared to stand-alone DER units. A comprehensive overview of the technological control capabilities of distributed generators and the resulting possibilities of providing ancillary services are analyzed. The technological potential is investigated by application of a new assessment approach that considers the grid-coupling converter separately with its particular capabilities. An enormous technological potential is identified. DSOs that use the TVPP concept can also be considered as Active Distribution Network (ADN) operators. An ADN operator can use ancillary services offered by DER units to optimize their network operation. On the other hand, an ADN operator can also provide ancillary services to other system operators. A hierarchical or parallel structure ofADNs may exist where the TVPP concept is applied, for instance according to different voltage levels or different network regions. Many examples of ADNs can be found in the Active Network Deployment Register. Some of the functionalities that have to be performed by TVPP are:

- Continuous condition monitoring - retrieval of equipment historical loadings.
- Asset management - supported by statistical data.
- Self-identification/self-description of system components.
- Fault location - automatically integrated with outage management.
- Facilitated maintenance.
- Statistical analysis and project portfolio optimization.

B. Commercial VPP (CVPP)

A CVPP has an aggregated profile and output which represents the cost and operating characteristics for the DER portfolio. The impact of the distribution network is not considered in the aggregated CVPP profile. Services or functions from a CVPP include trading in the wholesale energy market, balancing of trading portfolios and provision of services (through submission of bids and offers) to the [transmission] system operator. The operator of a CVPP can be any third party aggregator or a Balancing Responsible Party (BRP) with market access; e. g. an energy supplier. The CVPP enables:

- Visibility of DER units in energy markets.
- Participation of DER units in the energy markets.
- Maximization of value from participation of DER units in the energy markets.

This allows market access of small units and reduces the risk of imbalance by portfolio diversity and capacity compared to stand-alone DER units. CVPPs perform commercial aggregation and do not take into consideration any network operation aspects that active distribution networks have to consider for stable operation. The aggregated DER units are not necessarily constrained by location but can be distributed throughout different distribution and transmission grids. Hence, a single distribution network region may have more than one CVPP aggregating DER units in its region.

Basic CVPP functionalities would be optimization and scheduling of production based on predicted consumers' demand and generation potential. When actual needs differ from predicted ones, DRRs (Demand Response Resources) are introduced to fill the gap between production and real consumption. In general, CVPP functions should also include:

- Maintenance and submission of DERs' characteristics.
- Production and consumption forecast.
- ODM (Outage Demand management).
- Building DER bids.
- Bids submissions to the market.
- Daily optimization and generation scheduling.
- Selling energy provider by DERs to the Market.

7.3.5 Conclusion

Through the VPP concept IndividualDERs can gain access and visibility across all energy markets, and benefit from VPP market intelligence to optimize their position and maximize revenue opportunities. System operation can benefit from optimal use of all available capacity and increased efficiency of operation. Benefits from the Virtual Power Plant concept have been identified for different stakeholders:

Main benefits for owners of DER units:

- Capture the value of flexibility.
- Increasing value of assets through the markets.
- Reduced financial risk through aggregation.

Improved ability to negotiate commercial conditionsMain benefits for DSOs and TSOs:

- Increased visibility of DER units for consideration in network operation.
- Using control flexibility of DER units for network management.
- Improved use of grid investments.
- Improved co-ordination between DSO and TSO.

Mitigate the complexity of operation caused by the growth of inflexible distributed generation Main benefits for Policy Makers:

- Cost effective large-scale integration of renewable energies while maintaining system security.
- Open the energy markets to small-scale participants.
- Increasing the global efficiency of the electrical power system by capturing flexibility of DER units.
- Facilitate the targets for renewable energy deployment and reduction of CO_2 emissions.
- Improve consumer choice.
- New employment opportunities.

Main benefits for suppliers and aggregators:

- New offers for consumers and DER units.
- Mitigating commercial risk.
- New business opportunities.

7.4 Accessing Flexibility of Electric Vehicles for Smart Grid Integration

7.4.1 Introduction

The rising awareness of the limitation of resources and the environmental impact of CO_2 emissions in combination with the technological progress in the last decades brought electric vehicles (EV) back on the market. Contrary to the still mainstreaming passenger cars with internal combustion engine, electric cars do not necessarily depend on a CO_2-emmitting process as power source if they are equipped with a charging interface for the internal battery. The fact that an infrastructure for electricity already exists is beneficial in terms of power availability and the possibility to use renewable energy sources for charging. But the challenging aspect is that the capacity of the electrical grid was originally neither designed for the additional load by EV nor for the simultaneously ongoing energy transition from conventional to renewable energy generation.

Grid integration of EV is discussed for several years now and since the beginning different aspects have been in the focus of published research papers. The increasing electricity demand of a growing number of EVs was identified as a challenge to the distribution system already in 1998 and lead to the early development of a load leveling strategy for vehicle batteries in order to avoid the creation of new peaks in the electricity demand curve. Nearly one decade later renewable energy generation is added to the problem. Most renewable electricity generation is direct dependent on its natural, intermittent and fluctuating supply and therefore less predictable and less flexible than conventional generation. With the growing share of renewable energy and the therewith decreasing reliability on the generating side the need for more flexibility was raised in the grid and on the demand side. In order to keep investments into grid reinforcements as low as possible and to make use of the current state of the art in information and communication technology (ICT) several smart grid approaches have been developed and are still under investigation. The basic idea is to exploit synergies and to optimize the overall efficiency by a higher grade of coordination due to more available information. This pulled energy storage systems (ESS) in general and thereby the management of the energy-flow between the grid and the batteries of EVs back into the focus of scientific investigations. The fact, that an average vehicle is most of the time parking and potentially connected to the grid makes an EV flexible regarding the charging time schedule. This flexibility can be utilized in a smart grid context.

With a smaller focus the motivation is often expressed by stating that CO_2-emission-free mobility can only be realized if electric cars are charged by renewable energy sources and should therefore be synchronized with renewable power generation. Even thou this argument seems more appealing and easier to access, a greater benefit can be reached in a smart grid approach. Instead of following the renewable power generation the flexibility of the battery can be used to level out the remaining fluctuations in the difference between generation and consumption in the accounting grid. In this case one part of the intermittent renewable generation and one part of the stochastic consumption has already canceled each other out. This increases the consumption of renewable generated electricity on the inflexible consumption side and reduces the need for conventional CO_2-emitting countermeasures.

This paper deals with the integration of electric cars into the smart grid by making their flexibility accessible to a third party. Researchers are involved with this topic for several years now and many findings have already been published. The objective of this paper is to give an overview over already investigated problems and developed solutions. Therefore a general map of aspects which outline the problem of grid integration and the utilization of the electric vehicles flexibility is presented.

Since the topic consists of several aspects the findings and approaches are pooled by their major aspects and commented in the following chapters.

7.4.2 Lexibility in the Context of EV

The term 'flexibility' in the context of a system can be described as a variable which is a degree of freedom of the system. This means that regardless which value is chosen for the variable the successful operation of the systems primary task is not affected. The term utilization of flexibility says that this variable is put into a different context, where its value is of importance.

In the case of a passenger car the flexibility comes from the fact, that the vehicle is used less frequently than it could be used. In other words, the availability is higher than necessary. In comparison to conventional cars EVs are equipped with a different type of ESS which is in most cases a battery storage system that is used to store electricity. This has the advantages that electricity is at least theoretically highly available, that the charging process can be automated and that storage is rare and needed in the power grid.

The fact that traction batteries cannot be charged as fast as a fuel tank can be refilled enforces a different behavior of the driver and a different technology for the recharging process. The installation of a power supply at parking spaces where a car is parked for the longest periods of time, for example at home and at work, represents not a great challenge in many individual cases.

The best example to reveal the flexibility of the charging process is an overnight charging process, when the time needed for charging is shorter than the parking time. Charging can be described mathematically as an integration of power over the parking time period. This implies that the distribution of the power over time is not relevant for the result. Consequently as long as the end of the charging time period is known, the shape of the charging processes power demand can be freely defined within the boundaries set by the rated power of the supply and the batteries charging capability.

The pure flexibility of an EV is thereby just the possibility to schedule its power demand within its boundaries. To utilize this flexibility the driver of the car needs to be encouraged by incentives to plug in the EV whenever the car is parked, to reveal to the user of the flexibility when the car will be used again and to define under what conditions the battery can be charged. It becomes obvious, that in this case the operation of the vehicle is not affected.

Other than this, the vehicle to grid (V2G) approach is more demanding. First of all additional hardware functionality is required inside of the vehicle. A functionality that is not needed for the driving operation but that is always on board as additional weight. The conceptual discharging of the battery leads to additional charging cycles and thereby to a higher degeneration of the battery. These costs need to be added to the incentives in a business case.

7.4.3 General Map of Grid Integration Aspects

The general map of grid integration aspects is depicted in Figure 7.22 Beginning in the car, the first major aspect is the battery system and its management. The next aspect is the charging technology, which consists of on-board and off-board charging systems. Power quality at the grid connection point builds the next aspect. The smart grid context is introduced with the enabling aspect of communications and it is focused in the aspect energy management.

7.4.4 Battery System and ITS Management

At the current state of the art the most widely used type of energy storage technology in EVs are batteries. This application area makes high demands on the technology by needing simultaneously high energy density, high power density, wide range of operation temperature, low self-discharge, high number of charging cycles and low acquisition costs. Some of the properties can be enhanced by additional systems. But every enhancement that can be achieved comes at the cost of other properties. The power density, for example, can be increased, not continuously but for periods of time, by the integration of capacitors into the ESS. The range of operation temperature can be optimized by thermal insulation material in combination with higher climatisation effort. The additional system rises in both cases losses and costs. The additional required space leads to a reduction of the full systems energy density.

Battery systems are generally composed of multiple single battery cells. The nominal cell voltage of a battery cell is depending on the technology between 1, 2 and 3, 8 V. In order to keep current values and conductor cross-sections low, a higher Voltage is reached by interconnecting cells in serial strings. Serial and parallel connections of cells build a modul and on a higher level the battery pack. Due to production caused deviations and parasitic effects the voltage along a string can become unevenly distributed. This makes voltage balancing measures necessary, which preferably should come with low losses.

During operation in the battery system occurring thermal dissipation losses needs to be removed in order to stay within the acceptable temperature range. Therefore in most cases an active climate control is part of the battery system in order to remove the influences of thermal dissipation losses and environmental temperature deviations.

The battery management system (BMS) has to keep the system within the intended operating range and to protect it from undesirable conditions such as overcharging, deep discharge or overheating. To do this, the BMS performs safety measures like the prevention of power flow in a damaging direction.

For an effective realization of the BMS the not direct measurable state of charge (SOC) is of high importance. Its determination is normally based on a battery model which is fed by measurable values. This task is together with the safety management the main task of the BMS.

This is still a matter of research because chemical processes are difficult to observe and every new battery technology needs a new model together with other not direct measurable quantities.

A good overview over research on BMS can be found.

7.4.5 Charging Technology

For the determination of general conditions under which charging can be carried out first an overview of the different operating modes of EV is given. In general two pairs of conditions can be distinguished for an EV. It is either moving or standing still and either

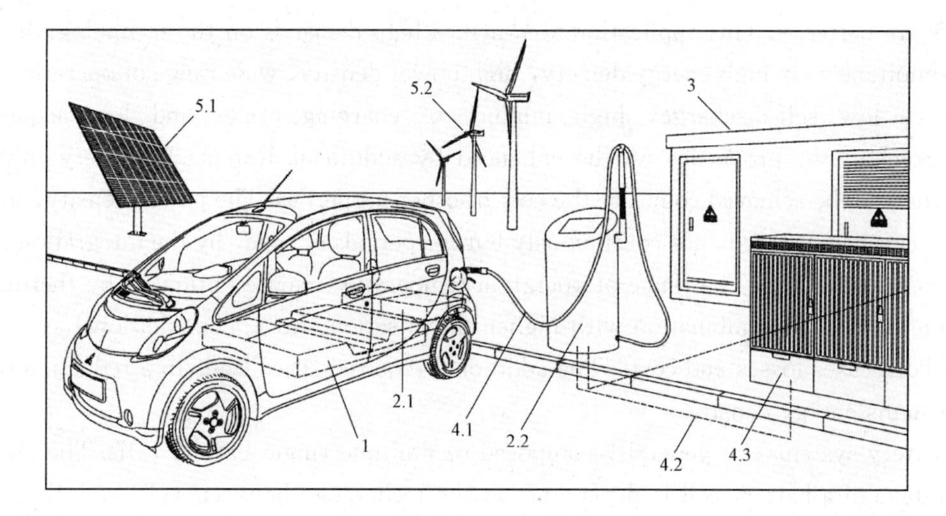

Figure 7. 22　General map of aspects which form the problem of grid integration and the utilization of the EV flexibility.

1. Battery system and its Management, 2. 1 On-board charger, 2. 2 Off-board charger, 3. Power quality at the grid connection point, 4. 1 Communication between vehicle and charging station, 4. 2 Communication between charging station and the central system, 4. 3 Communication grid connection point, 5. 1 Photovoltaic power generation (Smart grid challenge), 5. 2 Wind power generation (Smart grid challenge).

connected to the grid or autonomous. The combinations of these different conditions lead to the following four possible states:

a) autonomous and moving: the primary task of a vehicle, the driving operation.

b) autonomous and standing still: the result of non reachable optimal efficiency, e. g. traffic jam, stop and go traffic, stopping or parking.

c) connected to the grid and moving: a state which makes great demands on the infrastructure, because the whole stretch or at least a long section of the way needs to feature the infrastructure for example a contact line or in the street embedded induction coils.

d) connected to the grid and standing still: a state which is less demanding on the infrastructure, since only parking places need to be equipped with new infrastructure.

Consequently for a cost saving integration of EVs into the smart grid only state d) is of relevance. In this state wired systems or wireless inductive systems can be employed for the power transfer. A worldwide overview over charging technology research can be found.

In addition to the necessity of a medium for the transmission of power, current and voltage need to be adjusted regarding frequency and amplitude according to the desired mode of operation. For charging the functionality of a rectifier and a chopper is needed. For V2G applications the chopper needs additionally to be able to work with a current flowing in the opposite direction in combination with the functionality of an inverted rectifier.

The power electronic for the bidirectional power transfer between battery and grid is called charger '. Three different ways how the charger can be positioned in relation to the other power electronic components in the car are defined: off-board charger, onboard charger and integrated on-board charger. The fundamental idea of the integrated on-board charger is to make use of synergies by combining the hardware that is primarily needed for state a) with the hardware that is needed for state d). The particular optimum depends on the charging power that has to be realized. On the basis of the charging modes, defined in IEC 61851-1, the following power ranges can be summarized:

A. AC, single phase: up to 3, 7 or 7, 4 kW, (8 or 4 estimated charging hours)

B. AC, three phase: up to 11 kW or 22 kW, (3 or 1, 5 estimated charging hours)

C. DC, up to 400A , (0, 5 estimated hours for 80%)

Thereby stated charging times have been estimated conservatively for a 16 kWh battery. The particularly towards the end of the charging process decreasing charging power is considered by a correction factor, which was chosen to be 1, 85 for smaller power (case A) and 2 for higher power (case B and C). The correction factor was multiplied with the time that results from the calculation: energy divided by power.

Case A and B can be realized as on-board charger and have the advantage that existing infrastructure does not have to be changed very much. In case C the integration of the required power electronic components into the car would be inappropriate because of its weight and extend. This case is realized as off-board charger, e. g. CHAdeMO-charging stations.

Case B exceeds case A regarding the temporal flexibility and is additionally not an unbalanced load. For case A argues just the higher availability and a higher flexibility if short ranges were driven and just a recharge is needed. Nevertheless, both cases have the potential to realize the connection to the smart grid for the majority of parked vehicles, due to their proximity to the existing infrastructure.

Case C could offer the highest flexibility but the more expensive infrastructure and the conceptual short charging time are likely to prevent a utilization for long parking periods which reduces the flexibility drastically.

7.4.6 Power Quality

The battery of an EV is a DC based system. In contrast the mains power supply in the distribution network is AC based. For the conversion of AC to DC or DC to AC switching power electronic devices are either integrated in the car or in the charging station. The simplest and cheapest circuit arrangements cause system perturbations which have a non-negligible effect on the network, especially in the power classes of EVs. Particularly when in the expected future EVs have reached a large-scale and dense distribution in the distribution grid, missing countermeasures would result in severe loss of quality. But this is not a typical EV problem, the same applies to all simple rectifiers and inverted rectifiers that are connected to the grid. From this follows that approved approaches like input filter inductances and direct current control can be adapted as countermeasures. A similar approach is followed, where a battery charger for EV is proposed that mitigates the power quality degradation.

In the past inverted rectifiers were rare in many grid areas. Their occurrence has increased by private photovoltaic systems and will increase even more with the growing number of EV. Due to the increase of the supply from inverters at the same time, the number of supplying synchronous generators decreases, whereby the proportion of reactive power deployment capability also decreases.

The norm VDE-AR-N 4105 defines for decentralized generating plants in Germany which are connected to the low voltage grid that they have to be able to provide reactive power up to a shift factor cos (φ) of 0, 95 for a rated power between 3, 68kVA and 13, 8kVA. For plants with a rated power above 13, 8kVA even a shift factor cos (φ) of 0. 90 has to be possible.

As soon as inverters of EVs or charging stations are operated in V2G-Mode they are decentralized generating plants and have at least to fulfill the above mentioned norm.

Even without injecting active power from the battery into the grid a smart grid V2G application case could be just the delivery of reactive power. One interesting aspect about this would be the highly distributed appearance of EVs in the grid, which could allow for a spatial more accurate voltage regulation.

The spatial aspect is also addressed, stating that large parking areas with simple charging stations should be connected at the strongest grid lines and smart charging stations, which support V2G including reactive power, can be connected to weak grid lines.

7.4.7 Communications

Putting the access to the energy grid in the context of information exchange is a primary characteristic of a smart grid. In the case of EV different information needs to be exchanged for the realization of smart grid application cases, for example: the SOC, energy prices, switching commands, billing data or authentication data for invoicing. The communication link can be subdivided into the section between the vehicle and charging station and in the section between the charging station and the central system. A data transmission channel can first be subdivided into the physical realization and the logical encoding. Regarding the physical realization wired and wireless systems are discussed. For the user interface in addition to car based or charging station based approaches also application based interfaces for mobile phones have been developed.

Wired systems require due to the mobility requirements connectors that meet the requirements of the power transfer as well as those of the information transmission. The different interests of the industry have generated in addition to the necessities for different applications a variety of connector types, some of which were included in the standard IEC

62196. Also for the logical implementation of the communication link standards have been developed, e. g. for the section EV - charging station the standard IEC 61851 was defined. As an addition for bidirectional communication and V2G capability the standard IEC 15118 is mentioned. For communication between the charging station and the central system, the Open Charge Point Protocol (OCPP) has been established, which is based on the mentioned IEC standards among others. An overview about the OCCP is given. The mentioned standards and protocols are still being developed, and in particular the security of communication is still an important research topic

The relevance of the security goals: confidentiality, integrity, authenticity and availability is explained in relation to EV and smart grids: confidentiality affects the privacy of the customer and the secrecy of the service provider, authenticity gives proof about the sender and content of messages, integrity protects from malicious modifications and most important the availability which makes the difference between a smart grid and a simple energy grid.

In order to satisfy end-to-end security requirements for the communication in the wired case the idea of an extension for the IEC 61850, which deals with the design of electrical substation automation, is proposed.

However, the communication between EV and charging station can also be realized wireless, e. g. if a standard connector is used for the energy transfer, which has no extra communication lines, or an inductive interface is to be used. Different types of communication protocols can be used to establish a wireless information transfer between vehicle and base station. For example a GSM (Global System for Mobile Communications) or CDMA (Code Division Multiple Access) based approach using SMS (Short Message Service) is proposed. A test bed that allows for the exploration of Zig Bee, Bluetooth, Z-Wave, Home Plug and GSM/GPRS SMS has been described. A system demonstrator combining the wireless standard IEEE 802. 15. 4 with OCPP is described.

In the case of wireless communication a destructive effect can occur, that affects the security aspects in terms of availability. It happens, when the wireless transmission is jammed by other transmitters. To cope with this possibility a jamming-resilient protocol for wireless networks is proposed.

7. 4. 8 Energy Management

Accessible flexibility needs to be managed in order to gain an additional benefit. For this task different approaches and objectives can be found in published papers, which are outlined in this chapter.

In general two perspectives can be distinguished in this context: approaches with a local scope, where just one up to a small number of EV are managed within a very small grid context, or approaches with a regional scope, where many EV are connected distributed in a larger grid context. Most common objectives are load shaping aiming for a higher degree of grid capacity utilization, ancillary services or a higher rate of renewable energy utilization like wind energy or PV. Other goals are the minimization of charging costs or power quality.

The main two management strategies are centralized direct control or a local management based on a time-of-day electricity tariff. A comparative investigation can be found that concludes that the investigated direct control had a better performance.

For the simulative analysis of the management approaches different models have been developed, for example:

- A model for the charging behaviour of a single EV based on lead-acid battery technology under consideration of the driving mode of the EV is proposed.
- A model for the prediction of multiple EVs charging demand is used.
- A Model of the power electronics and control needed for V2G capability can be found.
- Grid models for the evaluation of charging impact on the grid are presented.

Methods used in this context are the following:

- a Markov model is used for the simulation of the vehicle use dependent change of the SOC.
- power flow calculation is performed on grid models for the analysis of the impact on the grid.
- hierarchical model predictive control (Hi-MPC) is used to realize primary, secondary, and tertiary control in the simulated grid.
- heuristic maximization is used for joint vehicle guidance and power control.
- mixed integer linear programming (MILP) is used for modeling a charging optimization problem and the description of a scheduling problem.
- microscopic discrete-time traffic operations are simulated for the investigation of the kinematic state of a EV fleet.
- activity recognition is applied to smart meter data and smart home events for the prediction of future human interactions with EV.
- and game theory is used for the optimization of a energy trading approach.

Many aspects of V2G related research have large overlaps with the integration of a stationary battery storage system in the network, be it power quality, the application as uninterruptible power supply (UPS), local energy management or even DC microgrid integration. However, for EV it should not be neglected that their number in the grid varies temporally and spatially. A fact that can also be an advantage, if it is considered that there is a causal relationship between the whereabouts of vehicles and the location of a subset of people. It might therefore be worthwhile to gear the V2G energy services more to the needs of nearby people, rather than to processes that are independent from the location of people. This would require a much stronger consideration of spatial and social aspects in the context of EV.

7.4.9 Summary

This paper deals with the integration of electric cars into the smart grid by making their flexibility accessible to a third party. After a discussion of the term flexibility in the context of EV, a general map of grid integration aspects has been drawn, consisting of the battery system and its management, charging technology, power quality, communications and energy management. In the following corresponding chapters an overview is given over already investigated problems and developed solutions that can be found in published papers. The report on the review of papers shows which topics have been investigated and what methods have been used.

7.5 Cyber Security for Smart Grid Systems: Status, Challenges and Perspectives

7.5.1 Introduction

The integration of electrical distribution system with communication networks forms smart grid where power and information flow is expected to be bi-directional. This transformation of traditional energy networks to smart grids revolutionizes the energy industry in terms of reliability, performance, and manageability by providing bi-directional communications to operate, monitor, and control power flow and measurements. Furthermore, smart grid is expected to automate the systems with the help of advanced communication systems. Along with several benefits the communication networks offers in smart grid, they bring the private power control systems to the public communication networks and associated security vulnerabilities. Smart grid can be a prime target for cyber terrorism because of its critical nature. As a result, cyber security for smart grid is getting a lot of attention from governments, energy industries, and consumers. There have been several research efforts for securing smart grid systems in academia, government and industries.

According to National Institute of Standards and Technology (NIST) conceptual model for smart grid, communication networks connect power system components as shown in Figure 7.23. There are seven logical domains: Markets, Service Provider, Operations, Bulk Generation, Transmission, Distribution and Customer. The first three deal with data collection and power management whereas the last four deal with power and information flows in the smart grid. These domains are connected with each other through secure communication links as shown in Figure 7.23.

Smart grid has different components and assets such as power generations, distributions, consumers, regional control centers, substations, field devices, communication and

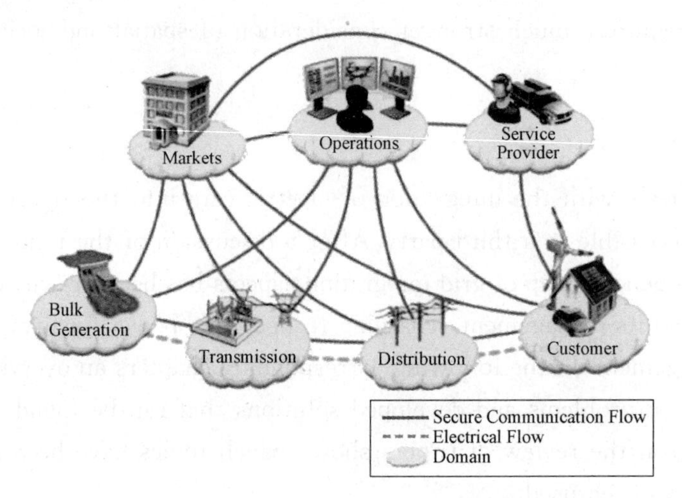

Figure 7. 23　The NIST conceptual model for smart grid.

networking devices, phase measuring units, protecting relays, intelligent electronic devices, remote terminal units, human machine interfaces, home appliances, circuit breakers, log servers, data concentrators, protocol gateway, tap changers, smart meters, etc. All of these components are connected in smart grid to operate, monitor, and control power flow and measurements. Thus, the legacy cyber-security techniques are not sufficient to meet the cyber security requirement of the smart grid and its assets. For instance, consumers are concerned about their privacy as their lifestyle could be exposed to malicious users because of compromised data communication networks. Similarly, Advanced Metering Infrastructure (AMI), commonly known as the smart meter, could be easily compromised however once they are compromised, it is almost impossible to change their passwords (PINs) as these devices do not have their own keyboards to change passwords/PINs. Thus, a controller may be needed to deploy new passwords automatically once it is compromised. Thus, the smart grid systems have unique features, goals and objectives to provide reliable power supply and robust communications.

This article provides a comprehensive study of challenges in smart grid security, which we concentrate on the problems and proposed solutions for cyber attacks and defense solutions. In this paper, specifically, we present the following issues:

- We outline the different requirements of smart grid systems and compare them with the traditional Internet. We then provide network vulnerabilities in the smart grid systems. It is noted that the most of the cyber attacks in smart grid systems are coming through malicious threats in communication networks.
- Because of the nature of the power grid and its impact in case of blackout, attack detection should be quick so as to have uninterrupted power supply. Thus, it is essential to deploy malicious attack prevention and defense solutions to protect the entire smart grid. Thus, we outline cyber attacks and respective defense solutions for the

smart grid.

Then we outline current state of the research and future perspectives. With this article, readers can have a more thorough understanding of smart grid security and the research trends in this topic.

The remainder of the paper is organized as follows: We present a background and smart grid architecture in Section 7. 5. 2 followed by the requirements and objectives of smart grid security in Section 7. 5. 3. Cyber attacks and defense solutions for smart grid systems are presented in Section 7. 5. 4. The challenges for securing smart grids are presented in Section 1. 2.

7. 5. 2 Background and Communication Network Architecture for the Smart Grid Systems

Electrical power networks are very complex since, for example, there are over 2000 power distribution substations, over 5500 energy facilities distributed throughout the country, and over 130 million customers all over the United States.

Typical communication framework for smart grid system is shown in Figure 7. 24. Home appliances of consumers are connected to Home Area Network (HAN) and they report their need and usage pattern of electricity in real-time to control and monitor the real-time power consumption. Smart metering system is composed of a micro-controller and a communication board to communicate with the gateway. Third party who provides electricity and manages operations relies on meter readings to provide value-added services for consumers. HAN covers single home or business. The HAN could use ZigBee, Bluetooth, WiFi, etc. HANs are connected to Neighborhood Area Network (HAN) through Neighborhood Area Network (NAN) gateway. NAN covers HANs, sub-stations and distribution systems. Note that a HAN could be treated as a single device at NAN as HAN gateway could aggregate the information received from HAN and forward that to NANs. The NANs are connected to Wide Area Network (WAN) through NAN gateways. WAN cov-

Figure 7. 24 A typical communication framework for smart grid systems.

ers power generations to transmissions as shown in Figure 7. 24.

Examples of WAN include optical fiber networks, WiMAX, and the latest cellular networks (such as 4G/LTE). Smart grid relies on wired and wireless communication networks, thereby inheriting both benefits they offer and security vulnerabilities. The smart grid has potential of introducing new security vulnerabilities into the power system and thus different cyber defense solutions are needed in different levels to safe guard the entire system.

A. Features of Smart Grid Networks

The smart grid network is expected to be larger than the existing Internet and share the somewhat similar architecture. However, there are significant differences between them.

1) Latency requirements: Internet (network of networks) is intended to provide data services (sharing files, surfing, etc.) to the users with high speed data rate. However, smart grid networks are intended for reliable, secure and real-time communication with low latency.

2) Data size and flow: Internet has generally bursty type communications however smart grid is expected to be bulky and has periodic data communications because of big size of the network and real time communication and monitoring requirements.

3) Communication model: In traditional power grids, the typical model for communication is one-way where electronic devices report their readings to the control center. But in smart grid, communication is bi-directional and real-time.

4) Password/PIN update process: In typical Internet, all end/networking devices have keyboards to enter/change PIN or password. However, in smart grid, end devices such as smart meter and/or some home appliances might not have keyboard to change or enter password/PIN. Thus smart grid needs some sort of automated process for this to deploy new policies and/or to change passwords.

5) Layered network architecture: There are different networks in smart grid at different levels such as HAN, NAN and WAN. The HAN works like a node for NAN and NAN works like a node for WAN through their respective gateways. Furthermore, for reliability, smart grid is expected to use variety of access networks including wireless, wired and fiber optical networks.

B. Power Systems Communication Network Protocols

There are various proprietary protocols and few open standard protocols. Two widely-used communications protocols in power systems are

1) *Distributed Networking Protocol* 3. 0 (*DNP*3): DNP3 was originally developed by General Electric Inc. that was made public in 1993. The DNP3 is the predominant standard used in North American power systems. Physical layer of the its initial version was based on serial communication protocols (such as RS-232, RS-422, or RS-485). However, its

current version is based on TCP/IP model which supports recent communication technologies with end-to-end communications.

2) *International Electrotechnical Commission* (*IEC*) 61850: IEC 61850 protocol is recently standardized with Ethernetbased communications for modern power substation automation by the International Electrotechnical Commission . IEC 61850 was designed to replace DNP3 in smart grid communications, however, current IEC 61850 is only limited within a power substation communications. Unlike DNP3, IEC 61850 built with a series of protocol stacks to support a variety of services which are time-critical and monitoring. Power substation communication deals with a number of time critical messages with end-to-end delay of 3 milliseconds to 500 milliseconds.

There are different types of messages with different requirements: 'Type1A/P1' and 'Type 1A/P2' messages have strict time delay requirements since they are used for fault isolation and power system protection purposes. The delay requirements for the 'Type 1A/P1' type messages is milliseconds and for the 'Type 1A/P2' type messages is 10 milliseconds. Next, the 'Type 1B/P1' and Type 1B/P2' messages are used for routine communications among automated components of power systems. The delay requirements for the 'Type 1B/P1' message has 100 milliseconds and for the 'Type 1B/P2 ' is 20 milliseconds. Similarly, 'Type 2' and 'Type 3' messages are used for monitoring and readings in substations which have less time-critical requirements. The delay requirement for 'Type 2' is 100 milliseconds whereas it is 500 milliseconds for 'Type 3' messages.

7.5.3 Smart Grid Sceurity Requirements and Objectives

Cyber attacks in smart grid systems depend on the several factors. Integration of bi-directional communication networks, incentives to attackers, socioeconomic impact of the blackouts, etc. Basically, the attack risk in the smart grid system relies on three factors as shown in Figure 7.25. Formally, the risk can be defined as

$$\text{Risk}=\text{Assets}\times\text{Vulnerabilities}\times\text{Threats} \tag{7.1}$$

where Assets are the smart grid devices (such as smart meters, substations, data, network devices, etc.), vulnerabilities allow an attacker to reduce a system's information assurance, and Threats are the attacks coming from outside or inside of the smart grid systems.

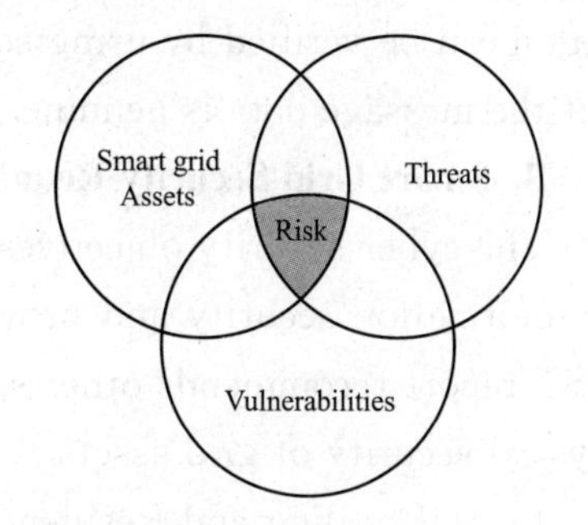

Figure 7.25 Evaluating the risks in smart grid systems.

In (7.1), the 'Risk' can be minimized or made zero if one the quantity on the right side is minimized or made zero. Note that in smart grid systems, Assets cannot be zero in the smart grid. Threats cannot be made zero as they are coming from unknown places or attackers. Thus, the main goal is to minimize the Vulnerabilities in the smart grid to minimize the overall 'Risk' in (7.1).

A. Security Objectives in the Smart Grid Systems

Objectives of smart grid security is to comply with policies while securing information using Confidentiality, Integrity and Availability, also known as the CIA triad. The CIA triad is a model designed to guide policies for information security in smart grid systems which is shown in Figure 7.26.

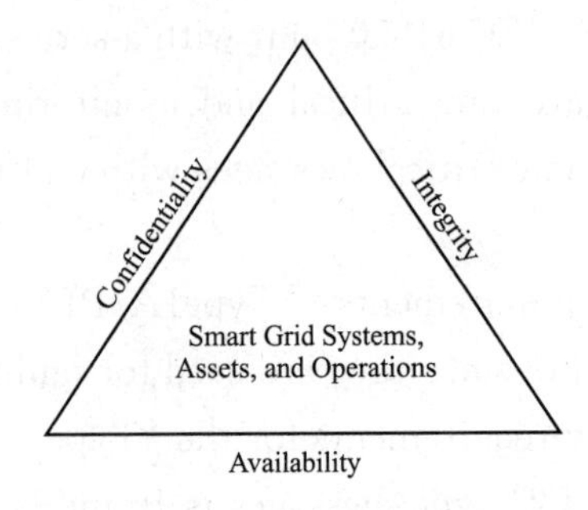

Figure 7.26　The CIA triad for smart grid security systems.

Confidentiality in the smart grid systems is needed to make sure that access to information is restricted to only authorized people and it is designed to prevent unauthorized access. In other words, it is equivalent to privacy. In smart grid systems where home appliances are connected to power grids for realtime bi-directional data communication and electricity flow, privacy is one of the important issues for the customers. If the information falls on wrong hand, they can keep track of the life style of the people, what appliances they use, whether the people are in there, etc. and misuse the information.

Integrity of information in smart grid is needed to maintain and assure the accuracy and consistency of data/information. The information should not be modified in an unauthorized or undetected manner. This feature helps the smart grid to provide robust real-time monitoring systems.

Availability in the smart grid implies that the information must be available to authorized parties when it is needed without any security compromise. Power systems is expected to be available 100% of the time, thus data availability also involves preventing denial-of-service attacks leading to blackouts.

Furthermore, Authenticity is also important in smart grid systems as it is important to validate that both parties involved are who they claim to be. Authenticity of the information can be verified by using some features such as "digital signatures" to give evidence that the message data is genuine.

B. Smart Grid Security Requirements

The cyber security objectives for the smart grid is to follow CIA triad to guide policies for information security and provide robust power supply. In addition to CIA triad, the NIST report recommends other specific security requirements for the smart grid including physical security of grid assets.

1) Self-healing and Resilience Operations in the Smart Grid: In smart grid systems, communication network is open as smart grid assets are distributed over large geographical area. Thus it is challenging to ensure the security in every single node in the smart grid to be invulnerable to cyber attacks. Thus the smart grid network must have some self-healing

capability against cyber attacks. Network must consistently perform profiling and estimating to monitor the data flow and power flow status to detect any abnormal incidents due to cyber attacks. To make data communication network available for power system operations, resilience data communication is essential.

2) Authentication and Access Control: There are millions of electronics devices deployed throughout the power systems and millions of home appliances are connected to smart meters, authentication is the key process of verifying the identity of a device or user to protect smart grid systems from unauthorized access. Furthermore, access control is used in smart grid to ensure that resources are accessed only by the appropriate parties that are correctly identified.

3) Communication Efficiency and Security: The smart grid communication needs to be efficient to support real-time monitoring and secure with self healing cyber defense solutions to protect from any security attacks. As these features are contradicting, trade-off between these parameters should be considered in smart grids.

C. Automated Policy/Password Update Process

Most of the nodes (e. g. , smart meters) in the smart grid do not have keyboards which makes the process of changing password/PIN more difficult. Furthermore, manually changing passwords in such million devices is not feasible, thus the smart grid needs an automated process to deploy the policies and/or passwords in real-time to prevent any attacks.

7.5.4 Cyber Attacks and Defense Solutions for Smart Grid Systems

In order to secure the sensitive data and smart grid, the Figure 7.27 shows five important layers that should be considered when defining cyber security schemes for the smart grid. Data security is in innermost layer and the WAN security is in the outermost layer. Note that cyber attacks could be at any layers to vandalize the entire smart grid.

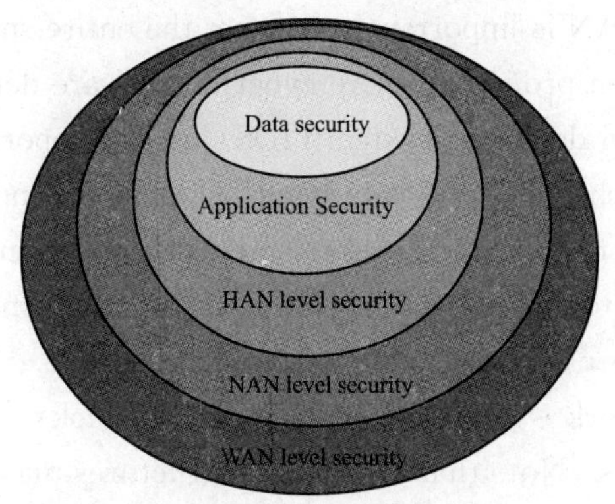

Figure 7.27 Smart grid security using multiple levels of protection against attacks.

In the following sections, we present classification of attacks in the smart grid systems.

A. Attack Classifications based on Networks

1) Home Area Networks (HAN) Attacks: Typical HAN attacks target the home appliances and smart meters. HANs use wired or wireless connections to provide interface to smart grid to support consumer awareness of energy consumption and to support demand response functionality in realtime. However, HAN is vulnerable to security attacks and the malicious users could use sophisticated attacks through easily accessible devices such as the smart meters and the associated communications hardware to interrupt the smart grid systems. Attackers could easily exploit the vulnerabilities available within the firmware of HAN devices and use reverse engineering of devices to attack the grid. Thus all of the known threats to such a network must be identified and addressed to avoid any damages caused by cyber vandalism to smart grid initiated at HAN level. Several approaches have been proposed in the literature to provide secure and reliable communications between the smart meter and consumer equipment in HAN to avoid security attacks. A security framework has been proposed which integrates HAN device registration and enrollment processes into a single network access authentication procedure. This scheme prevents unauthorized access to HAN by malicious users. A freshness counter based session key exchange scheme has been proposed to ensure defense against replay attacks (also known as playback attack) between the smart devices inside a house and the smart meter. This scheme helps prevent smart grid from valid data being delayed or transmitted repeatedly by malicious or fraudulent users. A scheme is proposed where security labels are used for data packets and enforce the information flow policy to avoid attacks in the grid through HAN. Other security mechanisms includes frequency hopping in wireless, dynamic security key management, advanced encryption schemes, intrusion prevention systems, intrusion detection systems, authentication and authorization.

2) Neighborhood Area Network (NAN) Attacks: Typical NAN attacks target the power sub-station and distribution centers. The NAN interfaces HAN with WAN in smart grid and attacks could be coming directly through power substations or through HAN gateways. Thus protecting NAN is important to protect the entire smart grid. A intrusion detection framework has been proposed where cyber attacks are detected at NAN level with the help of NAN intrusion detection system (IDS) using support vector machine (SVM) and artificial intelligent schemes. A 4-way handshaking mechanism has been proposed to establish secure links before smart grid assets start their communications. Note that HAN devices should be limited to communicating only with a HAN manager application within the meter to minimize cyber attacks.

3) Wide Area Network (WAN) Attacks: WAN attacks are targeted to power generation and control devices. Note that WAN infrastructures may be utility owned or public access depending on the business model of utility offices. There are different WAN standards including ANSI C12.21 and ANSI C12.22. In ANSI C12.21 based WAN

access protocol uses two-way authentication using DES encryption of a randomly generated toke. The ANSI C12. 21 protocol is session based thus a timeout can be implemented to release the session which reduces chances of cyber attacks based on the permanent key. ANSI C12. 22 adds another layer of security by having data encryption using AES-128 bit on top of WAN authentication. In ANSI C12. 22 standard each communication must be authenticated before any other operations. Note that the security schemes used in smart grid should provide highest level of security with minimal system and device performance impacts related to encryption, decryption, re-keying functions, intrusion detection, intrusion prevention, etc.

B. Attack Classifications based on Network Layers

Typical cyber attacks in the smart gird based on CIA Triad can be classified as below:

1) Cyber Attacks Targeting Confidentiality These attacks allow access to information to unauthorized users in the smart grid. Malicious users misuse the information to harm others or take advantages from it.

2) Cyber Attacks Targeting Integrity Main motivation of these types of attacks is to disrupt the data exchange in the smart grid by illegally modifying or inserting false information.

3) Cyber Attacks Targeting Availability Attackers' main motivation is to block or delay the communication in the smart grid so that the power delivery could be interrupted. This type of attack is also known as denialof service attack. A typical attacks are listed in Table 7. 8.

Table 7. 8 Typical Cyber Attacks Targeting Availability in Power Grids.

Network Layer	Attacks in Smart Grids
Application layer	CPU exhausting
Network and Transport layer	Data flooding Buffer overflow
MAC layer	Man-in-the-middle attacks
Physical layer	Jamming channels

C. Cyber Defense Solutions for Smart Grid Systems

To provide defense to cyber attacks targeting confidentiality in the power grid, network coding has been presented to maintain data privacy in the grid where all aspects of privacy such as anonymity, unlinkability, unobservablity, and undetectablity have been achieved.

Defense against cyber attacks targeting integrity, several approaches have been proposed. To prevent integrity attacks, a power fingerprinting technique has been proposed, a volt-var control (VVC) based scheme has been proposed, and a Trusted Network Connect (TNC) base approach had been studied.

There are several cyber defense solutions against attacks targeting availability. Solutions to channel jamming attacks include frequency hopping based on preshared sequence or uncoordinated rendezvous methods where transmitter and receiver meet to a common channel for communication while avoiding the jammer.

In man-in-the-middle attacks, the cyber security solutions include use of in-depth packet analysis (e. g. , IDS) since packets sniffed by the attacker have unmatched MAC and IP address pairs.

To deal with buffer-over flaw attacks in the smart grid, a flocking-based models of power system operation for the grid has been proposed and a Discrete-Time Markov Chain (DTMC) model has been studied.

A top-down analysis has been performed to prevent data flooding attacks including other attacks. A defense solution to puppet attack that results in denial-of-services in AMI network has been studied.

In order to overwhelm networking devices which have limited computing resources, attackers can flood computationally intensive requests using application layer attacks. A DDoS Shield based suspicion assignment mechanism and a DDoS-resilient scheduler have been proposed to prevent application layer attacks.

7.5.5 Challenges and Future Research Directions

There are several challenges to secure smart grid systems from multitude of attacks and meet the security requirements and objectives of smart grid as smart grid assets are distributed over large geographic areas. Because of the critical nature of power systems and socioeconomic impact of blackouts, smart grid can be a prime target for cyber terrorism. The cyber defense solutions should protect all aspects of smart grid systems. The defense solution integrate multiple defense techniques including proactive real-time intrusion prevention/detection systems (IPS/IDS) using machine learning and artificial intelligence, network segmentation, controlled wireless propagation, authentication, authorization, certification. The proposed solutions should comprise of scalable, resilient, and adaptive cyber security/defense techniques for smart grid operation without affecting any legitimate smart grid operations.

7.5.6 Summary

Communication networks in smart grid bring increased connectivity to revolutionize the energy industry in terms of reliability, performance, and manageability by providing bidirectional communications to operate, monitor, and control power flow and measurements. However, communication networks bring severe security vulnerabilities with them. Furthermore, smart grids can be a prime target for cyber terrorism because of their critical nature and socioeconomic impact of blackouts. In this paper, we have provided a compact survey of cyber security attacks and defense techniques in smart grid systems that are targeted at different networks and protocol layers. With this article, readers can have a more thorough understanding of smart grid security, its requirements and objectives, and the future research directions in this topic.

8

专业英语语法(4)

8.1 数量的增加和增加的倍数

(1) 表示增加意义的实意动词或词组 increase (enlarge, go up …) + by + n/n%/n times。

在这种表示数量、百分数或倍数增加的结构中，介词 by 后表示的是净增加数、百分数或增加倍数，所以在作汉译时照译不减，应译成"增加（了）n"/"增加（了）n%"/"增加（了）n 倍"。例如：

1) The production of transformers has been increased by 200. 变压器的产量增加了 200 台。

2) World nuclear power generation increased by 4% last year. 全球去年核能发电量增加了 4%。

3) The production of higher rating transformers has gone up by 20 to 30 percent since last year. 去年以来，大容量变压器的产量增加了 20%～30%。

4) With a temperature rise of one degree centigrade *rises* the electric conductivity of a semiconductor by three to six percent. 温度每提高 1℃，半导体的电导率就增大 3%～6%。

5) This year the generation of this power station has increased by three times as compared with that of 2008. 今年这个电站的发电量比 2008 年增加了三倍。

6) This year the value of our industrial output has increased by half as compared with that of last year. 我们今年的工业产值比去年增加一半。

7) The adaptively controlled drill outperforms a conventional drill by 15%. 自适应控制的钻床，比一般的钻床操作效率高 15%。

(2) n+表示增加的形容词或副词的比较级 (more、larger …) + than …

这种结构表示净增加，即净增加 n，由于英汉两种语言的习惯相同，故照译不减，故译为"增加（了）n"，或"比……大（多……）n"、"比……增加了 n"。例如：

1) A is three more than B. A 比 B 多 3。

2) The first quarter's output is three hundred thousand tons more than that of the same period of the last year. 第一季度的产量比去年同期增加 30 万 t。

(3) an increase of n% (of + 名词), a n% increase (of + 名词), n%+表示增加的形容词或副词的比较级 (more、larger …) + than …, n% + 表示增加的形容词的比较级 (more、larger …) + 名词。

这种结构也表示净增加，即净增加 $n\%$，由于英汉两种语言的习惯相同，故照译不减，故译为“增加（了）$n\%$”或“比……增加了 $n\%$”。例如：

1）There is an increase of 50% of insulators as compared with last year. 绝缘子产量比去年增加了 50%。

2）There is a 50% increase of transformers as against 2008. 变压器产量今年比 2008 年增加了 50%。

3）The power generation this year has increased 15% more than last year. 发电量今年比去年增加了 15%。

4）This month we'll produce 20% more high voltage electricity installations than we did last month. 我们这个月生产的高压电气设备将要比上个月多 20%。

5）The local coal output was 16% higher than that of 2008. 当地的煤产量比 2008 年增加了 16%。

6）The machine works 13% faster than that one. 这台机器比那台的工作速度快 13%。

7）Wheel A turns twenty percent faster than wheel B. A 轮转动比 B 轮快 20%。

8）The β values of the implanted samples were about 30～50% larger than those of unimplanted samples. 植入过的样品的 β 值比未植入过的样品大约大 30%～50%。

9）Whereas high pressure die castings have porosity problems，low-pressure parts have none，and are consequently 25% to 35% stronger. 高压铸件有气孔问题，而低压铸件则无。因此，后者强度比前者高出 25%～35%。

（4）表示增加意义的实意动词或词组 increase（enlarge、grow、go up、raise …）to ＋ n/$n$times。

这种表述在汉译时照译不减，即译为“增加到 n”/“增加到（了）n 倍”。例如：

1）The members of the association have increased by 200 to 500. 这个协会的成员人数增加了 200 名，已达 500 名。

2）The electric energy production has increased to six times since then. 自那时以来，发电量增加到了 6 倍。(或译为：增加了 5 倍。)

（5）表示增加意义的实意动词或词组 increase（rise、raise、grow、go up、exceed、multiply …）＋ n times/fold。

在这种表示数量增加倍数的结构中，英语均表示增加后的结果，所以应译为“增加（了）$n-1$ 倍”“增加到 n 倍”（译出增加部分与原有部分之和）（是）“为……的 n 倍”“n 倍于……”。例如：

1）In 2008，the output of nuclear plants increased 3.8 times as against 2004. 2008 年与 2004 年相比，核电站的发电量增加了 2.8 倍。(或译：是 2004 年的 3.8 倍。)

2）Total output value of Shanghai's light industry between 2007 and this year grew 13 fold，that of the textile industry went up 4.3 fold. 从 2007 年到今年，上海轻工业总产值增长了12 倍，纺织工业总产值增长了 3.3 倍。

3）This kind of equipment improves the working conditions and raises efficiency fourfold. 这种设备改善了劳动条件，并使功效提高了 3 倍。

4）This will increase its strength 10-fold. 这将使其强度增加 9 倍。（增加到原来的 10 倍）

5）The price of this device has increased fourfold since last year. 这种仪器的价格自去年以来增加了 3 倍。（增加到原来的 4 倍）

6）It is estimated that the production of fluid power components has been raised three times as vs 2000. 据估计，液压传动元件的产量比 2000 年提高了两倍。

但是，be multiplied by *n* 和 multiply *n* times 都表示“增加了 *n*−1 倍”。例如：

7）The sales of transformers have been multiplied by four since last year. 自去年以来，变压器的销售量增加了三倍。

8）During the past two years the output of the high voltage installations of our factory *has multiplied four times*. 最近两年中，我厂高压设备的产量提高了三倍。

9）The efficiency has been multiplied several times. 效率成倍提高。

（6）… a *n* times（*n*-fold 或 *n* fold）＋ increase…。

这种数量增加结构表示增加到 *n* 倍，即表示增加后的结果，所以应译为“增加了 *n*-1 倍”“增加到 *n* 倍”（译出增加部分与原有部分之和）、“（是）为……的 *n* 倍”“*n* 倍于……”例如：

1）A record-high four - fold increase in value was reported.（A record-high increase in value of four times was reported）. 据报告，价值破纪录地增长了三倍。

2）Multiband transmission permits a reduction in error probability in exchange for at least a two fold increase in bandwidth and carrier power. 多频带传输能降低误差概率，但其代价是带宽及波功率至少要增加一倍。

3）They produced 3000 high voltage surge arresters this year，a fourfold increase over last year. 他们今年生产了 3000 台高压浪涌保护装置，是去年的 4 倍。

4）The temperature of the furnace showed a two-fold increase over that of another furnace. 该炉子的温度比另一只的温度高一倍。（或译为：该炉子的温度为另一只的温度的两倍。）

5）There is a seven times increase of installed capacity. 装机容量比过去增加了 6 倍。（或译为：装机容量为过去的七倍。）

（7）表示增加意义的实意动词或词组 increase（rise，raise，grow，go up，exceed，multiply …）＋ by ＋ a factor of ＋ *n*/*n* times …。

这种数量增加结构表示净增加 *n*-1 倍，即表示增加后的结果，所以应译为“增加了 *n*−1 倍”“增加到 *n* 倍”（译出增加部分与原有部分之和）、“（是）为……的 *n* 倍”，“*n* 倍于……”。例如：

1）Both the power generating capacity and supply increase by a factor of four. 发电能力和供电能力都增加了 3 倍。（增加到原来的 4 倍）

2）The speed exceeds the average speed by a factor of 3.5. 该速度超过平均速度 2.5 倍（2.5 倍）。

3) The sales of transformers has increased by a factor of 1.1. 变压器的销售量增加了10%。

(8) *n*/*n*% + too long/short/many/few。这个结构表示“长出/短了/多了/少了 *n*/*n*% +”。

1) The insulator is five centimeters too short. 这个绝缘子短了 5 cm。

2) He has given me five too few (or many). 他少（或多）给了我 5 个。

3) There are forty workers in our factory, and they were given fifty tools, so we have ten tools too many. 我们厂有 40 个工人，而他们收到了 50 件工具，因此多出了 10 件。

(9) … by + *n* + 数量 + 表示增加的形容词的比较级（more、larger …）。

这一结构表示净增加，数字可以照译。例如：

This wire is by 4 meters longer than that one. 这根导线比那根长 4 m。

8.2 数量的减少和减少的倍数

(1) 表示减少意义的实意动词或词组 reduce（fall、drop、lower、decrease …）+ by + *n*/*n*%。

在这种表示数量、百分数或倍数减少的结构中，介词 by 后表示的是净减少数、百分数，所以在作汉译时照译不减，应译成“降低、减少（了）*n*/*n*%”。例如：

1) The price for certain high voltage electricity installations has dropped by one third to two thirds compared with last year. 一些高压电气设备的价格比去年降低了 1/3～2/3。

2) The cost of this kind of transformer decreased by 10%. 这种变压器的成本降低了 10%。

3) An increase in the oxygen content of a coal by 1 percent reduces the calorific value by about 1.7 percent. 煤的含氧量增加 1%，其热值下降约 1.7%。

4) The loss has been reduced by five-sevenths (reduced by 75. 6%). 损耗减少 5/7（减少 75.6%）。

5) The installations of this type have fallen by 10% in price. 这种类型设备的价格下降了 10%。

(2) 表示减少意义的实意动词或词组 reduce（fall、drop、go down、lower、cut、decrease …）+ to + *n* /*n*%。

在这种结构中，介词 to 表示“到”的意思，to 后的数字为减少后的结果，所以通常译成“降到 *n*”/“减少到 *n*%”。例如：

1) The employees have decreased to 200 this month. 这个月雇员人数减少到二百名。

2) By using the new process, the loss of metal was reduced to less than 20%. 采用这种新工艺，金属耗损降到不足 20%。

3) Hot gas is fed to the cooler, where its temperature drops to 20℃. 热气体加到冷却器，其温度降低到 20℃。

4) Last year the expense fell to two million dollars. 去年支出降低到了 200 万美元。

（3）表示减少意义的实意动词或词组 reduce（fall，drop，go down，lower，decrease …）＋ n times。

在英语中还可以用具有减少意义的词加倍数来表示减少的数量，这称为“成 n 倍地减少”。句中的倍数 n 是指原量（减少前的数量）为现量（减少后的数量）的倍数。由于汉语不能说“减少多少倍”，而说“减少了几分之几”或“减少了百分之几”，这是因为在汉语里，“减少”和“几倍”（表示增加）这两个概念是矛盾的。因此在译成汉语时应将原文中的倍数化为分数或百分数，即换算成分数或百分数来表示。这种由倍数换算成分数有两种方法：

a. 把倍数 n 作分母，用 1 作分子，即 $1/n$，表示减少后的结果，译成“减少到 n 分之一”或“减为 n 分之一”（指剩下部分）；

b. 把倍数 n 做分母，用 $n-1$ 做分子，即 $\frac{n-1}{n}$，表示净减少 $\frac{n-1}{n}$，译成“减少了 n 分之（$n-1$）”（指减去部分）。例如：

1）Since the introduction of the new technique the switching time of the new type transistor has been shortened three times. 采用这项新工艺之后，新型晶体管的开关时间缩短了 2/3（或缩短到原来开关时间的 1/3）。

2）The automatic assembly line shorten the assembling time 5 times. 这条自动装配线使装配时间缩短了 4/5。

3）The equipment under development will reduce error probability three times. 正在研制的设备将把误差减少 2/3（即减少到原来的 1/3）。

4）The weight of the transformer has decreased four times. 该变压器的重量减轻了 3/4（即减轻到原重量的 1/4）。

5）The cost decreased four times. 费用减少了 3/4（减少到 1/4）。

6）The wearing away of the heavily loaded components has been slowed down 10 times. 重载部件的磨损速度降低到原来的 1/10（降低了 9/10）。

如果减少的倍数里有小数点，则应进一步换算成不带小数点的分数。例如 reduce 4.5 times，由于汉语里不说“减少到四点五分之一”，所以首先应换算成整数分母，即 $\frac{1\times 2}{4.5\times 2}=\frac{2}{9}$，再译为“减少了九分之七”或“减少到九分之二”。如果换算成分数仍不符合汉语习惯，则应进一步换算

成百分数。如 shorten 8.2 times，换算成整数分母，即“缩短到四十一分之五”“缩短了四十一分之三十六”，仍不符合汉语习惯，应进一步换算成“缩短了 87.8%”或“缩短到 12.2%”。

尽管一般说来，英语常用 times 来表示数量（值）的减少，但“减少一半”却是个例外，在英语中从来不说 decrease 2 times。英语中表示“减少一半”的句型很多，汉译时须分别对待。

（4）表示减少意义的实意动词或词组 reduce（fall，drop，go down，lower，decrease …）＋ by ＋ a factor of ＋ n …

这种结构与“reduce（fall，drop，go down，lower，decrease …）＋ n times”意义

相同，也表示“减少到 n 分之一”或“减为 n 分之一”或“减少了 n 分之 $(n-1)$”。例如：

1）The new equipment will reduce the error probability by a factor of 7. 新设备将使误差概率降低 6/7（或降到 1/7）。

2）The collector series resistance is reduced by a factor of 5. 集电极串联电阻降低了 4/5（或降到 1/5）。

3）The water level in winter falls by a factor of three as against the average level. 冬天水位比平均水位下降 2/3。

4）With the technology improved，the rate of spoiled products has decreased by a factor of four. 技术改进之后，废品率下降了 75％。

5）The resistance is reduced by a factor of four. 电阻减少了 3/4（减少到原来的 1/4）。

6）The loss of electricity was reduced by a factor of two. 电的损耗减小了 1/2。

若上述结构中的形容词或副词是表示减少之意，例如 lighter，smaller，less 等，这称为“减少了 n 倍”，汉译应为“减少到 $1/(n+1)$”或“减少了 $n/(n+1)$（当 $n \leqslant 9$ 时）”。

（5）n％ + 表示减少的形容词的比较级（less、smaller …）（+ 名词）。这种数量减少的结构表示净减少 n％，译为“减少（了）n％”。例如：

1）An increase in lubricant temperature results in 25 ％ lower viscosity. 由于润滑剂温度增高，引起黏度降低 25％。

2）The new-type transformer wasted 10％ less energy supplied. 这台新型变压器少损耗所供能量的 10％。

3）The constructional cost was 15％ lower. 造价降低了 15％。

（6）n％+表示减少的形容词或副词的比较级（less，smaller …）+ than …。

这种数量减小的结构表示净减少 n％，译为“减少（了）n％”。例如：

The coal output was 12％ less than that of 2009. 煤产量比 2009 年减少了 12％。

（7）… a n－fold/n fold + reduction（decrease）…

这种结构与“reduce（fall，drop，go down，lower，decrease …）+ n times”意义相同，也表示“减少到 n 分之一”或“减为 n 分之一”或“减少了 n 分之 $(n-1)$”。例如：

1）The principal advantage over the old -fashioned machine is a three-fold reduction in weight. 与旧式机器相比的主要优点是，其重量比原来减轻了 2/3（为原来的 1/3 重）。

2）The principal advantage is a four-fold reduction in volume. 主要优点是体积缩小了 3/4（或缩少为 1/4）。

（8）… by + n + 度量名词 + 表示减少的形容词的比较级（less、smaller …）+ than …。

这一结构表示净减少，数字可以照译。例如：

This wire is by 5 meters shorter than that one. 这根导线比那根短 5m。

8.3 倍数的直接表示法

在科技英语中可以用表示倍数的动词 double 等或名词 times 以及百分数 $n\%$来表示一物为另一物的几倍或 $n\%$，构成倍数的直接表示法。下面分别议论如下：

（1）表示倍数的动词 ＋ 宾语在科技英语中，常见的倍数动词有 double（两倍）、treble（三倍）、quadruple（四倍）等。double 表示加倍，通常译成“为……的两倍”“两倍于”“增加一倍”“翻一番”；treble 译成“为……的三倍”“增至三倍”“增加两倍”；quadruple 译成“为……的四倍”“增至四倍”“增加三倍”。例如：

1）If the resistance is doubled without changing the voltage，the current becomes only half as strong. 如果电压不变，电阻增加一倍，电流就减小一半。

2）Reducing the frequency by one-half will double the period. 频率减半，周期加倍。

3）As the high voltage was abruptly trebled，all the capacitors were broken down. 由于高压突然增加了两倍，所有的电容器都被击穿了。

4）The power utilities will need to quadruple their capacity. 电力事业必须使它们的能力增加三倍。

5）The output of precision machine tools has quadrupled since 2005. 2005 年以来，精密机床的产量增加了三倍。

6）By the end of 2010，the annual power generation of hydroelectric power stations nearly quadrupled that in 2003. 到 2010 年底，水力发电厂的年发电量差不多为 2003 年的四倍。

7）The number of high voltage installations in the USA more than tripled. 美国高压设备总数比原有的多两倍以上。

（2）be n times …。这种结构直接译为“n 倍于……”“是……的 n 倍”“be twice …”译为“二倍于……”“是……的 2 倍”。例如：

1）The volume of gas in tube A is three times that in tube B. 管 A 内气体体积是管 B 内气体体积的三倍。

2）The peak value of an alternating current is 1.414 times its effective value. 交流电的峰值为其有效值的 1.414 倍。

3）The area of Electric Circuits Lab in this school brigade is twice that of Power Electronics Lab. 这个学院的电路实验室面积比电力电子实验室的大一倍。

4）The earth is 49 times the size of the moon. 地球的大小是月球的 49 倍。

5）The newly built highway is three times the width of the original. 新建的高速公路是原先公路的 3 倍宽。

6）The velocity of sound in water is nearly four times the velocity of sound in air. 声音在水中的速度差不多是在空气中的 4 倍。

7）The number of installations to be used is about nine times that of installations in the past. 所要采用设备的数量比过去约增加 8 倍（约为 9 倍）。

8）This distance can be estimated at twice the depletion depth X_d. 这段距离估计可能为耗尽深度 X_d的两倍。

9）The velocity of sound in water is 4，900 feet per second，or more than four times its velocity in air. 水中声速为每秒 4900 英尺，比空气中的声速大 3 倍多。

10）Two quantities are of the same order of magnitude if one is no larger than 10 times the other. 两个数量，如果其中的一个是在另一个的 10 倍的范围之内，那么这两个数量属于同一数量级。

11）The collector load resistance is about 40 times this value. 集电极负载电阻为此值的 40 倍左右。

12）In addition，arrester rating should be at least 1.25 times the maximum continuous line-to-ground system voltage. 此外，避雷器的额定值至少应为系统持续相电压的 1.25 倍。

（3）*n* times ＋ 示量名词（the size，volume，length，breadth，width，level，value，velocity）或 that ＋ of ＋ …。

此结构表示净增“*n*－1 倍”，即增加到原来的 *n* 倍，可以译为“是……的 *n* 倍”“为……的 *n* 倍”“比……（大、长）*n*－1 倍”。如果倍数是一个近似值，则照译不减。例如：

1）This device is twice the width of that device. 这台仪器比那台仪器宽一倍。

2）Uranus and Neptune have each about four times the diameter of the earth. 天王星和海王星的直径约是地球的 4 倍。

3）The new generating unit is about two times the height of the old one. 这台新机组的高度是那台旧机组的 2 倍。（这台新机组大约比那台旧机组高 1 倍）

4）The mass of an electron is 1/1840 that of hydrogen atom. 电子的质量为氢原子的 1/1840。

5）The weight of the proton is about 1837 times that of the electron. 质子的重量约为电子的 1837 倍。

6）The volume of the sun is 330，000 times that of the earth. 太阳的体积为地球的 330，000 倍。

7）The depth of gasoline in the first tank is four times that in the other. 第一个油箱内汽油的高度是另一个油箱中的 4 倍。

8）In this workshop the output of July was 1.5 times that of January. 这个车间 7 月的产量是 1 月的 1.5 倍。

9）In 2009 the export value of machine tools was 8 times that of 2008. 2009 年机床的出口额为 2008 年的 8 倍。

10）The service life of the cable is expected to be four times that of most cables now available for similar service. 这种电缆的有效寿命约为目前其他同类电缆的 5 倍。

（4）*n*% ＋（of）＋ 名词或代词这一结构表示的增减包括底数在内，因而译为“是……的 *n*%（0.0 *n* 倍）”。例如：

1）The steel output is 500%（5 times）that of the prewar level. 钢产量为战前的 5 倍。

2）This year we have produced 130％ the number of high voltage electricity installations of last year. 今年我们制造的高压电气设备是去年的 130％（1.3 倍）。

3）The cost of their power production is about 80％ that of ours. 他们的发电成本大约是我们的 80％（0.8 倍）。

8.4 倍数的间接表示法

在科技英语中还可以用 as … as 等结构来表示一物为另一物的几倍，由于它们没有用到表示倍数的词，故可以称为倍数的间接表示法。

（1）… half ＋ as ＋ 形容词或副词原级（many、much、large、fast、high、long …）＋ as ＋ …。

这种结构是将“*n* times ＋as ＋形容词或副词原级（many、much、large、fast、high、long …）＋ as ＋ …”结构中“*n* times”改为 half（半倍）而得，所以此结构表示程度相差一半，即“减少一半”的意思。汉译时，可译成“是……的一半”“有……的一半”或“比……一半”时，要把原文的形容词或副词译成相反的意思。例如：

1）The leads of new condenser are half as long as those of the old，yet their performances are the same. 新型电容器的导线比老式的短一半，但作用相同。（或译为：新型电容器的导线是老式的一半长，但作用相同。）

2）They have produced half as many power semiconductor devices as we have. 他们生产的电力半导体器件比我们生产的少一半。

3）The output power of this engine is half as great as that of the other one. 这台发动机的输出功率是那台的一半大。（这台发动机的输出功率比那台的小一半）。

4）Since the lever arm of E is twice as long as the lever arm of R，the effort is only half as large as the resistance. 因为杠臂 E 比杠臂 R 长一倍，所以作用力只有阻力的一半大。

在英语中表示“减少一半”的句型还有很多。例如：

5）If resistance of both resistors are of equal value—say 1，000 ohm—the output voltage is just one half less than the input. 如果这两个电阻具有相同的电阻值——比如说，都等于 1，000Ω——那么，输出电压正好比输入电压减少一半。（… be one half less than …）

6）Inspection time for the installation has been reduced by 50％. 检查设备的时间已减少一半。（表示减少的动词 reduced 等 ＋ by 50％.）

7）This sort of membrane is twice thinner than ordinary paper. 这种膜的厚度比普通纸张要薄一半。（be twice thinner than）

8）It takes half an hour less to accomplish the whole task. 完成整个任务可少花半小时。（用 half ＋ 名词 ＋ less 表示）

9）Halving the repeater spacing made it possible to quadruple the bandwidth. 把增音器间隔缩小一半，就能使频带宽度增加三倍。（动词 halve）

10）Only half of the products are up to the specification. 产品只有一半合乎设计（技术）要求。(half of …)

11）Reducing the data rate by one-half will double the duration of each symbol interval. 数据率减小一半，将使每一符号间隔的持续时间延长一倍。(表示减少的动词 reduce 等 ＋ …＋ by one half)

(2）… as ＋ 形容词或副词原级（many、much、large、fast、high、long …）＋ again ＋ as ＋ …。

这是表示倍数的一种特殊形式，表示净增加一倍，其中副词 again 表示"再一倍"之意，用以修饰前面的形容词或副词，简记为 as … again as，通常译成"是……的两倍""两倍于……"，"比……多（大、快、高、长……）一倍"。例如：

1）The leads of the new condenser are as long again as those of the old. 新型电容器的引线有老式的两倍长。

2）This year we have produced as many insulators again as they. 今年我们生产的绝缘子是他们的两倍。

3）These insulators cost me as much again as the ones I bought last year. 这些绝缘子比我去年买的贵 1 倍。

4）This substance reacts as fast again as the other one. 这种物质的反应速度是另一物质的两倍。

5）Wheel A turns as fast again as Wheel B. A 轮转动得比 B 轮快一倍。

6）The peak power is as great again as the carrier power. 峰值功率比载波功率大一倍。

7）Four is as much again as two. 4 比 2 大一倍。

8）The laboratory under construction will be as large again as this one. 兴建中的实验室的面积是这个实验室的两倍。

9）The driving gear turns as fast again as the driven gear. 主动齿轮的转速为从动齿轮的两倍。

显然，在这一结构中，所用的形容词或副词都应是表示增加的形容词或副词，而不会用到表示减少的形容词或副词，例如 slow，less 等。

(3）… again ＋ as ＋ 形容词或副词原级（many、much、large、fast、high、long …）as ＋ …。

这是前一结构 as … again as 的另一形式，两者的意义和译法均相同，简记为 again as … as。例如：

The frequency range of signal generator model A is again as broad as that of model B. A 型信号发生器的频率幅度比 B 型宽一倍。

(4）… half ＋ as ＋ 形容词或副词原级（many，much，large，fast，high，long …）＋ again ＋ as ＋ …。

这一结构是由"half as … as"结构演变而来，其中 half 表示一半，again 表示再一倍，故合为一倍半，即表示程度相差一倍半，即净增加一半，通常译成"是……的一倍半”“有……一倍半那么多”“比……多（大、快、高、长……）半倍”等。这种结构简记为“half as … again as”。例如：

1）This transformer tank is half as long again as the other one. 这台变压器油箱的长度是那台的一倍半。（或译为：这台变压器油箱比那台长一半）。

2）This machine turns half as fast again as that one. 这台机器转动得比那台快半倍。（或译为：这台机器的转速是那台的一倍半）。

3）The antenna is half as high again as that one. 这根天线比那根高一半。（或译为：这根天线是那根的一倍半高。）

4）Three is half as much again as two. 3 比 2 多一半。

5）The substation being built will be half as large again as this one. 兴建中的变电站的面积比这个变电站大一半。

（5）… half ＋ again ＋ as ＋ 形容词或副词原级（many、much、large、fast、high、long …）＋ as …

这是前一结构 half as…again as 的另一形式，两者的意义和译法均相同，简记为 half again as … as。

The antenna is half again as high as that one. 这根天线比那根高一半。

（6）… half ＋ as ＋ 形容词或副词原级（many、much、large、fast、high、long …）＋ as ＋ … twice thinner than …，reduce … by one half …，halve …等结构，表示程度相差一半，即“减少一半”的意思。汉译时，可译成“是……的一半”“有……的一半”或“比……一半”时，要把原文的形容词或副词译成相反的意思。例如：

This pipe is half as long as that one. 这根管子是那根的一半长。

（7）… ＋ 分数（百分数）＋ as ＋ 形容词或副词原级（many、much、large、fast、high、long …）＋ as …”。

这种结构可以表示一方的数量是另一方的几分之几，也可以表示一方的数量比另一方的少几分之几。前者与汉语的表达方式相同，分数照译，译成“是……的几分之几”。但后者则差别很大，往往需要把 as 之间的形容词或副词译成相反的意思，分母的数照译，但分子的数译成分母减分子的数。如 one third as fast as 译成“比……慢 2/3”。例如：

1）The new motor is 40% as heavy as the old one. 这台新电动机的重量是旧的 40%。（或译成：这台新电动机比旧的轻 60%。）

2）The effort is one-third as great as the resistance. 作用力的大小是阻力的 1/3。（或译成：作用力比阻力小 2/3。）

3）This substance reacts one-tenth as fast as the other one. 这种物质的反应速度是另一物质的 1/10。（或译成：这种物质的反应速度比另一物质慢 9/10。）

在这种结构中，分数（百分数）作状语。

8.5 表示数量比较的倍数或百分数

数量比较可以通过倍数或百分数来表示，这主要有以下四种句式。

（1）n times ＋ as ＋ 表示增加或减少的形容词或副词的原级（big、great、many、much、large …）＋ as ＋ …。

这是英语中表示“n 倍于”常用的结构，其中 n 为任何数量（值）。该结构表示一方为另一方的若干倍，即净增加“（$n-1$）倍”或减少“（$n-1$）/n”，其倍数包括比较对象的原有数。因此，在表示增加时可译为：

1）对其倍数照译不减，译为“是……的 n 倍”“为……的 n 倍”或“n 倍于……”；

2）译出双方比较相差的倍数时，将其倍数减一，译成“比……（$n-1$）倍”“增加了（$n-1$）倍”。例如：“… 10 times as large as …”，既可译成“大小是（为）……的 10 倍”或“比……大 9 倍”。

在这一结构中，当相比的对象十分明显时，第二个 as ＋ …可以省略，数量（值）n 或倍数（n times）在句中作状语。例如：

1）A yard is three times as long as a foot. 一码的长度为一英尺的 3 倍。(或译成：一码比一英尺长两倍。)

2）The oxygen atom is nearly 16 times as heavy as the hydrogen atom. 氧原子的重量几乎是氢原子的 16 倍。(或译成：氧原子几乎比氢原子重 15 倍。)

3）Nylon is nearly twice as strong as natural silk and much less affected by water than it. 尼龙的强度几乎是天然丝的两倍，而不像天然丝那样易受水的影响。

4）The coefficient of expansion of air is about twenty times as much as that of mercury. 空气的膨胀系数约为水银的 20 倍。

5）The speed of sound in water is about four times as great as in air. 水中的声速约为空气中的 4 倍。

6）The machine, after modified, can produce four times as many products as before in a given period of time. 这台机器改装之后在所给的时间内能够生产的产品为以前的 4 倍。(省略句，应为 as could do before)

7）On the earth everything is six times as heavy as on the moon. 地球上每件东西的重量为月球上的 6 倍。

8）The resistivity of iron is almost six times as much as that of copper. 铁的电阻率几乎是铜的 6 倍。(或译成：铁的电阻率几乎比铜的大 5 倍。)

9）The thermal conductivity of metals is as much as several hundred times as that of glass. 金属的导热率是玻璃的数百倍。

10）Jupiter is 5 times as far from the sun as is the earth. 木星离太阳的距离相当于地球离太阳的距离的 5 倍。

11）Cork is 0.25 times as heavy as water, mercury weighs 13.5 times as much as water. 软木的质量是水的 0.25 倍，水银的质量是水的 13.5 倍。

12）Magnesium itself is only 1.7 times as heavy as water. 镁本身的密度只是水的 1.7 倍。

13）Were the earth's mass twice as great as it is, it would attract an object twice as strongly as it does. 如果地球的质量是它现在的两倍大，地球就以它现在两倍的引力来吸

引物体。

14）When you get to the moon you can jump about six times as high. 当你到达月球时，你能跳到大约在地球上的 6 倍高。（在 as high 之后省略了 as you can jump on the earth）

15）Every square mile on the sunny side of Mercury's surface receives seven times as much heat as a square mile on the earth. 水星朝太阳的一面每平方英里上吸收的热量是地球上每平方英里的 7 倍。

16）A proton and a neutron in conjunction make up the nucleus of a type of hydrogen atom twice as massive as ordinary hydrogen. 一个中子和一个质子结合，构成一种氢原子核，其质量是普通氢原子核的两倍。

若倍数是一个很大的近似值，则一、二倍的变化差异没有太大影响，可以照译不减。例如：

17）Presence of the back oxide does not affect MOS capacitance measurements, because the back contact area is about 1,000 times as large as that of any front contact. 背面存在的氧化物并不影响 MOS 电容的测量，因为背面接触面积比正面接触面积约大 1000 倍。

18）An experimental fusion device has achieved a temperature 10, 000 *times as hot as* the surface of the sun. 一种聚变实验装置已获得比太阳表面高 10000 倍的温度。

这一结构所表示的减少称为"成 n 倍地减少"，它是指减少前的数量为减少后的数量的 n 倍，故译为"减少到 $1/n$"或减少了"$(n-1)/n$"。例如：

19）A is four times as light as B. A 比 B 轻 3/4。（或译为：A 减少到 B 的 $\frac{1}{4}$ 或 A 是 B 质量的 1/4。）

20）The education budget of 1938 was 30 times as small as that of 1959. 1938 年的教育预算是 1959 年的 1/30。

21）Thus the rate at which long-wave photons leave the earth, for example, is roughly 20 times as large as the rate at which visible and near-infrared photons arrive, because the energy per photon is around 20 times as small as that. 因而，长波光子脱离地球的速度约是近红外可见光子到达地球速度的 20 倍，因为每个光子的能量只是原先能量的 1/20 左右。（或译为：因而，长波光子脱离地球的速度约是近红外可见光子到达地球速度的 20 倍，因为每个光子的能量大约减少了 19/20。不可译为：因为每个光子的能量约要小 20 倍。）

这种结构可以简记为：倍数 ＋ as … as。

（2）分数/n% ＋ as ＋ 表示增加或减少的形容词或副词的原级（big, great, many, much, large …）＋ as ＋ …。

这种结构可以表示一方的数量是另一方的几分之几/百分之几，也可以表示一方的数量比另一方的少几分之几/百分之几。前者与汉语的表达方式相同，分数/百分数照译不减，译成"是……的几分之几/百分之几"，但后者则差别很大，往往需要把 as 之间的形容词或副词译成相反的意思而且分母的数照译，但分子的数译成分母减分子的数，如 one third as fast as 译成“比……慢 2/3”。例如：

1）This substance reacts one tenth as fast as the other one. 这种物质的反应速度是另一种物质的 1/10。（或译为：这个物质的反应比另一个慢 9/10。）

2）The electron weighs about 1/1850 as much as an atom of hydrogen. 电子的质量约为氢原子的 1/1850.

3）Pure magnesium weighs only sixty-five per cent as much as aluminium. 纯镁的质量仅仅是铝的 65%。

4）The new motor is 40% as heavy as the old one. 这台新电动机比旧的轻 60%。（这台新电动机的质量是旧的 40%。）

5）The effort is one-third as great as the resistance. 作用力比阻力小 2/3。（作用力的大小是阻力的 1/3）

（3）*n* times + 表示增加或减少的形容词或副词的比较级（more、higher、better、longer、faster、greater、broader …）+ than …。

在科技英语中，可以用这种结构表示数量比较，其中，倍数词（*n* times）置于形容词或副词的比较级前作状语，后接以 than 引起的状语从句（代表相比的对象）来表示数量（值）的增减或对比。若相比的对象很明显，than 引起的部分可以省去：

n times + 表示增加或减少的形容词或副词的比较级（bigger、greater、more、larger、longer …）…。

在使用表示增加或减少的形容词或副词的比较级时，这种结构中 *n* 可为任何数量（值）。由于这一结构所表示的倍数与上一结构：*n* times + as … as 相同，所以汉译时同样译成“是……的 *n* 倍”“为……的 *n* 倍”“*n* 倍于……”或“比……（*n*−1）倍”，“较……（*n*−1）倍”，比……高（大……）。例如：

1）Iron is almost three times heavier than aluminium. 铁的质量几乎是铝的 3 倍。（或译成：铁几乎比铝重两倍。）

2）Water conducts heat about 20 times better than air does. 水的传热能力约为空气的二十倍。（或译：水的传热能力约比空气强 19 倍。）

3）The volume of the earth is 49 times larger than that of the moon. 地球体积是月球体积的 49 倍。（或译：地球体积比月球体积大 48 倍。）

4）A meter is more than 3 times longer than a foot. 1m 的长度为一英尺的 3 倍以上。（一米比一英尺长两倍多。）

5）The frequency range of signal generator model A is twice broader than that of model B.

A 型信号发生器的频带宽度为 B 型的两倍。（或译：A 型信号发生器的频带比 B 型的宽一倍。）

6）In the universe hydrogen is ten times more abundant than helium. 在整个宇宙中氢含量是氦的 10 倍。

7）Kuwait oil wells yield nearly 500 times more than U. S. wells. 科威特油井产油几

乎500倍于美国油井。

8）This diode produces ten times more radiant power than that one. 这只二极管的辐射功率比那只的大9倍（或译为：是那只的10倍）

9）Sound waves travel through fresh water at about 1410m/sec., that is rather more than four times faster than through air（331.4m/s.）. 声波在淡水中的传播速度大约是1410m/s，这比在空中的传播速度（331.4m/s）快3倍还要多一些。

10）The circumference of an circle is about $3\frac{1}{7}$ times longer than the diameter. 任何圆的圆周长约为其直径的$3\frac{1}{7}$倍。

11）The resistance of aluminium is approximately $1\frac{1}{2}$ times greater than that of copper for the same dimensions. 尺寸相同时铝的电阻约为铜的一倍半。

12）The new method was ten times more efficient, while unit cost was 20% lower. 新方法的效率为旧的10倍，而单位成本降低了20%。（非倍数20%表示净增减，此句相比对象不言而喻，than的部分省去）

13）The new hobs have diameter two or three times larger than conventional hobs and have two or three starts. 这种新滚刀比普通的滚刀直径大1～2倍，头数为2或3。

14）If we place the 1lb. weigh 5 times farther away from the fulcrum than the 5lb. weight, then the two will balance. It means that the 1lb. weight is 5 times as far from the pivot as the 5lb. weight. 如果使一磅的重物离支点的距离为5磅重物的5倍那么远，那么两者会处于平衡。这就是说一磅重物离支点的距离是5磅重物离支点的距离的5倍。

需要注意的是，“*n* times ＋ 表示增加的形容词或副词的比较级 ＋ than …”这种结构表示“*n*倍于”的用法，有些语法家认为不对，因为从语法上讲，它应该比“*n* times ＋ as … as”结构多一倍。尽管这种说法在语法和逻辑上不合理，但现已广泛使用。由于这种倍数的表示法在英美尚未完全统一，还有人用“*n* times ＋ 增加的形容词或副词的比较级 ＋ than”结构来表示净增加的倍数，所以对于采用这种句子的翻译，除了一般译成“是……的*n*倍”之外，也不排除有译成“比……*n*倍”的可能性。比如“A is 3 times larger than B”，通常是表示“A是B的3倍”的意思，但也可能表示“A比B大3倍”即“A是B的4倍”的意思。又比如：

The frequency range of signal generator model A is twice broader than that of model B.

也可以译为：A型信号发生器的频带比B型的宽两倍。

但根据事实应译为：A型信号发生器的频带比B型的宽一倍。

因此，翻译这类句子时，必须特别注意，要根据科学道理和逻辑常理来判断。例如：

Mount Jolmo Lungma is 8843 meters high, about 2.3 times higher than Mount Fuji.

应译为：珠穆朗玛峰高8843m，约是富士山的2.3倍。

不可译为：珠穆朗玛峰高是富士山的3.3倍。

也不可译为：珠穆朗玛峰比富士山高2.3倍。（因为富士山高3776 m）

同理，在上述结构中使用表示增加或减少的形容词或副词的比较级时，也可能有两种情况：

a. “成 n 倍地减少”，故译为“减少到 $1/n$ ”或“减少了 $(n-1)/n$ ”($n \leqslant 10$)；

b. “减少了 n 倍”，这是指减少前的数量比减少后的数量大 n 倍，此时应译为：“减少了 $n/(n+1)$ ($n \leqslant 9$) ”或“为……的 $1/(n+1)$ ”或“减少到” $1/(n+1)$ 。译为“减少到 $1/n$ ”或“减少了 $(n-1)/n$ ”。例如：

15) A body of a given object would be six times less on the moon than on the earth. 已知物体在月球上的重量将是其在地球上重量的六分之一。

16) The substance reacts four times slower than the other one. 这种物质的反应速度是另一物资的 1/4。

17) If you double the distance between two objects, the gravitational attraction gets four times weaker. 如果两个物体之间的距离增加一倍，这两个物体之间的引力将减小到原来的 1/4。

译为“减少了 $n/(n+1)$ ”或“为……的 $1/(n+1)$ ”或“减少到 $1/(n+1)$ ”的例子：

18) A is 3 times less than B. A 是 B 的 1/4。

19) This transformer is nine times lighter than that one. 这台变压器比那一台轻 9/10。

20) The wire is 4 times thinner than that. 这根导线比那根细 4/5。

因此，对于这类表示减少的句子汉译时，也必须特别注意根据科学道理和逻辑常理来判断究竟应采用哪种译法。

这种结构可以简记为：倍数 + 比较级 + than。

(4) 表示增加或减少的形容词或副词的比较级（bigger、greater、more、larger、longer …）+ than … + by + n times。

这一结构只是将“n times + 形容词或副词的比较级 + than”中表示倍数的词由 by 引出，置于相比较的对象之后。因此它应译为“是……的 n 倍”“为……的 n 倍”“n 倍于……”或“比……（$n-1$）倍”，“较……（$n-1$）倍”“比……高（大……）”。例如：

1) A liter of oxygen is heavier than a liter of hydrogen by 15.88 times. 一升氧比一升氢重 14.88 倍（一升氧的质量是一升氢的 15.88 倍）。

2) With a 10∶1 turns ratio, the secondary voltage will be greater than that of the primary by ten times. 若变压器的变比为 10∶1，则二次电压比一次电压高 9 倍。（或译为：则二次电压为一次电压的 10 倍。）

除了上述四种句式外，还可以采用下述句式来表示倍数：

形容词或副词的比较级（bigger、greater、more、larger、longer …）+ than … + by a factor of + 数量词。

此结构译为“比……（$n-1$）倍”，“较……（$n-1$）倍”。若相比的对象很明显，than 引起的部分也可以省去。例如：

1) The error probability of binary AM is greater than for binary FM by a factor of at

least

二进制调幅的误差概率比二进制调频至少大 5 倍。

2）In case of electronic scanning the beam width is broader by a factor of two. 电子扫描时，波束宽度展宽一倍。

在以上各种句式中，如果倍数是一个相当大的近似值，差一倍没有多大意义时，往往可以直接照原数译。例如：

3）The sun is 330，000 times as large as the earth. 太阳比地球大 33 万倍（太阳的大小是地球的 33 万倍）。

4）In 1898，the Curies obtained a new element whose radioactivity was several million times stronger than that of uranium. 1898 年，居里夫妇获得了一种新的元素，其放射性比铀（的放射性）强几百万倍。

8.6 表示“原因”的 because，as，since，for 之间的差异

表示“原因”的 because，as，since，for 之间存在比较明显的差异，分别说明如下：

（1）Because 表示原因的语气最强，它回答 why 的理由，because 所连接的原因状语从句是整个主从复合句的重心，而主句所表述的结果却并不一定是主要的，但主句和从句所表示的是必然的因果关系。从句一般放在主句后。例如：

1）The material first used was copper because it was easily obtained in its pure state. 最先使用的材料是铜。因为纯净状态的铜是很容易得到的。

2）Steel is widely used in different branches of engineering because it has a number of good properties. 钢在各种工程部门中广泛使用，因为它有许多优良特性。

3）A rotating body possesses kinetic energy because its constituent particles are in motion. 旋转的物体具有动能，因为其所有的组成微粒都处于运动状态。

（2）as（=now that）用来表示原因时，在语气上不如 because 重，仅仅说明一般的因果关系。其用法和下面的 since 相类似，从句一般位于主句前。例如：

1）As water vapor is extremely light，it often rises high in the sky . 由于水蒸气极轻，故时常在天空中上升得很高。

2）As electricity is a kind of energy，it can be changed into heat energy or mechanical energy. 由于电是一种能，所以它能变为热能或者机械能。

3）As air has weight，it exerts force on any object immersed in it. 由于空气有重量，所以它对处于其中的任何物体都要施加作用力。

4）Now that we have discussed the meaning of a graphical solution of a system of simultaneous equations and the method of plotting a line . we are in a position to find graphical positions of systems of linear equations. 由于我们已讨论了联立方程组的图解含义及画线方法，现在就能求出线性方程组的图解了。(now that =as)

（3）since（=in that）表示的原因已为人所知，无须加以说明，或不如句子的其他部分重要，故一般把它译为“既然”，所以全句的中心自然落在主句上。. since 比 as 稍微正

式一些，从句一般位于主句前，科技文中也常位于主句后。例如：

1）Since light travels faster than sound，we see lightening before we hear the thunder. 因为光比声走得快，所以我们在听到雷声之前先看到闪电。

2）Since metal is fusible，it will change into liquid when the temperature is high enough. 既然金属是可熔的，因此在温度够高时，它就变成液体。

3）Since k and m are both constants，the ratio k/m is constant. 既然 k 和 m 均为常数，所以 k/m 比值是恒定的。

4）Obviously，no current can flow since there is no circuit. 显然不可能有电流流动，因为根本就没有电路存在。

5）Sunstroke differs from heat exhaustion in that one of the heat regulators I affected; namely，the sweat glands. 中暑不同于热疲惫，因为热调节机构之一——汗腺，受到了影响。(since = in that)

(4) for 与前三个从属连词：because、as 和 since 不同，它是一个并列连接词，for 连接的不是一个原因状语从句，而是与它前面的分句并列的一个分句。for 表示原因的语气最弱，只是根据前一分句所述的内容作一推理，或只是对某一问题或现象提供一些情况或作一点解释补充说明，附加说一下，翻译时可译为“因为，其理由是”等。根据情况需要 for 甚至可以不译出来。for 引导的从句通常位于后面，偶尔也有单独成为一句的情况。例如：

1）The fuel must be finished，for the engine stopped. 燃料一定用完了，因为发动机停了。(译出来)

2）The Chinese people are working with a will，for they know what they are working for. 中国人民正在坚持不懈地劳动，他们知道为什么劳动。(未译出来)

3）In previous chapters we did not use the trigonometric，inverse trigonometric，exponential，or logarithmic functions，for the derivative of each of these is a special form. 在前面几章我们并没有使用三角函数、反三角函数、指数函数或对数函数，因为其各自的导数均为一种特殊的形式。(译出来)

从说明原因的程度或语气来说，because、as、since、for 四者的关系可以表示为

because>as>since>for

8.7 形容词 no 后名词的单复数

no 可用作副词，也可用作形容词。当用作形容词与名词连用时，其后面的名词有时是单数，有时则是复数。基本规则如下：

(1) no 后接不可数名词时，不可数名词不受 no 的影响，应是单数。例如：

1）She has no time. 她没有时间。

2）They have no money for machines. 他们没有钱买机器。

(2) no 后接可数名词接时，应根据实际情况决定“数”的问题即意思上应是单数就用单数，意思上该用复数就用复数。例如：

1）This teapot has no lid. 这个茶壶没有壶盖。(茶壶一般只有一个盖。)

2）There are no (=not any) electric motors in our workshop. 我们车间没有电动机。(通常一个车间不止一台电动机。)

I have no camera. 我没有照相机。(通常一个人只有一架照相机。)

需要注意的是，no 在 have 后及 there be 句型中比 not 的否定语气要弱得多，但相反，它在其他动词后的否定语气又比 not 强得多，因此若不在动词 have 之后或在 there + be 句型中，而是用在系词 be 之后作表语或其他动词之后，可数名词用单数表示强调。

1）He is no engineer.（强调）他决非工程师。

He is not an engineer. 他不是工程师。

2）It is no joke. 这可不是开玩笑。

3）Neutrons carry no charge. 中子根本不带电荷。

4）He is no mathematician. 他根本不懂数学。

（3）no 后接泛指的具有抽象概念的名词，虽然有时也用复数，但通常只用单数。例如：

Liquids have no definite shape. 液体没有固定的形状。

There is no short cut in the study of science. 科学研究没有捷径。

It is no easy matter to make this experiment. 做这个实验并非易事。

（4）在许多情况下，no 后跟可数名词单数和复数，意思上没有很大的差别。例如：

1）We have no machine tool (or tools) left. 我们没留下机床。

2）There are no books (or There is no book) on the table. 桌上没有书。

3）Without dust there would be no clouds and rain.（rain 没有复数形式）没有尘埃，就不会有云和雨。

8.8 介词 of 的常见用法

（1）"of+某些抽象名词（绝不能是表示具体事物的名词）"等效于与这些名词相对应的形容词，但语气更强，例如：

1）This measurement is of great precision. 此测量非常精确。

2）This book will be of help in exposing computer scientists to the latest technology. 本书将有助于计算机专家们了解最新的技术。

3）It is of great interest to observe electrical phenomena。观察电现象非常有趣。

4）Engineers may find the book of value as a reference on basic electrical problems. 工程师会发现这本书作为一本有关基本电气问题的参考书是很有价值的。

5）It is (of) no use grounding this point.（在此句中 of 可以省略）将这一点接地是没有用的。

6）What is described in this paper is of great importance. 本文所讲内容极为重要。

（2）of 相当于 among，表示"在……之中"，其后只能接可数名词复数，由 of 引出的前置词短语在句中作状语，既可用作为最高级的比较范围，也可用于一般的范围，例如：

1）Of all these instruments，this one is the most expensive.（最高级的比较范围）所有这些仪表中，这一台最贵。

2）Of all the computers in this laboratory，this one works best.（最高级的比较范围）该实验室里的所有计算机中，这一台性能最好。

3）Of the four parameters，three can be disposed of rather quickly.（一般范围）在这四个参数中，有三个可以相当迅速地加以处理。

4）Of these three new chapters，the first treats overvoltage，the next covers radar，and the last one introduces television.（一般范围）在新增加的三章中，第一章论述，第二章讲解雷达，第三章介绍电视。

5）ECL III has the smaller propagation delay of the two ECL series.（一般范围）在这两种 ECL 系列中，ECL III 的传播延时比较短。

6）Of all engineering materials，only metals are truly weldable and repairable. 在所有的工程材料中，只有金属是能焊接，能修补的。

（3）of 表示前后两者处于同位关系，例如：

1）The concept of potential difference will be introduced in the next section. 电位差这一概念将在下一节介绍。

2）The science of Electricity is very useful in the modern world. 电气这门科学在当今世界非常有用。

3）In this case. the failure point corresponds to a factor of safety of 1. 在这种情况下，失效点相应的安全系数为 1。

4）This device can supply 4 mA of output drive current.

OR：This device can supply an output drive current of 4 mA. 这个器件能提供 4mA 的输出驱动电流。

5）The lighter machine part has a mass of 7 kg. 那个较轻的机器零件的质量为 7kg。

（4）of 表示“对…来说（而言）”，例如：

1）It is generally true of all bodies that an increase in size accompanies a rise in temperature.（of all bodies 是“对所有物体而言”）体积随着温度上升而增加，这种情况对所有物体来说一般是正确的。

2）This rule is true of all cases. 这条规则无论哪种情况都适用。

（5）of 后面的名词是其前面名词（来自于不及物动词）的逻辑主语。

1）Fig. 2 shows the variation of the output with the input.（the variation of A with B）图 2 画出了输出随输入的变化情况。

2）The dependence of y on x is expressed by $y=f(x)$.（the dependence of y on x）对于 x 的依从关系用 $y=f(x)$ 来表示。

3）Ellipses are used to describe the motions of the planets around the sun.（the motion of A around B）我们用椭圆来描述行星绕太阳的运行情况。

4）The response of a body to a net force F is an acceleration a proportional to F.（the response of A to B）物体对净力 F 的反应就是正比于 F 的加速度 a。

5）The rotation of the earth on its own axis causes the change from day to night. 地球绕轴自转，引起昼夜的变化。

（6）of 后面的名词是其前面名词（来自于及物动词）的逻辑宾语。

1）The resolution of a force into X-and y-components is possible.（the resolution of A into B）把一个力分解成 X 分量和 y 分量是可能的。

2）The definition of speed as the ratio of distance to time is familiar to all of us.（the definition of A as B）我们都很熟悉把速度定义为距离与时间之比（这一点）。

3）The separation of gold from its ore is not easy.（the separation of A from B）把金子从矿石中提炼出来并不容易。

4）Exposure of the body to potentially toxic substances should be avoided.（exposure of A to B）应避免人体接触有潜在毒性的物质。

8.9 enough ＋ 不定式的用法

enough 这个词在英语中是常用的词之一。它可以被用作形容词、副词、名词，甚至感叹词。虽然基本含义都是“足够”，但因上下文的影响，它在不同的句型中会有不同的含义。“enough ＋ 不定式”这个结构可能用在以下几个句型中：

（1）用在“一个表示否定的词 ＋ enough ＋ 不定式”这个句型中，意思是“甚至不足以”“甚至…都不行”等意。例如：

1）It was only a toy plane , with scarcely buoyancy enough to lift a cat.. 它只是个玩具飞机，其浮力甚至载一个猫都不行。

2）The heat was not enough to melt the metal. 热量甚至不够熔化这种金属。

3）At that time , to confess the truth I know not enough of circuits to tell resistor from capacitor. 那时候，老实说，我对电路知识之浅，甚至连电阻和电容都区分不清。

4）I found this lab. room not big enough to hold so many students. 我觉得此实验室房间不够大，甚至坐不了这么多学生。

（2）在“形容词或副词 ＋ enough ＋ 不定式”这个句型中，表示“简直够…”“已经够…”等意。例如：

1）This book is hard enough for me to read. 这本书让我阅读简直是够难的了。

2）The patient is strong enough to sit up. 病人结实得已经能坐起来了。

3）During combustion the oxidation takes place rapidly enough to be accompanied by light and heat. 燃烧过程中氧化作用快的会发出光和热来。

（3）在“动词 ＋ enough ＋ 不定式”句型中，有时表示“只要…就够了”之意；在一个表示“气愤”或“懊悔”的句子中，“enough ＋不定式”表示“竟然……”之意。如：

1）It is enough to remember these basic rules. 只要记住这些基本规则就够了。

2）They know enough to know what they learn for and how to learn. 只要他们知道为什么学和怎样学也就够了。

3）He was fool enough to agree with this method. 他真傻，竟然同意这个办法。

8.10 that 引导的定语从句修饰 time，moment 等名词

当 that 引导的定语从句修饰 time，moment，way，direction，distance，reason 等名词时，往往相当于“in which”“through which”“by which”等“介词＋which”构成的短语作状语。在这种情况下 that 可以被省掉。例如：

（1）One light year is the distance（that）light travels in one year. 一个光年是光在一年内走的路程。

（2）The way（that）this matter can be used depends upon its properties. 怎样能够使用这种物质取决于它的性质。

（3）The reason the two gases form as they do will be discussed later in this chapter. 这两种气体形成的原因将在这一章的后面讨论。

（4）With television we can see things happen almost at the exact moment（that）they are happening. 有了电视，我们差不多能够在事情发生的那一刹那看到它们发生。

（5）The direction a force is acting can be changed. 力的作用方向是可以改变的。

8.11 动名词的复合结构与“名词＋现在分词”的区别

由“物主代词、名词所有格、代词宾格或名词通格＋动名词”所构成的结构称为动名词的复合结构。这种结构在句中可以作主语，表语、宾语（特别是作介词宾语）等。例如：

（1）1Air leaking into the pump from any source is likely to cause erratic running.（主语）空气从任何来源漏入泵中似乎都会引起运行的不稳定。

（2）We do not object to you doing this test.（宾语）我们不反对你做这个试验。

（3）Any engine can give us an example of heat being turned into work.（介词宾语）任何发动机都能给我们提供一个热被转化为功的例子。

由“名词通格＋动名词”构成的动名词复合结构，由于形式上和由“名词＋现在分词”构成的结构完全一样，所以非常容易引起误解。分辨这两种不同结构的方法是看它们中间哪个词是该结构的中心词，最重要的要看它们在句中起什么成分作用。如果以带-ing 词为中心词，在句中起主语、宾语、特别是介词宾语及表语作用时，就是动名词复合结构；如果以名词为中心词，甚至没有后面的带-ing 词，句子的结构仍很完整，那就是现在分词修饰名词的结构；如果虽也是以带-ing 词为中心词，但整个结构则是起状语作用，那就是独立分词结构了。例如：

（1）Here was a clear case of magnetism being converted into electric current.（动名词复合结构）这里就有一个磁转化为电流的明显例子。

（2）The plano flying from England to American carry 300 passengers.（现在分词作定语）从英国飞往美国的那架飞机能载客300名。

（3）This attitude of flight may be caused by the pilot moving the controls.（动名词复合结构）这种飞行姿态可以由飞行员移动操纵机构引起。

（4）The speed of light being extremely great，we cannot measure it by ordinary method.（独立分词结构）由于光的速度非常大，我们不能用普通的方法度量。

参 考 文 献

[1] ALI K, MOHAMMADN, MARWALI M D, et al. Integration of Green and Renewable Energy in Electric Power Systems. USA: WILEY, 2010.
[2] GELLINGS W C. P E. The Smart Grid: Enabling Energy Efficiency. The Fairmont & CRC Press, Inc. , USA , 2010.
[3] SEN K K, Mey Ling Sen. Introduction to Facts Controllers Theory Modeling and Applications. USA: WILEY, 2010.
[4] GRIGSBY L L. Power System Stability and Control. CRC press. Taylor & Francis Group, USA, 2007.
[5] RYSZARD S, GRZEGORZ B. Power Electronics in Smart Electrical Energy Networks. Germany: Springer, 2008.
[6] MUHAMMAD H R. Power Electronics Handbook. Academic Press, USA, 2001.
[7] BIMAL K B. Modern Power Electronics and AC Drives. Prentice Hall, Inc. , USA, 2002.
[8] SABA M M F , SCHULZ W, et al. High-Speed Video Observations of Positive Lightning Flashes. 30th International Conference on Lightning Protection—ICLP 2010 (Cagliari, Italy—September 13th-17th, 2010).
[9] ALMEIDA M E, Correia De Barros M T. Modelling of Long Ground Electrodes for Lightning Studies. High Voltage Engineering Symposium . Conference Publication, 1999, 2 (467).
[10] DELLERA L, GARBAGNATI E. Lightning Stroke Simulation by Means of The Leader Progression Model I: Description of the model and evaluation of exposure of free-standing structures. Power Delivery, 1990, 5 (4): 2009-2022.
[11] PAUL D. Low-Voltage Power System Surge Overvoltage Protection. IEEE Transactions on Industry Applications, 2001, 37 (1): 223-229.
[12] 黄荣恩．科技英语翻译浅说．北京：中国对外翻译出版公司，1981.
[13] 董国忠．科技英语翻译初步（修订版）．北京：商务印书馆，1982.
[14] 魏新强，刘桂华．科技英语语法．上海：上海交通大学出版社，2009.
[15] 魏汝尧．科技英语篇章翻译．北京：外语教学与研究出版社，2009.
[16] 闫文培．实用科技英语翻译要义．北京：科学出版社，2008.
[17] 叶云屏．Reading Your Way into Technical Writing：方法与实践．北京：高等教育出版社，2007.
[18] 赵萱．科技英语翻译．北京：外语教学与研究出版社，2006.
[19] 戴文进．科技英语翻译理论与技巧．上海：上海外语教育出版社，2003.
[20] 秦荻辉．实用科技英语写作技巧．上海：上海外语教育出版社，2001.
[21] 翟天利．科技英语突破．北京：外文出版社，1999.